Olfa Kanoun, Christian Viehweger (Eds.)
Frontiers of Science and Technology

Also of Interest

Frontiers of Science and Technology
Automation, Sustainability, Digital Fabrication
Gabriela Celani and Olfa Kanoun (Eds.), 2017
ISBN 978-3-11-053623-2, e-ISBN (PDF) 978-3-11-053768-0,
e-ISBN (EPUB) 978-3-11-053626-3

Energy and Sustainable Development
Quinta Nwanosike Warren, 2021
ISBN 978-1-5015-1973-4, e-ISBN (PDF) 978-1-5015-1977-2,
e-ISBN (EPUB) 978-1-5015-1328-2

Advanced Materials
Theodorus van de Ven and Armand Soldera (Eds.), 2020
ISBN 978-3-11-053765-9, e-ISBN (PDF) 978-3-11-053773,
e-ISBN (EPUB) 978-3-11-053779-6

Proceedings of International Conference Building Services and Energy Efficiency
Modernizing and Increasing Performance of Building Services
Ciocan Vasilică (Ed.), 2020
e-ISBN (PDF) 978-83-957204-1-3

Energy Harvesting for Wireless Sensor Networks
Technology, Components and System Design
Olfa Kanoun (Ed.), 2019
ISBN 978-3-11-044368-4, e-ISBN (PDF) 978-3-11-044505-3,
e-ISBN (EPUB) 978-3-11-043611-2

Frontiers of Science and Technology

Reports on Technologies for Sustainability – Selected extended papers from the Brazilian-German Conference on Frontiers of Science and Technology Symposium (BRAGFOST), Potsdam 5-10 October 2017

Edited by
Olfa Kanoun and Christian Viehweger

DE GRUYTER

Editors
Prof. Olfa Kanoun
Technical University Chemnitz
Faculty of Electrical Engineering and Information Technology
Reichenhainer Str. 70
09126 Chemnitz
Germany
olfa.kanoun@etit.tu-chemnitz.de

Dr. Christian Viehweger
Technische Universität Chemnitz
Reichenhainer Str. 70
09126 Chemnitz
Germany
christian.viehweger@etit.tu-chemnitz.de

ISBN 978-3-11-058407-3
e-ISBN (PDF) 978-3-11-058445-5
e-ISBN (EPUB) 978-3-11-058414-1

Library of Congress Control Number: 2021932861

Bibliographic information published by the Deutsche Nationalbibliothek
The Deutsche Nationalbibliothek lists this publication in the Deutsche Nationalbibliografie; detailed bibliographic data are available on the Internet at http://dnb.dnb.de.

Cover image: Structural scale models developed at Henn Architectural Office, Berlin. Photograph by Gabriela Celani, 2014.
Typesetting: VTeX UAB, Lithuania
Printing and binding: CPI books GmbH, Leck

MIX
Papier aus verantwortungsvollen Quellen
FSC www.fsc.org FSC® C083411

www.degruyter.com

Preface

Modern technology continuously changes our life, usually for the better, but always with a social-economic impact. Information technology and digitalization are changing the way we communicate, modern buildings save energy, new means of transportation make us mobile in a more sustainable way. Sensors help us to constantly gather information about our surroundings, making it easier to understand the environment, processes and relations. Low power systems and nanotechnology open completely new perspectives. Among the most important emerging trends nowadays, a particularly important one is sustainability and the need for efficiency. Systems nowadays don't just need to work, they have also to be efficient in use, time and costs. Each new development has to be understood and accepted.

This book summarizes outstanding manifold contributions structured in four sections, dealing with modern building and smart cities with modern urban infrastructure and urban farming and agriculture, renewable energies and smart materials and nanotechnology.

In the first section, Lopes et al. report first on urban occupations in Belo Horizonte, Brazil, presenting projects for a new idea of society. Within the second contribution, Lenz et al. discuss old buildings and how to renew them, to fit the needs of modern structures. In the next chapter, Schmidt et al. describe simulation methods for modern buildings with respect to the special needs of the Brazilian climate zone. The following contribution from Mota et al. presents a monitoring technology to meter the electrical energy consumption within urban areas. Freire et al. concentrate their work on the hygrothermal behavior of buildings, especially the presence of moisture.

The next section of the book starts with Serra et al. investigating the challenges and opportunities of urban agriculture as improvement for future cities. This is enveloped by the contribution of M. Serafim; she focuses especially on urban agriculture in Brazil. A. Pavesi contributes with a chapter on the recultivation of the Brazilian Savannah within cities to fight the destruction of landscape.

Energy sustainability is becoming very important nowadays and touches at the same time several technological sectors on a global and small scale. Leite and Riffel provide a global insight on the recent challenges of energy decentralization as the new energy revolution. From a scientific and technological perspective, C. Birkel continues with an overview about inorganic energy materials and their synthesis.

The last section of the book focuses on efficient smart materials themselves, starting with A. Alves providing an exciting overview about the field of nature inspired materials and their use. Koerdt et al. contribute with their work on fiber-reinforcement in polymer composites with included sensors and their efficient production. S. Titotto reports about 4D printing to manufacture animated structures as a new production process. Matsubara et al. continue with their work on different CNT types for lightweight structures to improve the mechanical properties of carbon fibers. In the following contribution, Paterno et al. discuss about the use of nanoparticles and their bioimpact, es-

https://doi.org/10.1515/9783110584455-201

pecially to human health. Fiber metal laminates and their improvement by polymeric interlayers is the content of the contribution of Rossini et al., with an extended numerical analysis. A. Benchirouf reports on the development of piezoelectric CNT material in polymers and its production technique. The improvement of fatigue strength of fiber reinforced polymers is content of the work of Hausmann et al., with a focus on energy storage systems for renewable energies. An experimental study on nanofluids, especially breakdown voltage, viscosity and thermal conductivity, in mineral oil for smaller and lighter transformers, carried out by Filho et al., concludes the book.

Contributions like these, provide insight into intelligent scientific investigations with a direct impact on our environment, help us to improve our future life and will result in a continuously improved sustainability. They are the fruit of a symposium of the German Humboldt Society and the Brazilian Federal Agency for Support and Evaluation of Graduate Education (CAPES) welcoming every year outstanding researchers of both countries to a special symposium, where they discuss innovative developments in a variety of fields of interest. These interdisciplinary debates on "Frontiers of Sciences" enable a fresh view on the latest developments in areas of social importance, discussions with an outstanding deepness and provide a global platform for exchange at the same time.

We thank the authors for their valuable contributions and for providing insight in their fields of research and beyond it, at the frontiers of sciences.

Professor Dr.-Ing. Olfa Kanoun and Dr. Christian Viehweger

Contents

About the editors

Olfa Kanoun

Olfa Kanoun is a full professor and leading the professorship Measurement and Sensor Technology at Chemnitz University of Technology since 2007. She conducts research in the fields of impedance spectroscopy, sensors based on carbonaceous materials and energy aware wireless sensors. She served as the co-chair on the German side for the Brazilian-German Frontiers of Science and Technology Symposium (BRAGFOST), which was organized by the Humbold Foundation together with CAPES in Campinas in 2016 and in Potsdam in 2017.
With her team, she has been promoting the field of the application of impedance spectroscopy and developed several embedded solutions. With her chair, she received six best paper awards on international conferences and published more than 400 papers in peer-reviewed scientific journals, book chapters and international conferences.

Christian Viehweger

Dr. Christian Viehweger is the leader of the work group on energy autonomous sensors and wireless sensor systems at the chair of Measurement and Sensor Technology at Chemnitz University of Technology. He supported the organization of the Pre-BRAGFOST Workshop in 2017 in Potsdam.
Since 2017, he is a postdoc researcher at the chair of Measurement and Sensor Technology at Chemnitz University of Technology. Together with his team, he works on Energy Harvesting, low power communication technologies and system design of wireless sensor nodes for a variety of applications in industrial and environmental environments. He led two research groups for young scientists and different research projects in cooperation with industry, within the field of autonomous sensor systems.

https://doi.org/10.1515/9783110584455-202

Part I: **Future cities**

Marcela Silviano Brandão Lopes, Luciana Souza Bragança, Marcus Barbosa Deusdedit, Mayumi Ikemura Amaral, and Natacha Rena

Biopotent social technology: occupations park and university extensions

Abstract: The present article aims to discuss the relationship between the everyday practices in self-built urban occupations in Belo Horizonte (Minas Gerais, Brazil), and concepts related to social technology, aiming to build a different narrative that contrasts with the hegemonic discourse and modifies the approach to those territories. In order to support the debate, the case of Parque das Ocupações do Barreiro will be presented as a project that involves both agendas: the struggle for housing and the struggle for environmental preservation. The project is an example of resistance against the neoliberal production of urban space led by the capitalist state.

Keywords: Self-built urban occupations, social technology, biopolitcs

1 Times of neoliberal reasoning

This article seeks to discuss the relationship between the everyday practices of self-built urban occupations and the concepts of social technology. In order to start the

Acknowledgement: The authors would like to thank the support given by Fapemig, CNPq, and Pro-Ex/UFMG.

Marcela Silviano Brandão Lopes, PhD in Architecture and Urbanism in the Federal University of Minas Gerais (UFMG), coordinator of the extension project "Artesanias do Comum" from the research group "Indisciplinar," involved with CNPq, coordinator of the research project "Tecnologia Social, Sustentabilidade Cultural e Economia Solidária," in the "Ocupações Urbanas Autoconstruídas," involved with FAPEMIG, Belo Horizonte, Brazil, e-mail: marcelasbl.arq@gmail.com
Luciana Souza Bragança, Master and teacher in the faculty of Architecture, Urbanism and Design at the Federal University of Minas Gerais (UFMG), coordinator of the extension research "Natureza Urbana" from the research group "Indisciplinar," involved with CNPq, Belo Horizonte, Brazil
Marcus Barbosa Deusdedit, Undergraduate in the extensions researches "Natureza Urbana" and "Artesanias do Comum" from the research group Indisciplinar, involved with CNPq, Belo Horizonte, Brazil
Mayumi Ikemura Amaral, Undergraduate in the faculty of Architecture, Urbanism and Design at the Federal University of Minas Gerais, scholarship in the research project "Tecnologia Social, Sustentabilidade Cultural e Economia Solidária" in the "Ocupações Urbanas Autoconstruídas," involved with FAPEMIG, Belo Horizonte, Brazil
Natacha Rena, PhD in Architecture and Urbanism in the Federal University of Minas Gerais (UFMG), leader of Indisciplinar group research, involved with CNPq, coordinator of the extension programs IndLab, Cartografia das Lutas e Urbanismo Biopolítico and coordinator of the INCT Technopolitics: urbans territories urbanos and digital network, Belo Horizonte, Brazil

https://doi.org/10.1515/9783110584455-001

discussion, it is important to consider the current political and economic context, or as the French philosophers Dardot and Laval [6] would say, the "rationality" currently in vogue. What is understood as "government rationality" is the form of government in central countries that, until recently, used to associate public goods to citizenship and political democracy, maintaining a market logic focused on consumer practices. Such a change has radically affected the democratic game in such a way that philosophers have affirmed we are now living in a "post-democratic age." These same thinkers have stated that the "entrepreneurial subject" is the central figure in this new rationality, since this individual is no longer someone whose willingness is affected by the appeal of consumerism and competition. This subject, however, is now also a producer of "being in the world" based on efficiency and individual accumulation: "The desiring being is not only the point of application of this power; it is the relay of apparatuses for steering conduct." [6, p. 327]. In order to counterbalance the neoliberal rationality, Dardot and Laval bet on the construction of a different world reason – a "counter-conduct," whose way of acting might be "indivisibly a conduct toward oneself and a conduct toward others" [6, p. 400].

De Certau [3] says that there is always something that escapes from the power and control of the dispositives identified by Foucault [9, 10]. They would be, according to De Certeau, the ways of the everyday practices, whose ability of subversion derives both from its silent operations and from a form of action that works in the gaps of the hegemonic system. Within this understanding, Deleuze [7] goes further and states that the resistance can be engendered in another direction, apart from the relation between power and knowledge, which works simultaneously to the hegemonic relations without necessarily annulling them, but rather preventing an impasse. Pelbart sums up this dynamic and finds a term that explains the positivity of this "biopotent" force, which is the power of life [16]. Facing those conceptions, we are interested in investigating the possibility of mechanisms that articulate the resistances (or powers) in order to produce a conduct, or a counter conduct.

Through such approach, this article aims to develop the concept of biopotent social technologies in the production of space in self-built urban occupations,[1] though not as something to be implemented by technicians or intellectuals onto vulnerable social groups, aiming at increasing their capacity, qualifications or even awareness. Instead, our approach will be a process of adjustments aiming at the production of

1 Nowadays, in Brazil, the occupations of idle lands or abandoned downtown buildings, organized by social movements for housing struggle, are a strong phenomenon and constitute a resistance to an excludent land organization and to institutional politics, which are unable to break with this power system. When the choice of the area is planned, it is usual to opt for tracts of land and/or lots with legal issues, with the clear intention of exposing a weakness in the urban land system, considering the primacy of housing rights over the right to property in the Brazilian Federal Constitution of 1988. By occupying the grounds, it becomes immediately evident that the land is not fulfilling its social function as it should, also according to the Federal Constitution [14].

new subjectivities, which target not only people from vulnerable territories, but also academic and technical public officials, here understood as agents in a network of relations.

There is an understanding that the favelas and self-built occupations are a consequence of the excluding modernization. In our opinion, despite not being totally false, such a statement simplifies and erases the subtleties of a more complex procedure than expected. It is possible to identify actions and subversions raised in these territories, which can also be characterized as biopotent actions within cities, considering that they are engendered in the gaps of established power, denouncing the very fragility and the contradictions within institutions [15].

Thus, even if the hegemonic ways of building space and housing are being reproduced, the everyday inventions are present, whether by forced circumstances or by the immanent power of life. This perception raises some questions: do not these people used to producing their own space have some relevant knowledge to solve their own issues? Are not the well-intentioned technicians, like the engaged individuals or even the critics, attached to a place of knowledge shaped by predefined values? Can the very place of precariousness and lack also be a place of biopotent invention and creation? Does academia always have a pedagogical project embedded in it, whether through science and humanism or through social-political consciousness, which tends to disregard local knowledge?

2 Biopotent social technology

To answer these questions, we mobilized two important concepts to the university[2] extension production: social technology and biopower. Reappliable social technologies, according to Lassance and Pedreira, can be defined as “a set of techniques and procedures, associated with forms of collective organization, that represent solutions to social inclusion and improvement in the quality of life” [12, p. 66]. Also according to Bava [2]:

> “Social technology is a set of transformative techniques and methodologies, developed through interaction with the population, which represent solutions for social inclusion. From this perspective, innovative experiences can be assessed and valued simultaneously for their dimension of building processes for new paradigms and new social actors, strengthening democracy and citizenship, and for the results they provide in terms of improvement in the quality of life.” [2, pp. 106–107]

2 The university extension programs in Brazil follow the constitutional principle of indissociability between education, research, and extension, and for that reason represent “an interdisciplinary, educational, cultural, scientific, and political process, which promotes the transforming interaction between the university and other sectors of the society.” (Forum of Extension Pro-Rectors of Brazilian Public Universities – XXVIII National Meeting, 2010).

Considering the main guidelines for the production of university extension programs and the idea of social technology, we believe that it is mostly through the exchange (without hierarchy) between erudite and popular knowledge, and between the university and the community, that social technologies can become biopotencies, according to the concept presented by Pelbart [16]: BAVA

> "Because it is always about life, in its dimension of production and reproduction, that the power invests, and that is, however, the background from which the counterpowers, the resistances, the lines of escape emerge. Hence the persistent presence of the prefix bio in that conceptual variety. Biopower as a general rule of life domination, biopolitics as a form of domination of life which can also mean, at the opposite side, an active resistance, and biopotency as a crowd's potency of life, including the vital work, the power of action, the potency of self-valuation that surpasses itself, the constitution of an expansive communiality (...)" [16, p. 86]

University extension processes can work as a counterhegemonic struggle strategy, through alliances that become possible because they are based on common denominators and goals, mobilized to produce positive and purposeful actions [18, p. 107].

> "More than the ability to implement solutions to certain problems, [those actions] can be regarded as methods and techniques that drive the empowerment of collective representations of citizenship, in order to enable them to fight in public spaces for the development alternatives originated from innovative experimentations, which are guided by the interests of the majority and through the distribution of income." [2, p. 116]

We thus propose the concept of a social biopotent technology, for which the daily inventions from the intervention territory are the starting point. Academic knowledge lies side- by-side with local nonacademic knowledge, and the role of the researcher or technical official is that of an actor whose actions are not neutral. This is why they require, at the same time, a transparent political position. This understanding is connected with the concept of an "actor-network," bringing a simultaneity: the actor is submitted to power forces present in the network, while also interfering and acting on those forces.

> "It is no coincidence that this expression, as a character, was taken from the stage. Far from indicating a pure and simple source of action, both refer to puzzles as old as the very institution of theater (...). To employ the word "actor" means that it will never be clear who or what is acting when people act, because the actor, on the stage is never alone in acting." [13, p. 75]

This concept is, therefore, a relational idea of what an actor is, recognizable through one's action in the network, that at the same time considers this action uncertain and displaced. Thus, in the construction of a social biopotent technology, it is necessary to recognize that all those people involved in the extension activities constitute a network in constant movement.

By admitting the value and the power of the counter conducts engendered in subversive everyday practices, including those concerned to the production of space, the challenge for researchers and intellectuals is building resources and instruments that do not reproduce the knowledge-power logic, in which knowledge is located at and restricted to the academic environment. This concern leads us to Foucault's [9, 10] warning of 30 years ago, which remains current:

> "Well, what the intellectuals have found out recently is that the masses don't need them in order to know; the masses know perfectly and clearly, much better than they do; and they can say it very well. But there is a power system that blocks, bans, and invalidates such discourse and such knowledge. A power not only found in the superior courts of censorship, but that penetrates very deeply, very subtly all of the social fabric." [9, 10]

Cartography as a methodology considers research as an intervention device, producing events open to the unpredictability of action. Therefore, the confluence between the researcher and the subject of research will, necessarily, bring destabilization, causing the production of new knowledge and subjectivities. Through this approach, the alternating movement of the observer/researcher, which sometimes goes toward the process that he or she intends to analyze, and sometimes moves away from it, destabilizes the divide between subject and object. This divide involves all the political subjects in the process, with voices and knowledge to be shared, and which are then open to transformation. Thus, it is about an availability to the unknown, which leads us to a process of constant deterritorialization.

3 Urban occupations of Barreiro

In order to develop the proposed concept of a biopotent social technology, we will rely on an action front of two associated extension projects included in the Grupo de Pesquisa Interdisciplinar,[3] "Artesanias do comum"[4] [Crafting the commons] and the "Natureza urbana"[5] project. This front refers to the "Parque das Ocupações do Barreiro" [Barreiro Park of Occupations], based in Belo Horizonte, Minas Gerais. It is relevant to present, first, how the process of self-built urban occupations has been happening in the municipality and how the Park (Parque das Ocupações do Barreiro) is included in this context.

In the city of Belo Horizonte, the process of organized urban occupations began from a "dissidence" within the participatory process for housing developed under the management of Mayor Patrus Ananias. Understanding the offer under discussion was insufficient, a group organized the first occupation, called Corumbiara, in 1996. It was

3 Interdisciplinary Research Group.

4 http://naturezaurbana.indisciplinar.com/artesanias-do-comum/

5 http://naturezaurbana.indisciplinar.com/

after 2010, however, that the occupations in Belo Horizonte began to happen in a more assertive way.

It is important to emphasize that, when an area is chosen by social movements for housing struggle, the chosen area often presents legal issues. This choice intents to expose the weakness within the urban land system, considering the primacy of housing rights over the right to property in the Brazilian Federal Constitution. The presence of technicians in these processes is not an indispensable requirement for occupations, but it is deemed positive by the members of the group. They can be present even before occupations become effective, when choosing the area to be occupied and dividing it into lots, as it was the case for the occupations, Eliana Silva Paulo Freire, in the region of Barreiro, in Belo Horizonte. These occupations were organized by the Movimento de Luta em Bairros, Vilas e Favelas (MLB) [Struggle Movement in Neighborhoods, Villages, and Favelas],[6] whose coordinator, Leonardo Péricles, sought the support of the School of Architecture of UFMG in 2012. Such support came initially through a partnership with project Diálogos [Dialogues], from the research group PRAXIS,[7] and in 2015 a new partnership was formed with the extension projects Artesanias do Comum and Natureza Urbana, both associated to the Indisciplinar research group (see Figure 1).

4 Parque das Ocupações do Barreiro

In large Brazilian cities, inserted in a neoliberal context, we can highlight two guidelines that aim at increasing justice in the city: the struggle for housing rights and the struggle for environmental preservation. In the hegemonic narrative, those questions are opposed, or at the very least, they do not join forces. This is because when the green has not been suppressed from the city by the formal urbanization logic, it is concentrated in the territory occupied by the upper middle class, often in response to the tension caused by environmental movements, organized, in general, by middle class groups. Because of such conflict, we pose the following question: could there be a relation of coexistence between man and nature, in which man cares for and benefits from nature at the same time? Although hegemonic discourse, which promotes market interests, is against this possibility, we started from the premise that those questions can be complementary and mutually reinforce one another. The real opponent of those movements is the voracity of real estate capital that, in a recurring way, creates justifiable grounds and strategies to prioritize policies that exclude and ignore not only the environment, but also culture, and any possibility of housing for the poor that is effectively included in the city. With that in mind, we started building the narrative for "Parque das Ocupações do Barreiro," in 2015. In that year, there

6 https://www.mlbbrasil.org/
7 http://praxis.arq.ufmg.br/

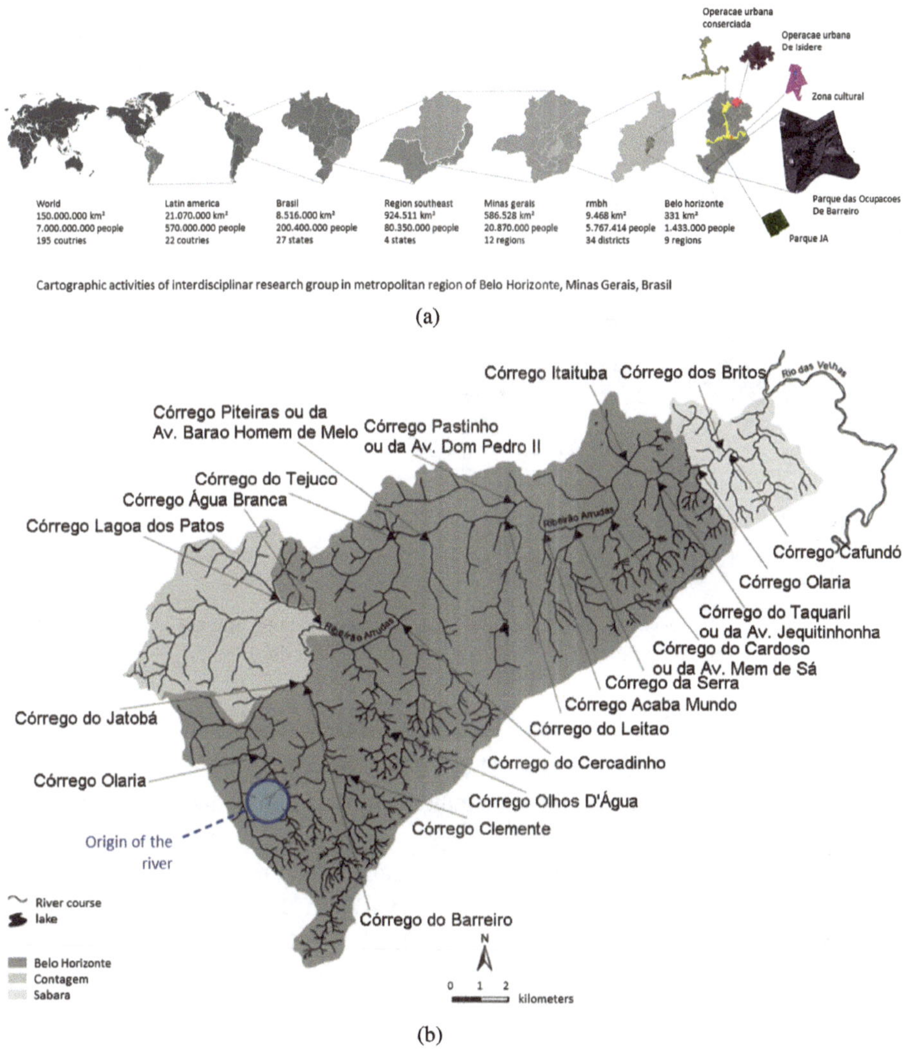

Figure 1: Geographic context of the urban occupations from Barreiro and its location in the Arrudas River's basin. (Source: Indisciplinar's collection and Belo Horizonte government base map (modified)).

was a meeting between Leonardo Péricles (MLB) and Professors Natacha Rena and Marcela Brandão – members of the Indisciplinar research group and professors at the School of Architecture and Urbanism – UFMG (Escola de Arquitetura e Urbanismo – UFMG) – in which the importance of inserting the environmental agenda in housing struggle was discussed, considering the proximity of Barreiro's occupations to a large environmental preservation area. In that meeting, the name "Park of Occupations" ("Parque das Ocupações") emerged, thus initiating a process of building the imagination around the idea of the park, considering it to be from the beginning the entire environmental preservation area and all the urban occupations (see Figure 2).

Figure 2: First visit to the occupation Paulo Freire, in 2015, when teachers and social movement leaders discussed the occupation process. (Source: Indisciplinar's collection).

One of the first repercussions from this was the participation of Leonardo and Poliana, leaders of the MLB, in a meeting hosted by Rede Verde,[8] which at the time was formed by the most relevant environmental movements in Belo Horizonte and surrounding area (see Figure 3).

The second repercussion took place in February 2016, during the event Verão Arte Contemporânea [Contemporary Art Summer] (VAC)[9] in Belo Horizonte. The event, which already belonged to the official city calendar and included Indisciplinar research group in the curation of the debates about architecture and urbanism had in

8 Rede Verde [Green Network], according to its official page, "[...] emerges in Belo Horizonte through the connection of different environmental, social and cultural movements that involve collaborative actions in defense of Mata do Planalto [Planalto Woods], Parque Jardim América [Jardim América Park], the Ficus trees on Bernardo Monteiro Avenue, Serra do Gandarela [Gandarela Mountains], Parque Lagoa Sêca [Lagoa Seca Park], Isidoro's Region."

9 The event "Natureza Urbana e Produção do Comum" [Urban Nature and Production of Commons] was one of the initiatives of the Indisciplinar research group, in partnership with VAC 2016, and joined collectives that struggle for Urban Nature and Heritage Conservation and groups that advocate for housing rights for all. The curation and organization of the event "Natureza Urbana e Produção do Comum" by Indisciplinar researchers (Ana Isabel de Sá, Luciana Bragança, Marcela Brandão and Natacha Rena) happened between days 01 and 02 of February 2016. Indisciplinar researchers, activists, and militants of many different movements were present. In addition, a visitation circuit to the Barreiro Occupations (location of the Park of Occupations) and the Parks Jardim América – BH and Mata do Planalto – BH, was held. https://www.facebook.com/naturezaurbanavac2016/?ref=br_rs

Figure 3: Leonardo and Poliana attend to Rede Verde's meeting, after a formal invitation from the group participants. (Source: Indisciplinar's collection).

that year, the theme "Natureza Urbana e a Produção do Comum" [Urban Nature and the Production of Commons]. Several social movements, with different agendas – struggle for housing, environmental preservation and historic and cultural heritage preservation – were invited to join a round table and a circuit at the urban occupations of Barreiro. This aimed precisely at putting into practice a reflection about the conflict between the struggles for housing and environmental preservation through the case of "Parque das Ocupações do Barreiro." One can say this was the first moment during the event in which the agenda of struggle for housing for a low-income population was included in the environmental discussions (see Figure 4).

Figure 4: Participants from the VAC event visits the Paulo Freire occupation as part of the activities and discussions promoted during the event. (Source: Indisciplinar's collection).

Afterwards, still in the first semester of 2016, the subject of the park was taken inside the School of Architecture and Urbanism of the Universidade Federal de Minas Gerais (UFMG), through the projects course “Parque das Ocupações do Barreiro.” The course began with collective mapping, in order to produce a shared cartography with local residents, to lend visibility to everyday practices in action at the territory of the occupations. Through that cartography, students will create proposals, incorporating the existing potential and providing answers to the weaknesses they found (see Figure 5).

(a) (b)

Figure 5: Illustrations of project proposals made by students during the course Parque das Ocupações, taught at the School of Architecture – UFMG. Students, teachers, and occupation residents worked together discussing urban and architectural projects. (Source: Indisciplinar’s collection).

As the course came to an ending, the park’s landscape and architecture project was further developed by the group of teachers and students who were part of the extension projects “Natureza Urbana” and “Artesanias do Comum,” resulting in a notebook named “Caderno Parque das Ocupações” [Park of Occupations Notebook]. The notebook was handed to the coordination of the MLB in March 2017 and used as an instrument during negotiations with municipal public officials (see Figure 6).

Figure 6: Parque das Ocupações notebook. (Source: Indisciplinar’s collection).

In 2017, the park project became the major action front for the mentioned extension projects. Their intention then became not only to reflect, but also to consolidate the counter-narrative under construction, having as their target audience the technical public officials, as the ones from the Companhia Urbanizadora de Belo Horizonte – URBEL [Urbanization Company of Belo Horizonte] and the Companhia de Saneamento de Minas Gerais – COPASA [Sanitation Company of Minas Gerais], in charge of sanitation at the municipality.

As an effect of the intersection between academia, social movements, and public authorities, it should be noted that, in February of the same year, the project was inserted in the agenda of the Subcomitê da Bacia Hidrográfica do Ribeirão Arrudas [Subcommittee of the Arrudas Stream Drainage Basin], thanks to the participation of Professor Luciana Bragança (coordinator of the "Natureza Urbana" project) and of Cristiano Abdanur (COPASA technician) as representatives in this subcommittee. Therefore, the park project could participate in two open bids, one for the preservation of one of the springs in the area and the other for planting trees on the occupations' streets.

It is our understanding that the discussion about the hybridization between the agendas of housing struggles and environmental conservation also needs to be extended in the academic environment. For that reason, Parque das Ocupações do Barreiro has been the subject of articles presented in conferences like contested cities, in Madrid (Spain) and Arquisur, in Santiago (Chile), as well as in events like 4ª Semana do Meio Ambiente do Instituto Federal de Educação Ciência e Tecnologia de Minas Gerais (IFMG), in Santa Luzia (Brazil); Transformar Cidades a Muitas Mãos at the Escola de Arquitetura e Urbanismo – UFMG, in Belo Horizonte (Brazil); Cidades Invisíveis, in Goiânia (Brazil) and in the meeting BRAGFOST 2017 (Germany) (see Figure 7).

Figure 7: Dr. Marcela Silviano Brandão Lopes presenting the Parque das Ocupações do Barreiro at the socio-environmental science class from UFMG.

In addition to that, entering the proposal into architecture and urbanism competitions, like the one hosted by the VI Bienal de Sustentabilidade José Lutzemberger (VI José Lutzemberger Sustainability Biennial) enabled Parque das Ocupações to enter into another territory of narrative dispute about a project methodology that aims to expand the experiences already present in the territory. The tree planting proposal, for example, did not follow the commonly used criteria for tree selection. Heading into another direction, the group in charge of the project opted for an openly political choice of trees, based on three criteria. The first refers to native trees for the Permanent Protection Area (PPA) in order to recompose vegetation that was suppressed in the park, not only by the occupations, but mostly by industrial plants located in the region. In second place, the so-called "hardwood trees" were included, considering that their felling is regulated by environmental legislation, thus hampering their suppression. They were planted on occupations' main roads, which allow for tall vegetation. Finally, the proposal included "affection trees" ("árvores dos afetos"), fruit trees planted on small streets in order to establish an affectionate bond with the residents through smells and flavors already present in their previously mapped everyday life and stories (see Figure 8).

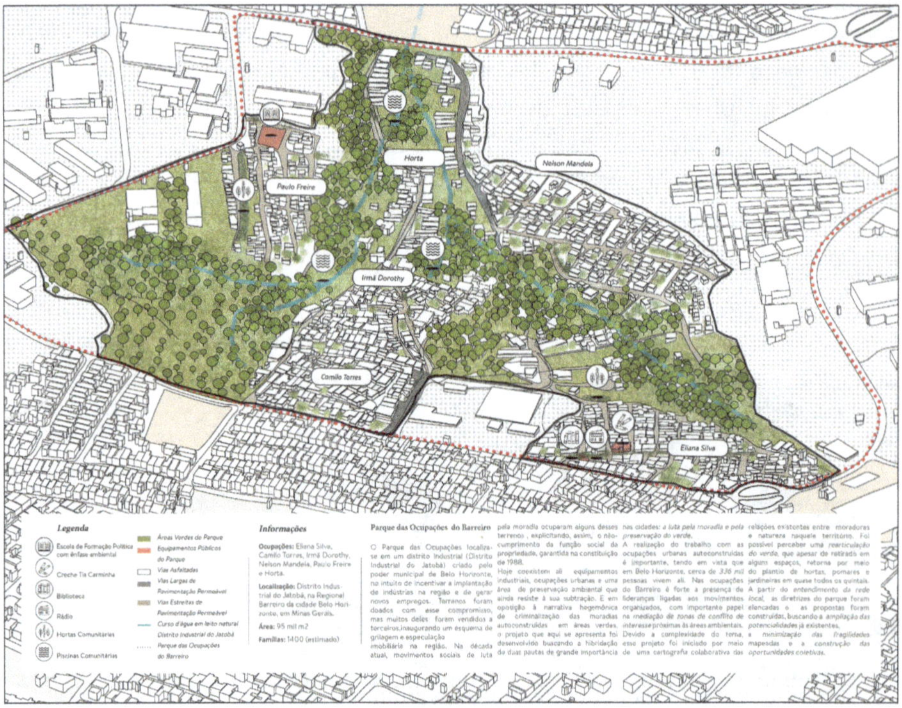

Figure 8: Proposal for the VI Bienal José Lutzemberger competition, in which sustainable strategies were brought together to the park's project and discussions. (Source: Caio Nepomuceno, Marcus Barbosa, Mayumi Amaral and Octavio Mendes).

The axis dedicated to the waters in the occupations presents similar diversity to the tree planting proposal, adopting different solutions for each different water issue. The preservation of the three water springs found in the area was prioritized and the creation of natural pools using the river courses that permeate the territory was meant as a way to promote recreation for children and adults alike through a public space of contact with nature. The proposal was inspired by the reports of a water pit created by one of the residents, that ended up being seized by the children as a swimming space on warm days (see Figure 9).

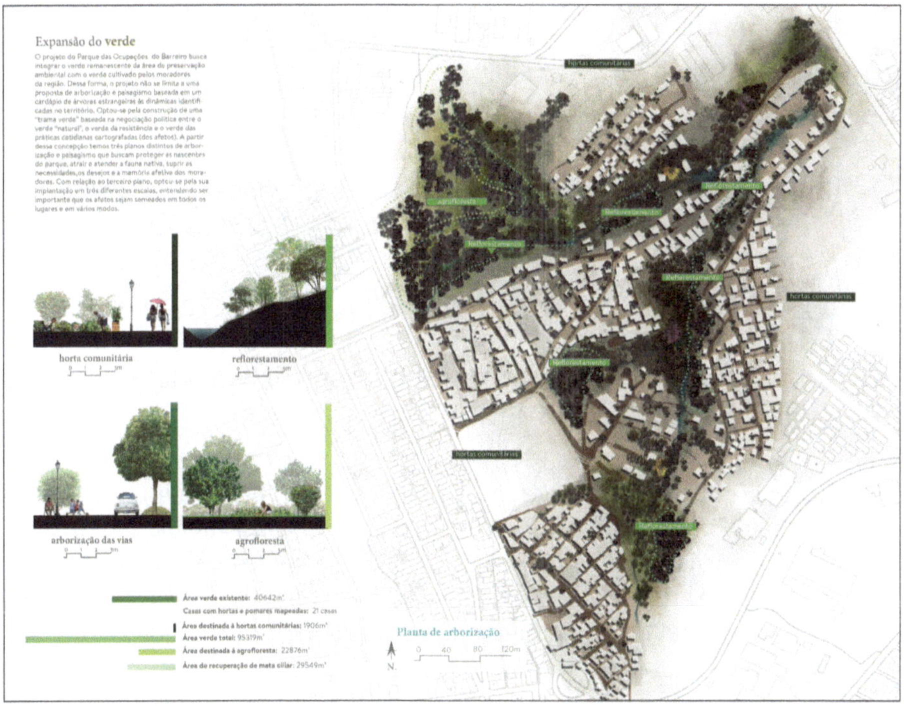

Figure 9: Landscape project for the Parque developed during the VI Bienal José Lutzenberger competition.

For the roads, different drainage, paving, and equipment solutions were adopted, depending on their specificities (gradient, width, and access). Following this logic, the permeability of interlocked paving was chosen for the streets that allowed for this type of material, while the usual asphalt pavement became restricted to the streets with a heavy flow of vehicles. This suggestion, along with the proposition of urban equipment to compose the sidewalks, aims to maintain the system of shared streets that exists in the occupations, and that is usually lost in formal cities, where motor vehicles are prioritized (see Figure 10).

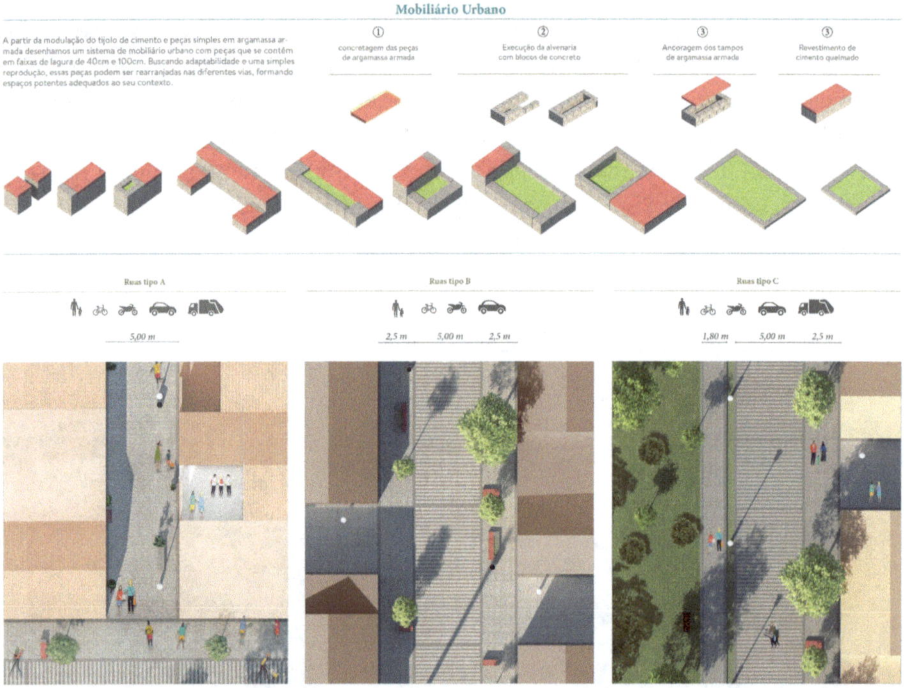

Figure 10: Urban Equipment project developed during the VI Bienal José Lutzenberger competition. (Source: Caio Nepomuceno, Marcus Barbosa, Mayumi Amaral, and Octavio Mendes).

In the course of "Visual Communication of Buildings and City" (Comunicação Visual do Edifício e da Cidade), in the second semester of 2017, house numbers for occupation Paulo Freire, along with a park sign, aiming to consolidate the park narrative, were developed. The narratives of nature were created from the elements of water, cultivation, fauna, and housing struggle. The sign will be inserted in the territory in 2018 (see Figure 11).

Empowerment of local residents was amplified in 2018, especially due to the arrival of a bus donated to the cause, through a partnership between Indisciplinar and a local company. The transformation of the bus into an itinerant space for events, lectures, and meetings strives for the production of caravans for educational and political intents, throughout the occupations coordinated by the MLB.

5 Concluding remarks

The concept of a biopotent social technology presented in this work assumes that everyday inventions present in the socially vulnerable territories have a subversive value, in the way that they are engendered in the gaps and in the absence of the power

Figure 11: Sign developed by Design and Architecture students from UFMG for Parque das Ocupações. The design was based on the work and discussions between students, residents, and members of social movements. (Source: Luciana Bragança).

and control devices enforced by the agents of a hegemonic space production, currently established by the neoliberal rationality. With that in mind, we propose for such inventions to be regarded as the starting point for any intervention in these territories and also that, along with the existing problems, they be mapped out and made visible, in order to avoid disregarding the already constructed solutions. In other words, it is not about a "romanticized view" of poverty, but rather about recognizing the existence of potent non-academic knowledge. Apart from mapping, construction strategies for new agents need to be planned. However, it is fundamental that such a plan remains open and flexible. Cartography was then presented not just as a method for investigation, but also, and maybe mostly, as an understanding that scientific neutrality does not exist, and that academic research is a political act.

> "A gesture is revolutionary not by its own content, but by the chain of effects that it engenders. It is not the intention of the authors, but rather the situation that determines the direction of the act. It is by the meaning acquired in contact with the world that an action is or is not revolutionary" [4, p. 175/176]

Therefore, under the presented assumption, the project "Parque das Ocupações do Barreiro" is being planned strategically. It is understood that every device in action from its naming, to the project courses in which guidelines and propositions are developed, the discussions in lectures and conferences, the participation in contests and its inclusion in the agenda of a subcommittee has produced different kinds of agency, between different units and instances of academia itself, between academia and social movements, between academia and sectors of political power. These interactions have had varied effects and ranges, articulating actors and territories in the most different ways.

> "A movement is only alive through the displacements it operates over time. It is at all times, therefore, a certain distance between its state and its potential (...) The crucial gesture is the one that

> lies one step ahead of the state of the movement and that, breaking through the status quo, gives access to its own potential. Such a gesture can be that to occupy, to crush, attack, or just to speak within truth; it is the state of movement that decides. The true question for revolutionaries is to enhance the living power of which they are a part, to care for revolutionary duties with the intention to get to a revolutionary situation." [4, p. 176]

This process brings us back to the idea of molar and molecular micropolitics, proposed by Guattari and Rolnik [11], for which it is necessary to enter the domain of subjective economy. This proposal is very up to date, once we consider the urgency in building new conducts.

> The micropolitical question, that is, the analysis of formations of desire in the social field has to do with the way in which the level of broader social differences (which I call "molar") intersects with the level that I call "molecular." (...) It is no longer a question merely of reappropriating the means of production or the means of political expression, but also of leaving the field of political economy and entering the field of subjective economy. [11]

However, the seams that have been sewn are dedicated and subject to constant ruptures and, for that reason, it is necessary to always be alert in order to promote new agencies, in a continuous and nonlinear movement.

Bibliography

[1] Agamben, G.: O que é um dispositivo? Florianópolis: Outra Travessia, n. 5, pp. 9–16 (2005). Available in: Access in 07/03/2018.
[2] Bava, S. C.: Tecnologia social e desenvolvimento local. Fundação Banco DO Brazil (org.) Social technology: a strategy for the development. Rio de Janeiro (2004).
[3] Certeau, M.: A invenção do cotidiano: 1.artes de fazer. Vozes, Petrópolis (1994).
[4] Invisivel, C.: Aos nossos amigos: crise e insurreição, n-1 edn.: São Paulo, (2016).
[5] Costa, A.: Tecnologia Social e Políticas Públicas. Instituto Pólis, São Paulo, Brasília, Fundação Banco do Brazil (2013).
[6] Dardot, P., Laval, C.: A nova rasão do mundo. Boitempo, São Paulo (2016).
[7] Deleuze, G., Guattari, F.: Mil Platôs. Cap Vol. 1. São Paulo: Editora 34 (2000).
[8] Foucault, M.: Vigiar e Punir. História da violência nas prisões. Editora Vozes, Petrópolis (1977).
[9] Foucault, M., Deleuze, G.: Os Intelectuais e o Poder. In: Foucault, M. (ed.) Microfísica do Poder. Edições Grall, Rio de Janeiro (1979).
[10] Foucault, M.: Microfisica do Poder. Edições Graal, Rio de Janeiro (1979).
[11] Guattari, F., Rolnik, S.: Cartografia do Desejo. Editora Vozes, Petrópolis (1986).
[12] Lassance Júnior, A. E., Pedreira, J. S.: Tecnologias sociais e políticas públicas. In: Tecnologia social – uma estratégia para o desenvolvimento. Fundação Banco do Brasil, Rio de Janeiro (2004).
[13] Latour, B.: Reagregando o Social. EDUSC/EDUFBA, Bauru/Salvador (2012).
[14] Lopes, M., Bragança, L., Sá, A., Rena, N.: Natureza Urbana e Produção do Comum, Indisciplinar Group: Method, Activism and Tecnopolitics in Defense of Urban Commom Goods (2016).

[15] Lopes, M. S. B.: Artesanias Construtivas e Urbanas: por uma tessitura de sa-beres. Doctoral Thesis in Architecture and Urbanism. School of Architeture and Ur-banism of UFMG, Belo Horizonte. Available at http://www.bibliotecadigital.ufmg.br/dspace/handle/1843/BUBD-9WRGLR. Access in 07/03/2018.
[16] Pelbart, P. P.: A Sociedade dos sem Sociedade. Editora Iluminuras, São Paulo (2011).
[17] Oliveira, F.: O vício da virtude. Autoconstrução e acumulação capitalista no Brasil. São Paulo: Novos Estudos CEBRAP, n. 74 (2006).
[18] Rena, N.: Extensão como Resistência: ativando a biopotência do coletivo. In: Estudos avançados de direito à moradia. Belo Horizonte: Arraes Editores.
[19] Santos, B.: A filosofia à venda, a douta ignorância e a aposta de Pascal. Critical Magazine of Social Science, número 80, pages 11–43, March 2008. Available at http://www.ces.uc.pt/myces/UserFiles/livros/47_Douta%20Ignorancia.pdf. Access in 27/03/2012.

Bernhard Lenz and Roberto Zanetti Freire

Performance potentials: the optimization of buildings in operation

Abstract: Many buildings – such as administrative, research and high school buildings which were designed and built in the 1960s deal with problems of thermal discomfort and a high level of power consumption. Due to the mentioned difficulties, this type of building is being observed closely. Since the buildings often show similarities in their inner structure, the following research project analyzes which problems can occur after a restructuring. Contrary to conventional restructuring concepts, which primarily deal with reducing the heat loss and to improve the comfort by upgrading the building technology, the following approach follows a much more integral and future oriented concept.

Keywords: Building energy consumption, renovation, air flow, water consumption

1 Exemplary building

The collegiate building of mathematics of the Karlsruhe Institute of Technology – KIT can be seen as representative for the type of building mentioned above. It is being intensively observed in the course of energetic monitoring with a following operational optimization. The building used to consist of an open ground floor with three raised (elevated) floors and a basement floor. The support structure was constructed as a ferro-concrete skeleton with concrete ceilings; the inner walls were completed as lightweight constructions.

The building demonstrates typical deficits considering the period of construction. Especially, the heating and power consumption were rated very high. The loss of heating energy resulted from a poor insulation standard of the main building shell, heat bridges and simply outdated building technology. Also, the building was not airtight. Heat bridges were found in the expected forms for this period of construction – uninsulated concrete supports in the facade surface. Also, uninsulated overhanging ceiling panels and beams were problematic issues. The share of heat bridges was at about 5 % of the transmitting building surface, but contributed to the total heat loss by about 12 % [1] (see Figure 1).

Acknowledgement: This project on which this report is based was funded by the Bundesministerium für Wirtschaft und Technologie (BMWi).

Bernhard Lenz, Hochschule Karlsruhe, Karlsruhe, Germany, e-mail: bernhard.lenz@hs-karlsruhe.de
Roberto Zanetti Freire, Pontifical Catholic University of Parana, Polytechnic School, Curitiba, Brazil; and Université Claude Bernard Lyon 1, Villeurbanne, France, e-mail: roberto.freire@pucpr.br

https://doi.org/10.1515/9783110584455-002

Figure 1: Outside view before and after the restructuring (left: [4], right: [5]).

In the course of the restructuring, the overhanging ceiling panels as well as the railing were scaled back to the interior. Like this, the supports and beams, which beforehand were in the facade surface were relocated to the inside. This extension not only enabled a gain in space, but also a costly and complex insulation could be left out.

Before the restructuring, the open ground floor was primarily to have access to the upper floors, which implicated a large area of the ceiling panels being in the open air. After the reconstruction, the whole ground floor is now indoors. As a consequence, the A/V-ratio (surface-area-to-volume-ratio) was improved and a complicated insulation of the ceiling slabs (floor panel of the first floor) and the ceiling panel of the basement floor was not necessary. The basement floor barely was provided with daylight. In the future, eight rooms in the basement will be improved by adding "light eyes" in the floor panels of the ground floor. As a result, less artificial lighting is needed (see Figure 2).

Figure 2: View of the atrium before and after the restructuring (left: [4], right: [5]).

The building technology was outdated by far and was not up to the state of the art. This caused a very restricted adjustment to the various needs and the thermal loads. The primary energy consumption for heating and electricity was approximately at 580 kWh/m^2, whereas the share for heating was at 350 kWh/m^2 [1]. Especially due to structural and planning deficiencies, there was a high thermal discomfort in the summer as well as winter, moderate room acoustics and a rather unsatisfactory air quality. Also, the aspect of daylight could be rated as unsatisfactory before the restructuring, which was especially traced back to the 7m long office rooms. This caused a low daylight factor in the rear half of the rooms. Also, the sun protection system was inadequate. The sun protection system consisted of an external, nonflexible horizontal blind system, which did not allow any adjustments depending on the outside lighting situation. As a result, neither a high heat input in the summer could be reduced nor a desired heat recovery in the winter could be optimized.

The restructuring of the building was supposed to offer a transferrable strategy to eliminate the deficiencies and reduce the primary energy consumption. Also, the load reserve was analyzed to examine whether an extension of the building is possible. As with many other similar buildings, this was the case, so within the architectural redesign the building was extended with a partial floor. The previously open atrium was enclosed with a roof (converted from a cold atrium to a warm atrium). The A/V-ratio was significantly improved and an additional inside area was created. The energetic standard of the building was also improved and the system technology was completely renewed. The goal of this measure was especially to increase the energetic performance of the building. A new energy efficient ventilation system for the seminar rooms should not only reduce the heating loss but also improve the air quality in the seminar and work rooms. The individual offices still receive the fresh air supply over the windows, whereas the window ventilation is part of the ventilation concept in the summer. An external controllable sun protection system shall improve the daylight quality and protect against high heat intakes. To offer thermal storage mass, which helps reduce an overheating in the summer, the inside ceilings were not designed as suspended ceilings.

With the restructuring, the legal requirements were supposed to be met and even dropped below them. The goal was to restrict the yearly primary energy consumption to a maximum of 100 kWh/(m^2 a). The legal binding reference value of the energy saving regulations (Energieeinsparverordnung – EnEV 2009) for modernized old buildings, which were valid at the time of the application, was at 225 kWh/(m^2 a). For new buildings, the reference value was at 160 kWh/(m^2 a) [2]. The threshold for modernized old buildings was going to fall below the legal values by more than 50 %. In the course of different simulations, a further improvement of the primary energy consumption was anticipated by adding further binding parameters (user profiles). With an individual adjustment of the profiles with regard to usage time, operating period of the ventilation, cooling and heating, room-temperature, the person related and area related air flow volume, as well as the times of full utilization and the specific internal loads a reduction of the primary energy consumption to only 75 kWh/(m^2 a) was predicted [2]. To assess the restructuring concept about 700 measurements were sorted and analyzed over a time period of 2 years. Some of these measurements were drawn from the central building control system, others over mobile temporary installed sensors. Long term measurements are checked in situ by meter readings and temporarily enhanced with measurement technology.

2 Energy concept

The restructuring concept was developed from an accompanying optimization of the primary energy consumption. In the course of the energetic assessment various variants for the heating and cooling supply were analyzed. The following variants were compared:

V1: Heating = district heat; Cooling = Absorption cooling

V2: Heating = heat pump (70 %); District heating (30 %); Cooling = well water V3:
Heating = district heating; Cooling = district cooling

The use of well water (V2) to cool the building could not be followed up, due to a slight soiling in the ground water of the property. The possible use of geothermal energy (V2) to operate a heat pump could not be realized because of confined space as well as the restricted drilling depth [2] (see Figures 3 and 4).

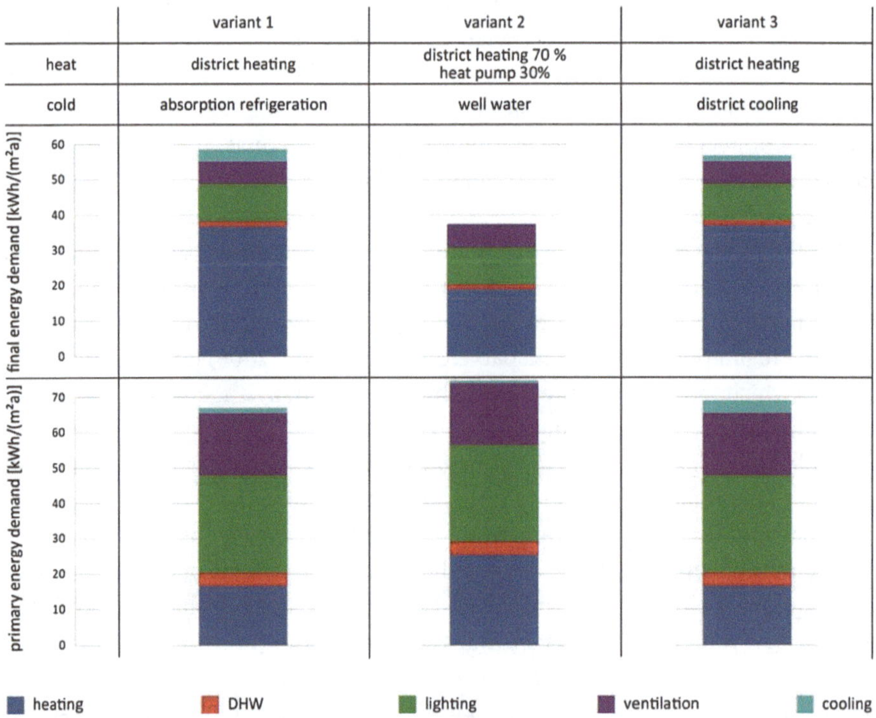

Figure 3: Comparison of the primary energy demand and the final energy demand of different energy supply systems [2].

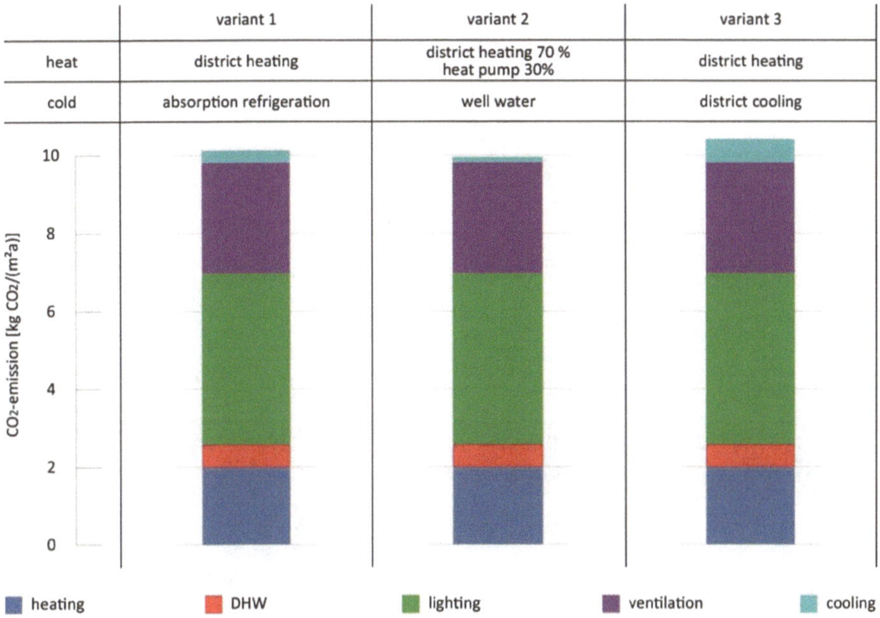

Figure 4: Comparison of the CO2 emissions of different energy supply systems [2].

The comparison of the remaining options V1 and V3 showed that both solutions show similar results concerning the primary energy demands and CO_2 emissions. Since the area already provided a district warming and cooling system and V3 showed economic benefits, this is the solution which was implemented.

3 Ventilation concept

The air exchange of the offices alongside the facade is enabled by a window. The inner offices, as well as the rooms of the ground and basement floor receive the supply air through a central ventilation system. Outgoing air from the upper floors is diverted to the atrium and extracted beneath the roof. So the atrium is used as an open overcurrent zone. This creates a system, which uses the room itself to transport air and there is no need for channels to transport the air. This substantially reduces the installation effort for ventilation channels. The exhaust air from the ground and basement floor are yet conventionally led over channels. The increased air exchange rate of the single rooms is achieved by a redistribution of the total air volume, without raising the actual total air volume (see Figure 5).

In the course of the energetic monitoring, it showed that this concept causes a massive loss of cooling recovery as the exhaust air from the seminar rooms heats up extremely, especially in the upper levels of the atrium. On a typical summer day, a

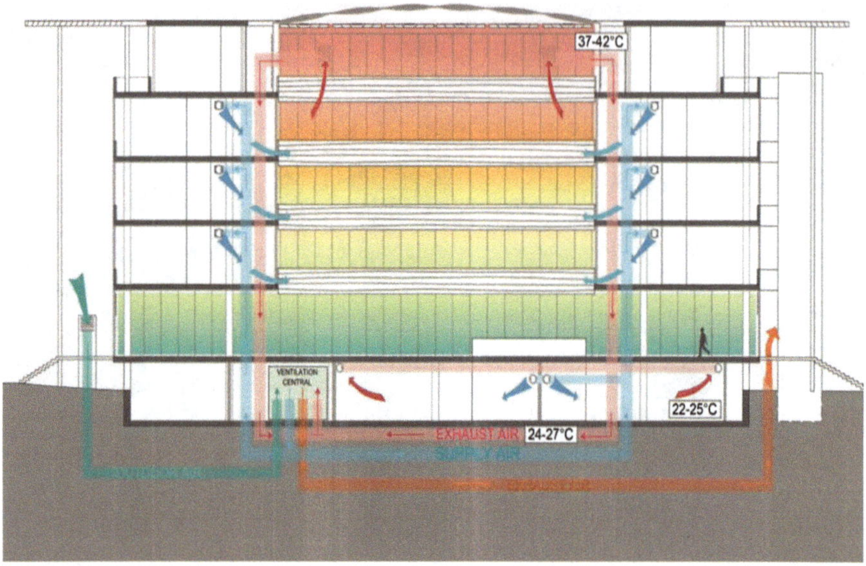

Figure 5: Ventilation concept in the summer with exemplary temperature values [3].

vertical temperature stratification builds up in the atrium. The extraction of the exhaust air takes place in the upper level of the atrium, just below the foil cushions of the ceiling. Before the air can be extracted, which is intended to precool the outside air, it heats up that much, so the temperature actually exceeds the outside temperature. This reheated exhaust air mixes with the significantly cooler air of the ground and basement floor before it can precool the warm outside air (supply air). Often this precooling is not possible due to the high outside temperatures (see Figures 6–8).

Resulting from the minimal cooling recovery (see Figures 7–8) is a very high need of district cooling during the summer time. The cooling recovery could be improved by redirecting the exhaust air from the atrium through the existing ventilation flaps. Like this, only the exhaust air from the ground and basement floor would be used to precool the supply air, which would help reduce the need for district cooling (see Figure 9).

Generally, the supply air volume flow matches the exhaust air flow. Since the air volume flow from the ground and basement floor is less than the total incoming air volume, the possible cooling capacity is reduced, but the need for total cooling is also reduced. The time frame for possible precooling is increased during the cooling period (see Figure 10).

4 Single zones

To ensure an optimal energetic performance, it was inevitable to implement customized climatic concepts in the various zones. The main challenge in planning this

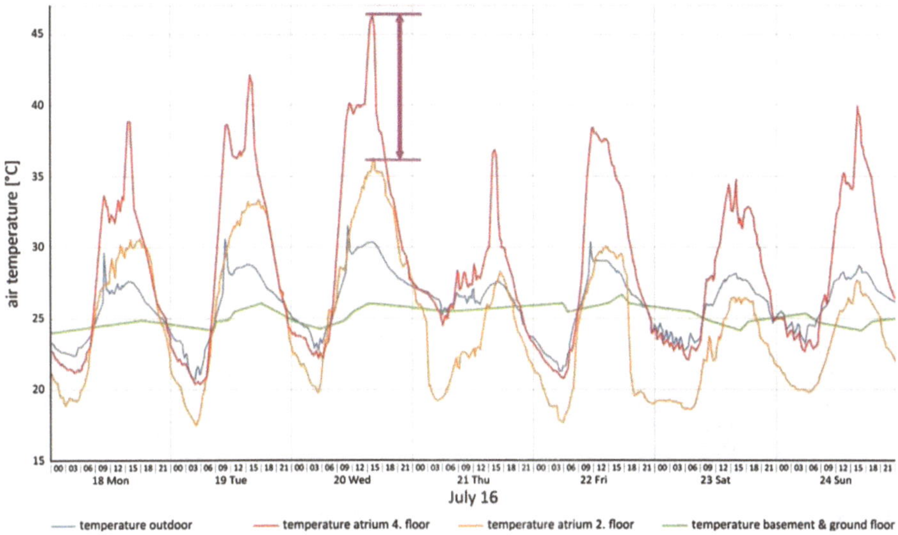

Figure 6: Air temperature profiles in different zones in the summer [3].

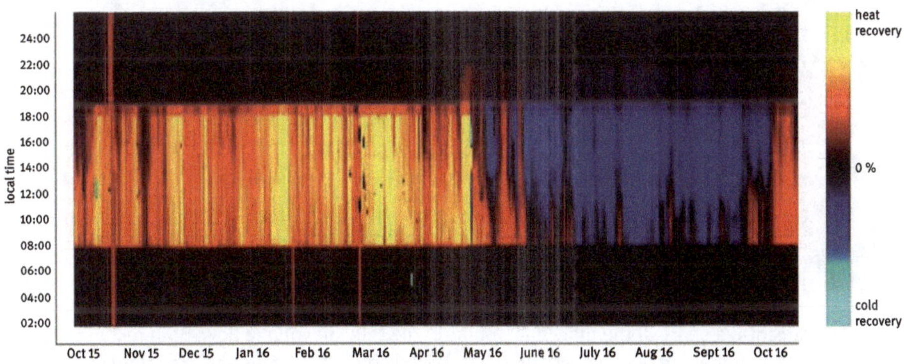

Figure 7: Proportionate energy recovery of the ventilation system over heat exchange [3].

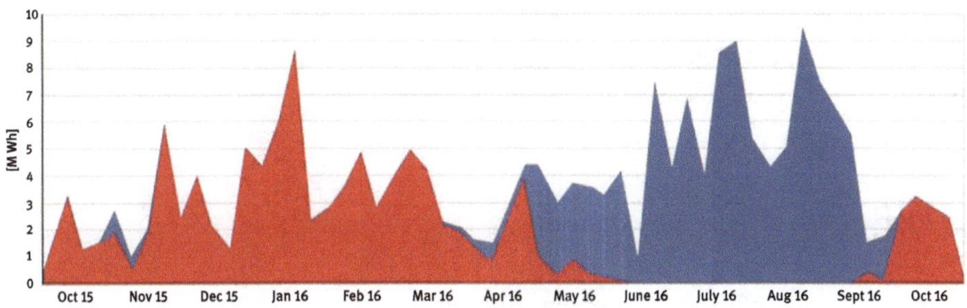

Figure 8: Consumption of district heating and cooling [3].

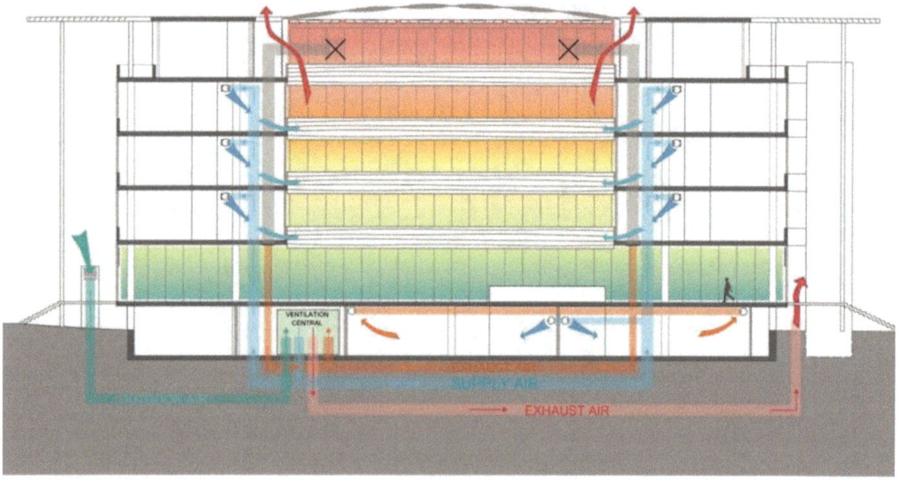

Figure 9: Suggestion for a better ventilation concept in the summer [3].

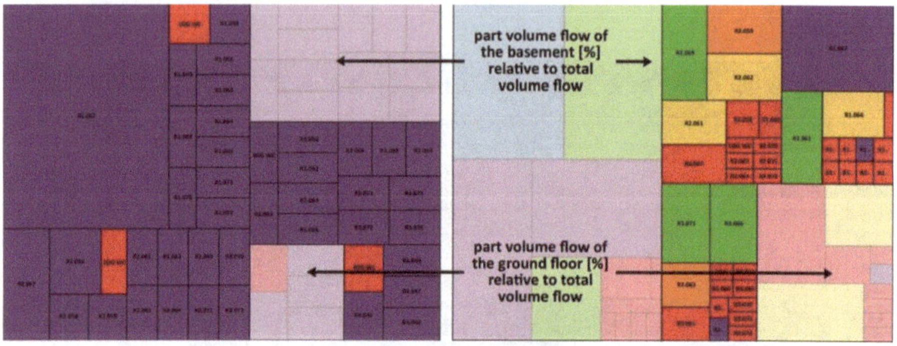

Figure 10: Percentage share of the individual rooms of the total air volume flow. Both rectangular areas show the supplied total air volume flow, which is divided differently. The grey shaded boxes show the share of the outgoing air from the ground and basement floor, which is used for precooling the supply air. Presentation of two different using times [3].

was to develop the dependencies of the subsystems in a way that they reinforce each other and do not have negative effects.

5 Offices

For the individual offices, a minimum solution was chosen with the lowest possible technical effort. The offices are neither actively cooled, nor supplied with additional air. The air exchange and the cooling of the offices takes place only over manually controlled opening flaps, since the restructuring exposed ceiling panels serve as thermal storage mass. Over a night-time ventilation, the absorbed energy can be emitted, which enables new storage capacities during the day time. The night air serves as a natural heat sink. This means that cooling does not actively have to be created, rather the ceiling panels function as a temporary storage, which can release the energy with a certain delay. A goal of this planning was also to enable a high thermal comfort for the usage time during extreme weather periods without having to actively cool the rooms (see Figure 11).

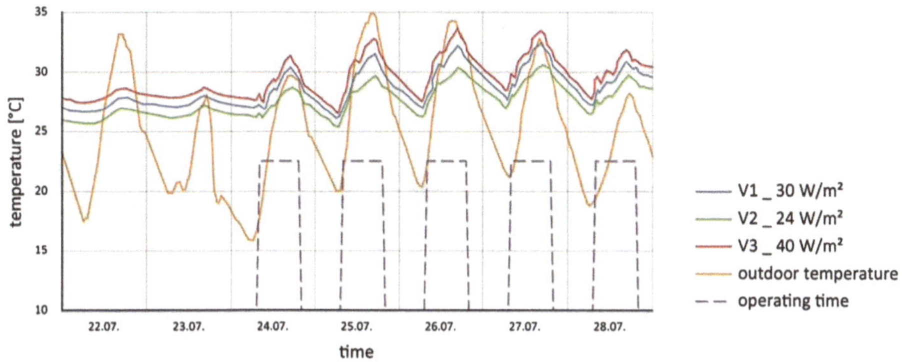

Figure 11: Felt temperatures at different thermal loads (technical equipment) during an extreme weather period [2].

Analysis of a southern office during extreme weather based on different loads:
Version 1: internal load with 30 W/m^2 (by AMEV*)
Version 2: internal load with 24 W/m^2
Version 3: internal load with 40 W/m^2 (by AMEV*)
* Work group machine technology and electrical engineering of state and municipal administration.

In the daily routine, it showed that the planned concept was not as practical as expected, since the office workers did not feel able to decide when to open and close the flaps. As a result, many poor decisions were made. If the flaps are not open long enough, the storage mass cannot emit enough heat to cool the room. If the flaps are open too long, it causes too extreme cooling of the room. This problem could be solved by installing a system supplement. A type of assistance for the ventilation strategy could be provided over the internet. This assistance could be based on predicted weather forecasts and under consideration of the spatial orientation. The number of poor decisions could be reduced.

As sun protection the offices are supplied with an outside system with a fc-value of 0.25 (cut-off position). This is controlled automatically. The system is activated when the solar radiation exceeds 200 W/m^2. The solar protection system consists of connected blinds. When the blinds are completely closed, additional lighting is necessary. Due to this, the energy consumption of the southern offices (higher need for sun protection) is much higher than in the northern offices (see Figure 12).

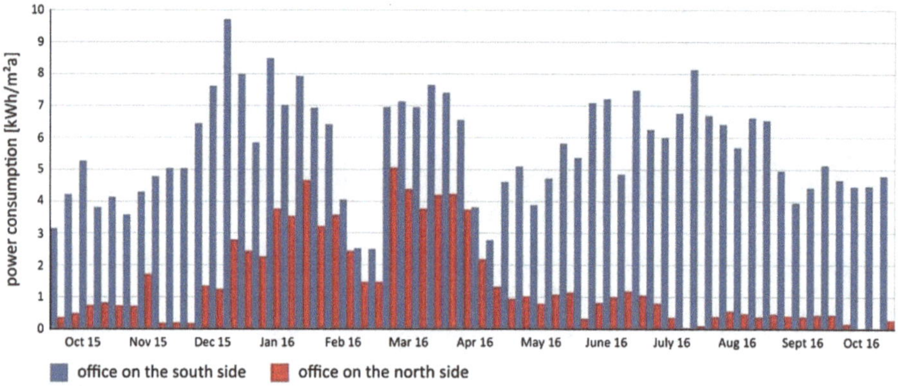

Figure 12: Electricity consumption for artificial lighting in the northern and southern offices [3].

If blinds with separated parts had been installed, the energy consumption of the southern offices would be lower than in the northern offices, because an overheating of the rooms would be prevented, but also natural lighting would be let in. The decision to install a connected blind system resulted in significantly higher energy consumption for artificial lighting.

6 Seminar rooms

The heating and cooling of the seminar rooms happens over the ventilation system. The room ventilation operates mechanical and should be controlled primarily over the CO_2-content and secondarily over the temperature. This means during low occupancy and, therefore, a low CO_2-content there is only a minimal air exchange. During high occupancy (high CO_2-content), the exchange raises (see Figure 13).

The size of the individual surfaces shows the percentage share of the total volume flow of the building. Over the total area at a 100 % performance of the ventilation system the maximum air volume that can be spread is at 68000 m^3. In graphic 13, it is evident that the volume flow regulator of the individual rooms let in nearly 80 % of the supplied volume flow, the leftover share is throttled. The colors of the in-

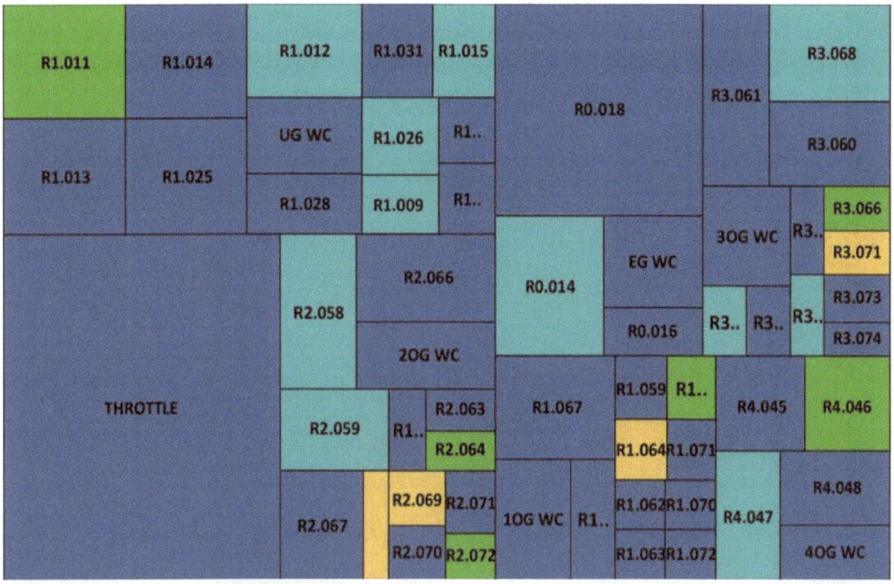

Figure 13: Graphic for indicating the different supply air volumes of individual rooms at a random time (snapshot). Different colors indicate the occupancy rate [3].

dividual rooms show the occupancy rate of the rooms (blue = not occupied, green = 50 %, yellow = 75 % occupancy). Although certain rooms are not occupied, a high air exchange rate is found. In direct comparison of the two identical rooms R-1.011 and R-1.013, it is seen that even at different occupancy rates there is a nearly identical air exchange rate. In room R-1.013, as well as other rooms a lower air exchange rate could be achieved by an optimized setting relating to the actual CO_2 value of the room (see Figures 14–16).

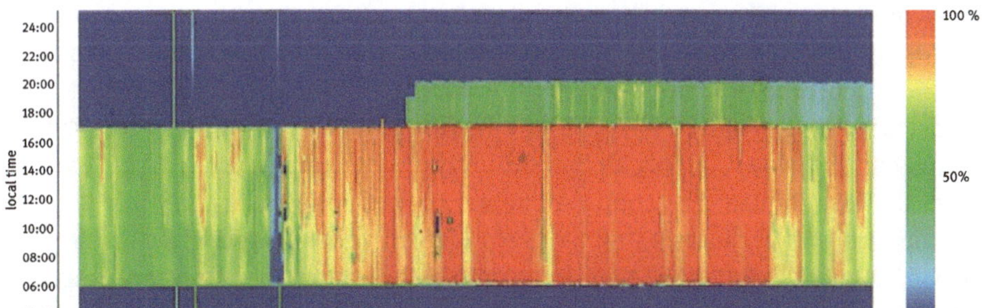

Figure 14: Performance of the ventilation system for the supply air – November 2015 to October 2016 [3].

Figure 15: Average outside temperature – November 2015 to October 2016 [3].

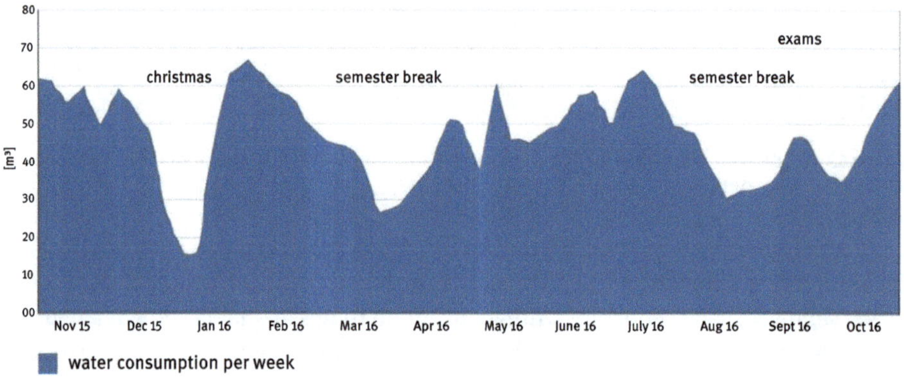

Figure 16: Water consumption as indicator for the occupancy of the building – November 2015 to October 2016 [3].

Comparing the operation conditions of the ventilation system (supply air) with the outside temperatures, it shows that an increase in the outside temperature and a rise in the supply air flow volume are congruent. Between the usage intensity, derived from the water consumption, and the supply air flow there is no correspondence. With a needs-based control of the supply air flow (regulated by the CO_2 value), there should be a significantly lower supply air flow during the nonlecture periods (august), which is not the case though. The supply air values are controlled especially by the room temperature, which causes an unnecessary high air volume flow, which results in a high energy consumption. To reduce the energy consumption of the ventilation system, the reference factor should be based more on the CO_2 value than on the room temperature.

7 Atrium

The atrium which was part of the outside (cold atrium) was enclosed with a foil roof and now functions as a thermal buffer zone with lower comfort needs to protect the inside from thermal discomfort. The sun protection closes when the insolation rises above 300 W/m^2. An overheating of the atrium shall be avoided by opening flaps. In the upper area, the opening surface is 55 m^2, in the lower part 45 m^2. Like this, the expected thermal lift can cool the atrium. Especially, a night time ventilation shall be used to remove the warm air.

Figure 17 shows that there are short opening periods of the ventilation flaps and the potential of the night-time cooling is not exhausted. The minimal temperatures in the third and fourth floor are still significantly over the outside temperature after several hours in the example.

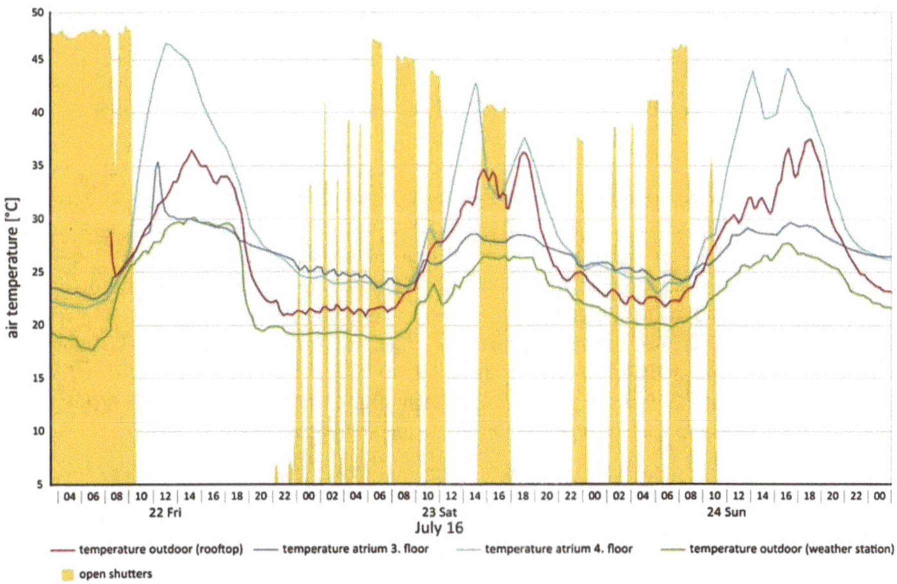

Figure 17: Opening times of the ventilation flaps and temperature levels of different areas [3].

Due to the settings for the threshold value of the ventilation flaps, they are only open for a short time during the night. It is clear that the temperature of the third and fourth floor does not continuously drop, rather it increases again after closing the flaps. This results from the stored energy in the structure, which is released to the air inside the rooms. An adaptation of the threshold settings in consideration of the stored energy potential would lead to a significant reduction of the temperature in the atrium in the night and also day time. Like this, an improvement of the thermal comfort (lower temperatures) could be achieved. If the temperature in the atrium was lower, also the adjoining seminar rooms would need less cooling due to lower thermal loads.

8 Conclusion

An energy optimized planning is the base for a high energetic building performance. As shown in the examples though, it becomes clear the theoretical conceptions alone are not a guarantee for a highly efficient management. Be the means of energetic monitoring, which is implemented within an accompanying research, a variety of parameters can be analyzed and used for an operation optimization. A very large conversation of energy is achievable.

Bibliography

[1] Selig, M., Wambsganß, M.: Energetische Bestandsanalyse, Kollegiengebäude für Mathematik, Technische Universität Karlsruhe, Fachbereich Architektur, Fachgebiet Bauphysik und Technischer Ausbau, Prof. Andreas Wagner (2002).
[2] Treiber, M., Hopf, J., Gama, R. O.: Abschlussbericht Energiemanagement Kollegiengebäude Mathematik, DS-Plan Ingenieurgesellschaft NL München (2009).
[3] Lenz, B., Abromeit, A.: Zwischenergebnisse des Forschungsvorhabens Monitoring und Betriebsoptimierung im sanierten Kollegiengebäude Mathematik des Karlsruher Instituts für Technologie, Hochschule Karlsruhe – Wirtschaft und Technik, Fakultät für Architektur und Bauwesen, Lehrgebiet Energieoptimiertes Planen und Gestalten (2017).
[4] Beuchert, P.: Technische Universität Karlsruhe, Fachbereich Architektur, Fachgebiet Bauphysik und Technischer Ausbau, Prof. Andreas Wagner (2008).
[5] Lenz, B., Abromeit, A.: Hochschule Karlsruhe – Technik und Wirtschaft, Fakultät für Architektur und Bauwesen, Lehrgebiet Energieoptimiertes Planen und Gestalten (2017).

Additional Information

http://www.enob.info/de/sanierung/projekt/details/sanierung-eines-universitaetsgebaeudes-aus-den-1960er-jahren/.
http://www.hs-karlsruhe.de/fakultaeten/fk-ab/personen/architektur/professorinnen.html.

Simon Schmidt and Klaus Peter Sedlbauer

Climate culture building: comparison of different computer generated building envelope designs for different Brazilian climate zones

Abstract: This contribution explains a way to simulate the performance of different building types with respect to the Brazilian climate zone. A high level of sustainability is necessary to maintain our way of living in the future. The search for suitable materials becomes therefore more and more important.

Keywords: Computer generated building design, Brazilian climate zones

1 Introduction

Right now we are at a point in history were major measures have to be conducted in order to prevent the way we are living. In addition, it also is a point in history where rather minor decisions have major impact on our future way of living [9].

In addition, in the coming years and decades our society will have to face global trends such as population growth, urbanization, rising energy requirements, changes in existing social structures and others. Questions like which influence these trends have on the buildings of the future and which innovations are necessary to meet them arise especially in the building sector.

The building sector is going to meet the major challenges for the construction industry. The fields of study in this discipline are closely interwoven with socials issues and are responsible for 35 % of energy consumption, for example, in Germany [1]. Moreover, also the resources needed for the building sector are bigger compared to all other industry sectors as, for example, the automotive industry. With respect to energy and resource shortage, we are and will be facing the building sector that has a major impact on the future way we are going to live worldwide.

Although it is known that there are many influences on buildings as, for example due to the location, the usage and the social context, a worldwide trend to erect a similar building type, and ignoring these influences can be seen.

Simon Schmidt, Klaus Peter Sedlbauer, Fraunhofer Institute for Building Physics, Stuttgart, Germany, e-mail: simon.schmidt@ibp.fraunhofer.de

https://doi.org/10.1515/9783110584455-003

2 Research field

Climate Culture Building (CCB) is a research field bringing together climate and cultural influences on buildings, in order to get the best possible results in terms of comfort, sustainability, resource efficiency and cost for buildings in their surroundings.

The aim of climate culture building is to improve the planning, construction, utilization and energy efficiency of the building, conversion process and the dismantling of buildings qualitatively and quantitatively to and in this way make the rooms available to the people that meet their cultural, climatic and health requirements to support their well-being and to support the local economic and ecological to meet your requirements in the best possible way also.

Build spaces in all climatic zones of the earth have to be built in a way such that they can be used for the well-being, the cultural, health and economic demands of the local people for the rest of their lives. The guiding principles of sustainability apply throughout the entire group, life-cycle of buildings as a basic principle.

Traditional, autochthone buildings all around the world are highly adopted according to their location, their usage and their social context. These buildings were designed by “try and error” over a very long period and because of availability and costs of local materials. According to their performance in terms of indoor comfort, only little is known but the assumption lies near that those autochthone buildings have reached the best possible comfort with the measures taken.

One aspect of the CCB research is to focus on possibilities to reinvent the “try and error” process using numerical simulation in order to find the best solution in terms of comfort, sustainability and cost, including also aspects of local materials.

Using computer simulation to reinvent “try and error” design is not a new topic. Especially in evaluating energy consumption, indoor comfort, moisture transport and others, computer simulation, in this context named “building performance simulation” (BPS), is in use since the early 1970s. Hauser 1978 [3], the developers of TRNSYS [10], the Berkley labs [5], the University of Strathclyde [11] of Equa [7] and many others working since and still at improving their simulation tools meeting actual requirements. But using just these tools the reinvent of “try and error” is not capable. Just with the beginning of combining BPS with mathematical optimization this is possible.

Since 1990, when both disciplines were likely to be joint, the new discipline “building performance optimization” (BPO), was created (Figure 1). Within this discipline BPS and mathematical optimization are used to find optimal solutions for one, or more set goals. Early representatives of this discipline are Wetter [12], Nielsen [4] and Palonen [6]. By using BPO, it is now possible to use a computer integrated “try and error,” optimization approach.

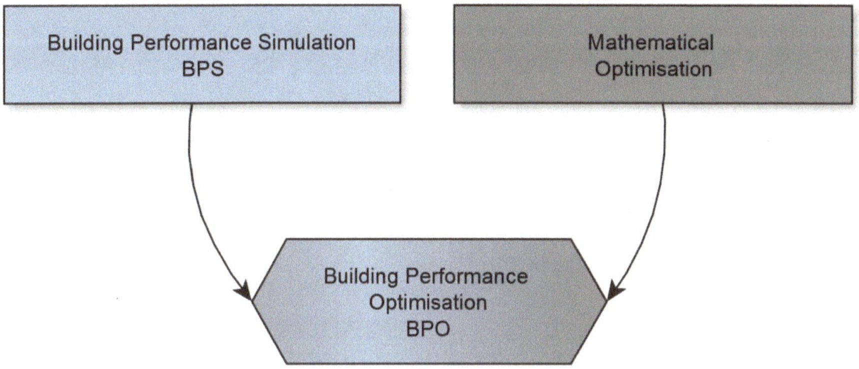

Figure 1: Schematic of the combination of BPS and mathematical optimization to BPO [8].

3 Background BPO-simulation

Each BPO basically consists of the six components (Figure 3), model development, parametrization, simulation, optimization, output and their combination. The method used for this work is characterized by the fact that all components except the model development were embedded in one environment.

To use a model within this framework, a predefined thermal model has to be created externally with all necessary variables. After the implementation into the BPO, the model is parametrized by an internal process and transferred into different variants automatically. These variants then are passed to the optimization, simulation and evaluation internally. The framework consists of an IDA ICE-engine [6] as simulation engine, combined within a Java framework with GenOpt [12] as optimization engine.

The possibilities of this method lay in an effective way for the calculation of different variables, to gain experience and also to evaluate the influence of different variables.

This method was also chosen because it allows a quick calculation of different variants. This method offers advantages, especially with regard to variants in which different building parameters have to be changed simultaneously, such as the weight of the building or the energy level. For example, different building parameters are automatically varied by predefined material possibilities and construction stratification, as well as internal processes, which considerably reduces the number of variables to be calculated (Figure 2).

This also makes it possible to carry out optimizations for different climatic conditions.

The method consists of the main parts, the model, the simulation, the optimization and the analysis, all linked together using different tools. Herewith, a basic computer model of a formal building is designed externally and implemented into the system. The model then is automatically parametrized and passed on to an optimiza-

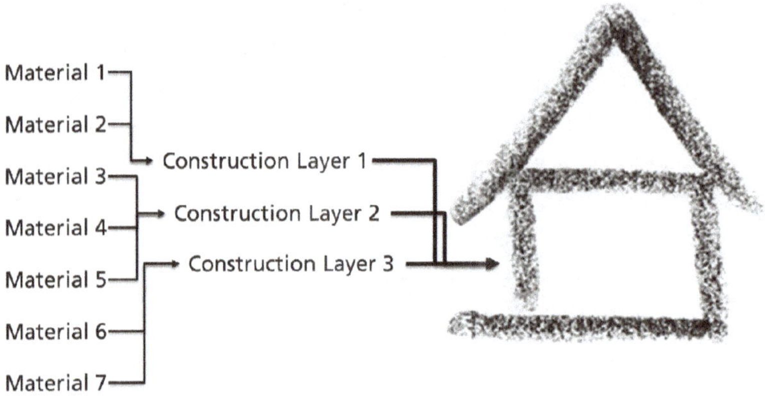

Figure 2: Schematic of variations in building construction implemented in the BPO method used for this evaluation.

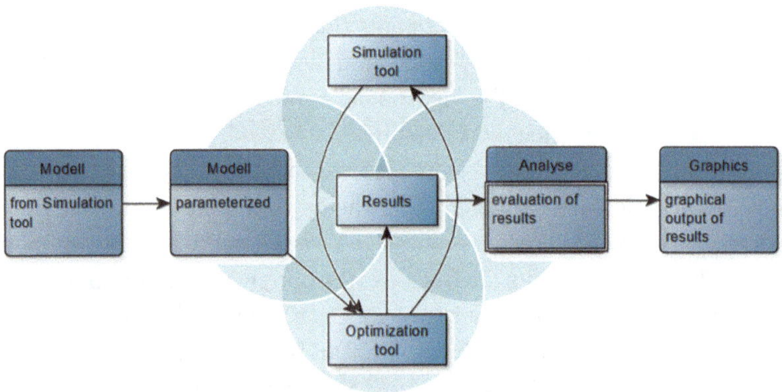

Figure 3: Schematic representation of the used BPO method [8].

tion tool. According to settings, implemented by the user, the optimization starts, using a BPS to evaluate the performance of each variant according to the set goals. All gained results are stored for further evaluation and later on analysis. The analysis is performed after the optimization tool converted to a stable solution (Figure 3).

4 Methodology

By calculation of a predefined model in the climate region of Munich and under German standard boundary conditions, a base case is created. These calculations are repeated for three destinations in Brazil with different climate conditions, which are

Belem, Brasilia and Sao Paulo. The differences and resulting influences are determined directly by comparing with the base case and are visualized afterward.

For these calculations, the overall net energy demand is set as the major goal for reduction.

This work is divided into three parts. The observation of climate regions forms the first part. The model and boundary condition description for simulation and the subsequent calculations using the BPO form the second part. The third part is formed by the representation and discussion of the results.

5 Model description and boundary conditions

5.1 Base case

For the base model, a simple residential building with a pitched roof, as displayed in Figure 4, is used. The building is a two-story massive built building with an roof-inclination of 45 (knee height 0.8 m), an aspect ratio of length to width of 2 and a living area according to DIN 277 [2] of 150 m^2 the building has two floors, a clear room height of 2.5 m each, with the rooms on the first upper floor partially bounded by the roof area. The energy standard of the building is set to the rules of the basic site and has a HT of 0.39 W/(m^2 K). The U-values of the base model are displayed in Table 1. The building is ideally heated and cooled so the calculated net energy demand directly can be evaluated. The set point temperature used for cooling is 25 C. For comparability purposes, these values are not changed.

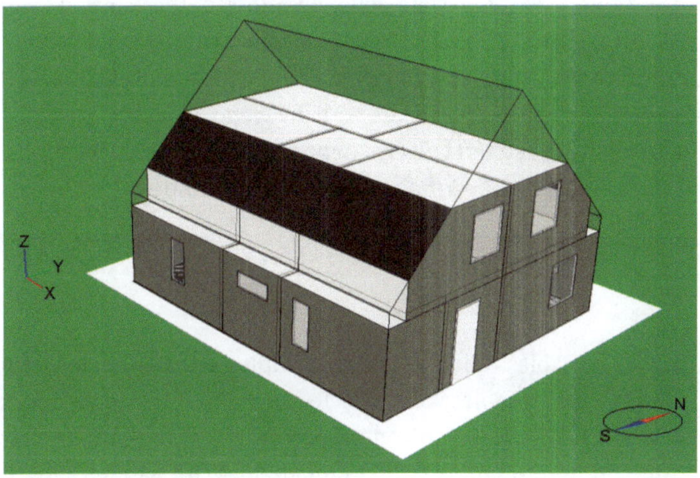

Figure 4: Graphical outline of the base model.

Table 1: U-values used for the different components of the basis model.

Description	Value W/(m²K)
U_{window}	1.3
$U_{outerwall}$	0.28
U_{roof}	0.20
U_{cellar}	0.35
H_T	0.39

The window area of the entire building amounts to 28.3 m² and is with 46 % centered to South. Direction West: 18 %, North 9 % and East 27 %. The frame portion of the modeled windows corresponds to 30 %. The glazing used has a U-value of 1.2 W/(m² K) and a total energy transmittance of 60 %. An automated shading device with an shading coefficient of 0.14 is modeled which is drawn according to the control level on the outside surface of 100 W/m².

5.2 Boundary conditions

The climatic boundary conditions for all sites under consideration were analyzed with regard to temperature, relative humidity, radiation and wind direction and speed. In Figure 5, the daily course over a year or the speed vector is presented. In addition, the radiation range was divided into direct (black) and diffuse (blue). By using only the climate parameters used for the simulation, the boundary conditions can be directly compared.

The analysis shows, Table 2, that the average temperatures at the selected locations in Brazil differ by only about 7.5 K, but by about 19 K compared to Munich. In the area of relative humidity, it can be seen that Belem has a relatively humid climate with an average of 83 % and Brasilia has an dry climate with an average of 68 %. The maximum values of the solar irradiation are relatively similar at all considered locations, however, differ strongly, up to approx. 75 W/m², with the average values.

5.3 Variables

For these investigations, 10 variables were selected and implemented in the method for the creation of variants. The variables were selected in such a way that control and regulating variables, such as ventilation control schedule and the set point for shading, are changed in addition to purely structural variables. All variables used, their acronyms, the default settings and the variable bandwidth are shown in Table 3.

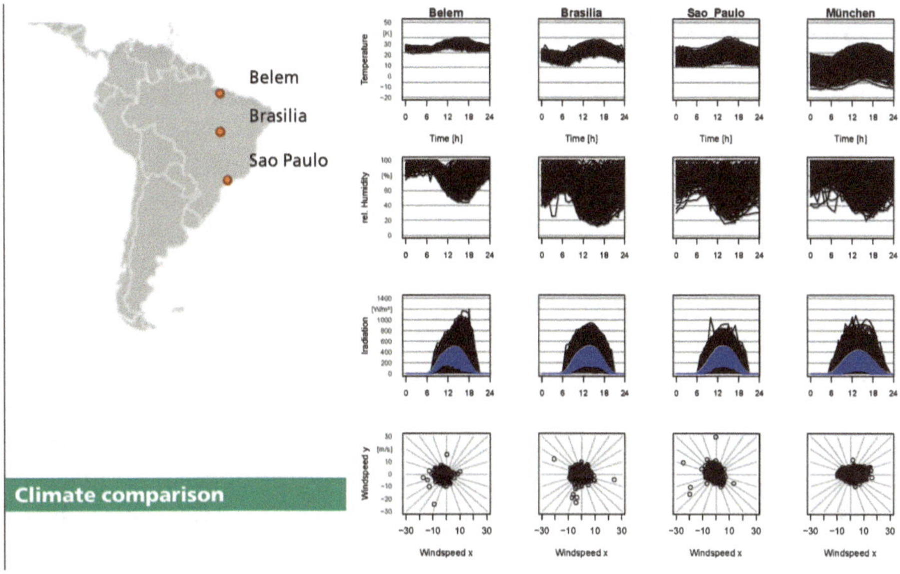

Figure 5: Graphical weather evaluation of Munich, Belem, Brasilia and Sao Paulo according temperature, relative humidity, irradiation and wind direction.

Table 2: Weather evaluation of Munich, Belem, Brasilia and Sao Paulo according temperature, relative humidity and irradiation.

City	Temperature °C	Rel. humidity %	Irradiation W/m^2
	min-average-max	min-average	max-average
Belem	21.0-27.8-35.5	44-83	1210-247
Brasilia	9.0-21.8-34.0	12-68	1068-263
Sao Paulo	8.0-20.3-36.5	15-75	1074-200
Munich	−12.7-8.8-31.1	19-78	1140-175

Table 3: Table with all observed variables and their ranges of values as the number of possible variants Variable acronym Value range variants.

Variable	Acronym	Base case	Value range
Window distribution	win. dist.	centered	Centred, uniform distributed
Thermal storage	stor. cap.	light	Light, mean, heavy
Aspect ratio	a to b	0.5	0.5 to 1.5
Heat bridge coefficient	heat bridge	0.01	0.01 to 0.5 $W/(m^2 K)$
Infiltration rate	infiltration	0.01	0.01–1.5 1/h
Shading control at	shading cont.	100	100–350 W/m^2
Shading coefficient	shading	0.14	0.14–0.82
Ventilation rate	ventilation	10	10-500 l/s
Ventilation control schedule	vent. cont.	16	From none to full time ventilation
Building energy Standard	build Std	4	EnEV, KfW70, KfW55, KfW40

6 Result analysis

For this study, all the variables listed in Table 3 are considered individually for each climate and their influence are evaluated qualitatively.

The qualitative evaluation of the results shows, that the change in the window distribution has but a little positive influence on the energy demand, when the windows are centered to the south. This can be observed regardless of the location.

This positive influence can be seen also by increasing the thermal storage capacity of the building. Only for the climate conditions of Sao Paulo the influence on the energy demand is slightly higher, which corresponds with an decreased heating demand. Overall, it can be stated that if the thermal storage capacity is increased, a decrease in the required demand is the result.

Changes in the aspect ratio of the building influences the energy requirements differently in each observed climate zone. As there is a maximum increase of the net energy demand in Belem and Sao Paulo, the influence in Munich is 6 times as high.

The influence of the thermal bridge coefficient is limited in climates where the cooling energy demand is very high compared to the heating demand. This applies for Belem and Brasilia. In climates with an high heating demand, the influence of the heat bridges is much higher, as for Sao Paulo and Munich.

The infiltration rate affects both, the heating and cooling energy demand, but dominates the heating energy demand. An increase of the infiltration rate therefore is reflected in a higher increase in energy demand in Munich and Sao Paulo.

By changing the closing criteria for the sunscreen from 100 to 350 W/m^2, the sunscreen is open longer so the energy demand for cooling is increased and the demand for heating decreased accordingly. Like the change in the heat bridge coefficient, the influence of the closing criteria on the heating demand is higher than on the cooling demand. In this case, a higher closing criteria results in a higher energy demand for cooling. Therefore, there is an increase in energy demand in Belem an Brasilia and an decrease in Sao Paulo and Munich.

The changes in the shading coefficient from a high level of 0.14 to a low level of 0.82 result in an increase of internal gains and, therefore, in an reduction of heating and increase of cooling energy demand. The changes are similar to the once found changing the closing criteria.

Significant differences arise when the influence of the ventilation rate is displayed. The higher the rate, the higher the energy demand, with a linear influence if just the heating or cooling demand is increased as for Belem and Munich and an exponential influence if both demands are influenced as for Sao Paulo and Brasilia.

Looking at the graphs with a change in the ventilation strategy, it can be seen that in most locations a increase of the energy demand can be seen. Just in Brasilia, a potential of an reduction by using a fixed ventilation strategy result in a lower energy demand. All Schedule 24: Brasilia: −1.4 (−11 %); Sao Paulo: 1.2 (12 %); Belem: 21.9 (99 %); Munich: 24.8 (44 %).

Changing the building standard mostly influences the heating energy demand but also has a positive effect on the cooling energy demand. So in climates with an high heating demand the influence is much higher, as for Sao Paulo and Munich, compared to climates with an lower or no heating demand.

Changing the perspective from a qualitative to a quantitative view, the results can be displayed as shown in Figure 6. The net energy demand per m^2 according to each variable shown in Table 3 is displayed in a radar graph. The demand is calculated relative to the first value of the value range of each variable. Figure 7 refers to the maximum reduction of the energy demand for each variant. Figure 6 specifies the maximum increase in energy demand.

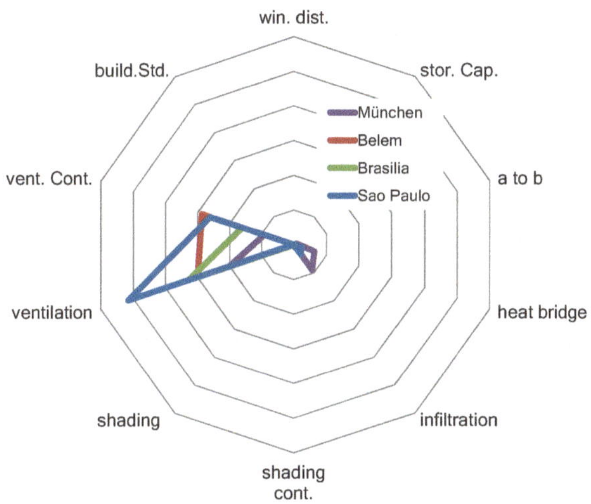

Figure 6: Display of relative maximum increase in net energy demand according to the base case.

The figure shows that the highest increase of the energy demand is strongly reliant on the variables for ventilation, its control and the infiltration rate but also a minor influence of the aspect ratio and the heat bridge can be seen. Contrary to this the highest decrease can be seen by changing the storage capacity and the variables for shading and its control, as well as the building standard. The ventilation control has, but just for the climate of Brasilia, a positive effect. Overall, it can be deduced that for each climatic zone observed, a different set of variables influence the net energy demand.

7 Discussion and analysis

Using this method, it becomes possible to evaluate different variables and a huge number of variations in order to find the most influencing set of variables for a given cli-

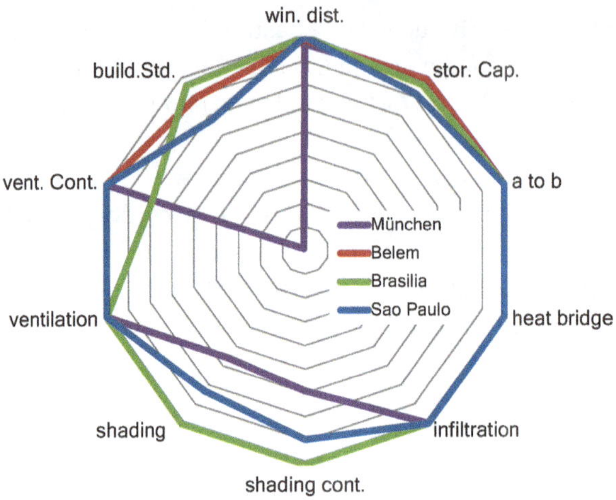

Figure 7: Display of relative maximum reduction in net energy demand according to the base case.

matic zone. With this, the first step in the direction of the reinvent of the "try and error" has been taken. The next in the direction of the goals set for the BPO then would be to evaluate these sets under different goals as, for example, comfort, sustainability and cost using an optimization algorithm as described in Section 3. But even with these calculations, the results just indicate possibilities and not a fixed design. This approach focuses on simplifying the work of finding the influence of different variables on the performance of buildings in different climatic zones.

But in order to build spaces in different climatic zones in a way such that they are beneficial for the well-being, the cultural, health and economic demands of the local people this approach has to be expanded. The results gained have to be implemented in a real surrounding and designed accordingly by designers and architects.

Bibliography

[1] Bundesministerium für Wirtschaft und Energie. Energiedaten: Gesamtausgabe: Stand: Februar 2017. Internet (2017).

[2] DIN 277-1. Grundflächen und rauminhalte von bauwerken im hochbau teil 1 begriffe, ermittlungsgrundlagen, Februar 2005.

[3] Hauser, G.: Rechnerische Vorherbestimmung des Wärmeverhaltens großer Bauten. PhD thesis, Universität Stuttgart, Stuttgart (1977).

[4] Madsen, K., Bruun Nielsen, H.: Introduction to Optimization and Data Fitting. PhD thesis, DTU, Darmstadt (2010).

[5] National Renewable Energy Laboratory. Energyplus programmsystem (2016).

[6] Palonen, M., Hasan, A.: Mobo programmsystem (2014).

[7] Sahlin, P., Grozmann, P.: Ida ice programmsystem. Digital (2014).

[8] Schmidt, S.: Development of a novel method for thermal – energetic and economical optimization of residential buildings. PhD thesis, Technical University of Munich, Munich (2016).
[9] Stallmann, M.: Pressemitteilung nr. 15 vom 24.04.2017: Internationale konferenz berät über konzept der plantetaren belastungsgrenze: Pressemittteilung, 24.04.2017.
[10] Transsolar: Trnsys: Transient system simulation tool (2016).
[11] University of Strathclyde. Esp-r programmsystem (2011).
[12] Wetter, M.: Genopt manual: Generic optimization program (2011).

Lia Mota, Ivan Lemos, Luiza Santos, and Renata Panseri

Electrical energy efficiency in urban infrastructure systems: nonintrusive smart meter for electrical energy consumption monitoring

Abstract: The following contribution shows the development of an IoT based smart meter for the monitoring of energy consumption. The system uses Arduino together with a special designed monitoring board and has a cloud functionality.

Keywords: Urban infrastructure, smart meter, IoT

1 Introduction

In the last decades, as a result of stimulus policies and investments, the Metropolitan Region of Campinas (Brazil) presented an intense urbanization process, which resulted in accelerated metropolization with great population growth and a disorderly urban expansion.

Having undergone this process, the metropolitan region of Campinas, as well as other metropolitan regions around the world, endured many of the typical imbalances of the great metropolitan regions, among them: in the area of information and communication technologies (ICTs), large deficiencies in metropolitan networks that impede a comprehensive and quality service and, in the area of energy, irregular supply, due to high consumption, energy inefficiency of distribution services, and widespread waste.

In this context, in the last years, research was developed, in the scope of the Urban Infrastructure Systems Graduation Program of PUC-Campinas and also in the scope of the Electrical Engineering Graduation Program of PUC-Campinas, related to Electrical Energy Efficiency in Urban Infrastructures Systems, more specifically the Telecommunications and the Electrical Power Systems. This research focused on the development of methods, techniques and devices for application in smart cities, aiming at monitoring the consumption/generation of electricity (consumption of different types of consumers and generation inserted in the cities as, for example, photovoltaic distributed generation), optimizing the use of electric energy and inserting renewable energy sources in the urban environment, including photovoltaic energy.

Lia Mota, Ivan Lemos, Luiza Santos, Renata Panseri, PUC-Campinas, Campinas, Brazil, e-mails: lia.moreira.mota@gmail.com, ivanpedrotti@yahoo.com.br, luizahiginoss@gmail.com, renata.panseri@hotmail.com

https://doi.org/10.1515/9783110584455-004

With regard to the monitoring of electricity consumption/generation, a nonintrusive energy meter was developed (using open source platforms), which meets the billing requirements of the white tariff modality. This tariff modality consists of different electric energy consumption tariffs, depending on the hour of the day, counting on three hourly tariffs, being: I – Peak period tariff, corresponding to 3 hours; II – Intermediate period tariff, consisting of 2 hours, one before and one after the peak period; III – Off-peak period tariff, consisting of 19 hours. The developed energy meter has the ability to communicate, being possible to use it in the context of smart grids and smart cities, contributing to the reduction of energy consumption (through the change of consumption habits) and to the management of consumption by the demand side.

Due to their broad conceptual scope, smart grids can be studied in four areas of concentration: advanced metering infrastructure, advanced distribution automation, distributed generation and storage resources and energy conservation [3, 4].

This work focuses on the area of advanced measurement infrastructure, based on electronic measurement, which requires the use of technologies such as intelligent electronic devices, telecommunications and information technology [1, 2, 4]. The electronic meters are fundamental in obtaining a more efficient network, allowing, among other benefits, the application of energy tariffs according to the hours of use, the so-called hourly rates.

This type of tariff policy is practiced to encourage consumers to reduce or transfer consumption at times of peak demand, where the cost of energy is more expensive for periods where demand is lower and, therefore, with lower tariffs. Policies such as this can allow the reduction of investments to expand generation, as well as transmission and distribution infrastructure, considering that the electric system is sized for periods of peak consumption. It is also possible that this hourly rate policy allows the reduction of the consumer's energy bill, without reducing its consumption, only by modifying its habits in order to use energy in the off-peak period. In this sense, the potential of gain with the adoption of differentiated tariffs according to the hours of consumption is directly associated to the number of users that adopt this type of tariff, justifying even the investments in meters [5].

Thus, it is clear that to fulfill its purpose; this type of tariff policy must be advantageous to the consumer, so it is necessary to know the profile/ behavior of the energy consumption of the residence. In this context, the load curve of the consumer can be lifted from the use of energy meters with recording of temporal information.

Therefore, this research intended to develop an energy meter using open source platforms that meet the billing requirements of the white hour rate modality, and have a communication capacity, being possible to use it in the context of smart grids, contributing to the reduction of energy consumption (by changing consumer habits) and consumption management by the demand side.

2 Methodology

A key feature for the operation of an intelligent network is the interoperability between components and equipment, so that any exchange of a device does not compromise the performance of the system.

Based on this premise, the selection of the components of the meter proposed in this work was performed, searching for technologies already consolidated in the market, in order to guarantee robustness to the project, as well as the use of open source platforms, due to the capacity of standardization, in view of the large amount of available materials and the possibility of dissemination of knowledge.

The proposed energy meter consists of a metering circuit composed of analogue-to-digital converters, a digital signal processing core, registers, a microcontroller responsible for collecting measured quantities and sending them to the communication module. All components are controlled by the meter module.

Still composing the system, but with partial control, are the access point (AP) and ThingSpeak data server. A block diagram of the design is represented by Figure 1.

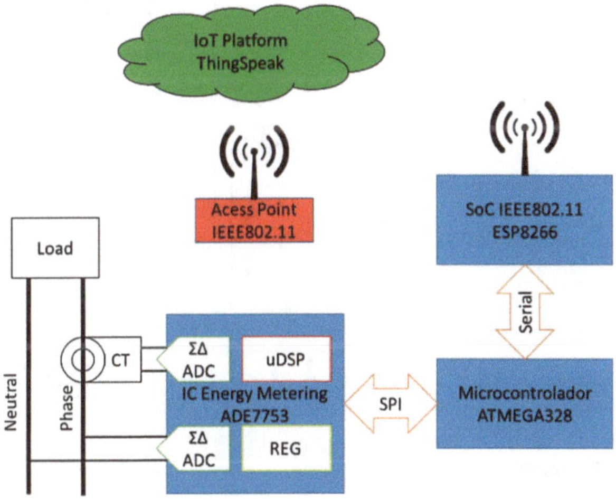

Figure 1: Design of the energy meter.

2.1 Energy meter circuit

The ADE7753 single-phase integrated circuit of Analog Devices was used to measure electrical energy, consisting of 2 second-order analog-digital converters with 16-bit resolution and a digital signal processing core capable of measuring active, reactive and apparent energy, voltage and RMS (root mean square) line current and frequency [6–9].

The option of an integrated circuit to perform the above functions is justified due to the reduced resources of the microcontroller, small amount of external components, miniaturization of the printed circuit board, in addition to the high precision guaranteed by the manufacturer, being specified in less than 0.1 % Still as a benefit, one can mention the cost, since this component, when purchased in large quantities, has a price of less than US$3.

Through a selectable digital integrator, internal to the chip, it is possible to measure electrical current by means of a (di/dt) type sensor such as a current transformer or Rogowski coil, thus eliminating the need for external analog integrator circuit, allowing stability and precise phase matching between the voltage and current channels. The functional block diagram of this IC (Integrated Circuit) is shown in Figure 2.

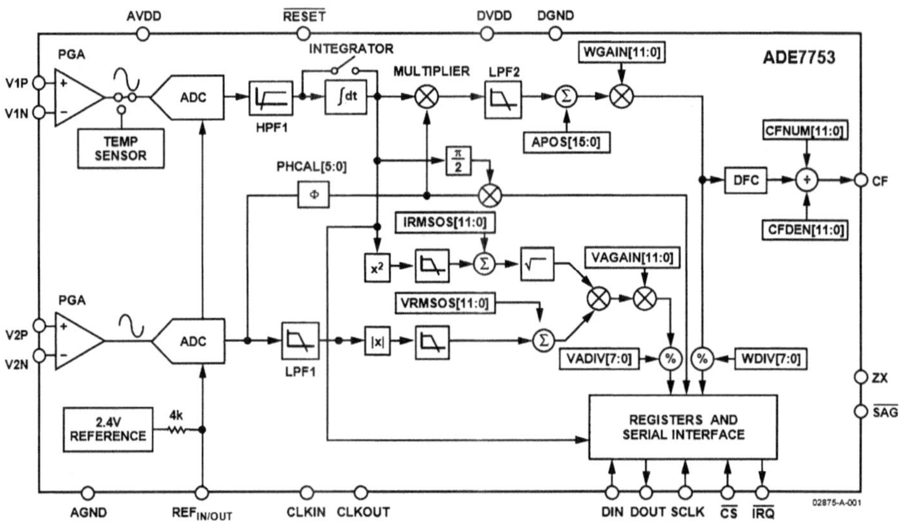

Figure 2: Internal block diagram of ADE7753 [9].

The AD7753 has two channels of analog inputs with differential voltage, limited to ±0.5 V over the analog ground signal (AGND). In channel 1, the current measurement is performed through a current transformer, using SCT-013 as shown in Figure 3 for measurements up to 100A.

The choice of this current sensor was due to the possibility of using this type of transducer in conjunction with the ADE7753, besides being a noninvasive and easy installation method.

Figure 3: Current transformer SCT-013 [10].

2.2 Microcontroller

The microcontroller circuit is responsible for collecting the measurements of the ADE7753 through Serial Peripheral Interface (SPI) communication. It performs the conversions in electrical measurement units and sends this information to the communication module. To accomplish these tasks, the ATMEGA328 microcontroller from Atmel Corporation [11], an 8-bit microprocessor with Reduced Instruction Set Computing (RISC) architecture, 2 kB RAM (Random Access Memory), 32 kB program memory (flash) and 1 kB EEPROM (Electrically Erasable Programmable Read Only Memory), was used. This microcontroller has 23 digital inputs and outputs, Pulse Width Modulation (PWM), 6-pin analog readout with 10-bit resolution, 3 counters with comparison modules, 2 external interrupt pins, USART serial communication (Universal Synchronous/Asynchronous Receiver/Transmitter), Serial Peripheral Interface (SPI), built-in programmable oscillator for watchdog and 5 low power modes. The firmware shipped on the microcontroller is programmed through the Arduino open source development platform, as shown in Figure 4, which uses C/C ++ programming language.

The option of using this hardware and software platform is justified due to the large amount of available development materials generated by the community, due to the fact that it is financially viable, as it does not require proprietary development tools. The main component of this platform, the ATMEGA328, when purchased in large quantities is priced at just under US$2. In addition to these benefits, it is necessary to emphasize the option of this low processing microprocessor (1MIPS – million instruc-

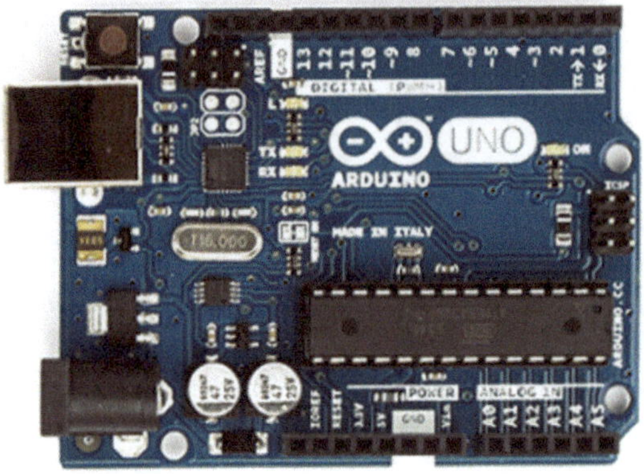

Figure 4: Arduino Uno board [12].

tions per second – per MHz) in order to reduce the cost of the equipment, and thus become a feasible option for scale production.

2.3 Communication module

In order to send the data, that was already sent to the microcontroller, to the internet, the intelligent connectivity platform of the Espressif Wi-Fi Systems (Wireless Fidelity) was used through the SoC (System on Chip) ESP8266 [13], as shown in Figure 5. This component is a stand alone and complete solution on Wi-Fi networks (IEEE802.11 standard), and can be used as an application host or by unloading network adapter functions from another application. It supports IEEE802.11b/g/n protocols with maximum transmission power of +20 dBm, reception sensitivity of −91 dBm (11 Mbps) and frequency range between 2.412 and 2.484 MHz.

The ESP8266 was chosen because it uses the IEEE802.11 standard, already widespread in the world and has a wide application in residential environments, as well as being a complete solution, reducing development time, and being low cost (about US$2) when compared to other solutions of similar robustness.

2.4 Access point

To make energy consumption measurements available on the Internet, a Wi-Fi router is required in order to connect information between the meter and the data server. The device used was the D-Link DI-524, as shown in Figure 6, also operating in the IEEE802.11 standard, with maximum transmission power of +14 dBm and manufacturer-specified range of up to 30 meters indoor.

The data transfer rate is variable according to the Received Signal Strength Indicator (RSSI). It can reach 54Mbps, in the condition of being close to the equipment.

Figure 5: Wi-Fi module [14].

Figure 6: Wi-Fi Access Point DI-524 [15].

The internet plan used was known as Broadband Popular, with a transfer rate of 1Mbps, created by Decree No. 7175 – National Broadband Plan (PNBL), with the purpose of expanding Internet access, mainly in needy regions of technology [16].

Despite being considered as broadband, this Internet plan is one of the most basic of the service providers in the country proving, therefore, that the communication infrastructure for this type of application does not need high transfer rates and, consequently, not financially burdening the user.

2.5 Internet of things platform

For a real-time storage and visualization of household consumption information, a file server is required and, for this purpose, the Application Programming Interface (API) of the open source platform ThingSpeak was used [17].

ThingSpeak is used to build Internet of Things (IoT) applications, allowing interaction between sensors and control of any device that supports the HyperText Transfer Protocol (HTTP), having as main characteristics: Open API; data collection in real time; geolocation data; data processing; data visualization; device status messages; plugins.

To use ThingSpeak, it is only necessary to create an account (user with login and password), add a channel and, through channel settings, it is possible to include up to 8 fields, which will be the variables to be monitored, as shown in Figure 7.

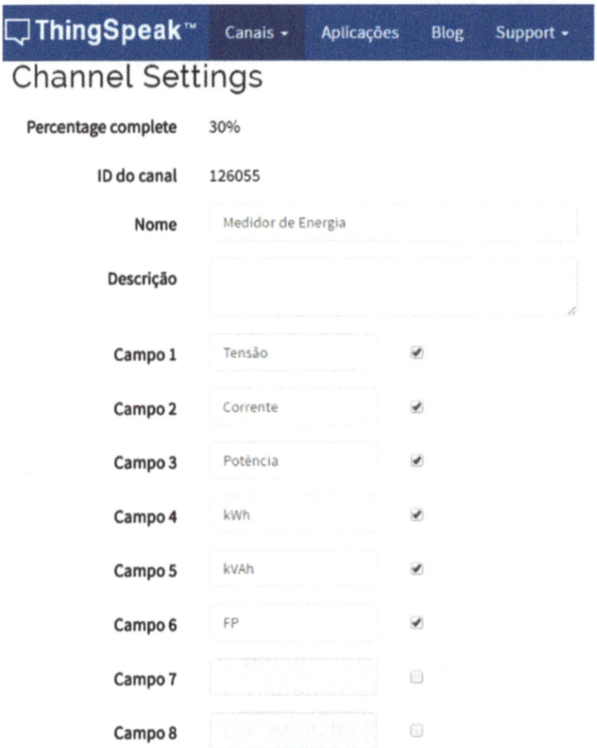

Figure 7: Channel settings of ThingSpeak [17].

3 Results

3.1 Hardware development

With the meter components selected, the first step was to consult the manufacturers' datasheets in detail, in order to verify all the considerations for the correct functioning of the circuits and, after that, an electrical schematic with the connections between the components that are part of the device could be proposed, as shown in Figure 8.

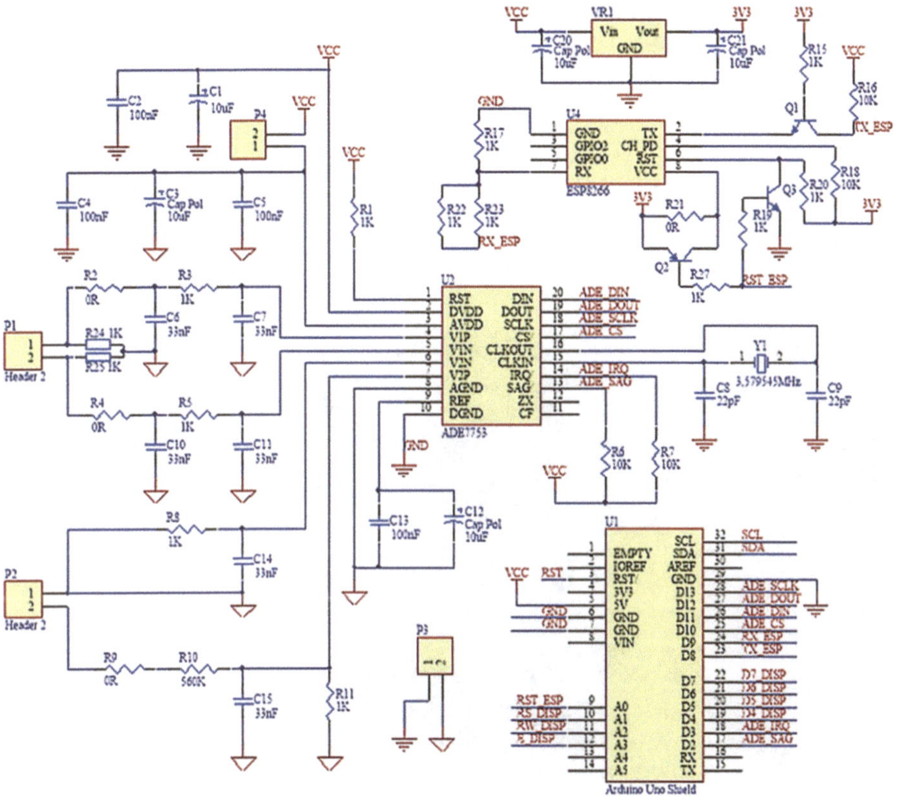

Figure 8: Electrical diagram of the energy meter.

For implementing a prototype, it was necessary to the develop the hardware, which consists of the designing the printed circuit board, where the components of the electric circuit were assembled. This procedure requires care to maintain the integrity of the signals, especially with regard to the way the reference planes of the analog and digital signals are connected, and once again the manufacturers' orientations have been taken into account, resulting in a two-layer plate (layers) as shown in Figure 9.

After the layout of the printed circuit board was finished, the manufacturing files were sent to the printed circuit board (PCI) manufacturer, who was able to make the prototype, and with the board ready, the components were assembled in it, as shown in Figure 10.

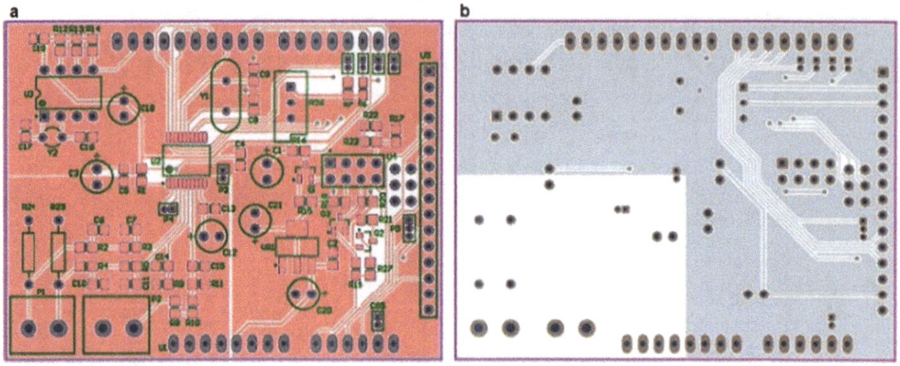

Figure 9: PCI layers: Top (red) and Bottom (grey).

Figure 10: Board with components.

3.2 Firmware development

For this task, the Arduino development environment (Integrated Development Environment – IDE) was used, being able to compile the code and also the recording of the ATMEGA328 microcontroller.

As a first activity, the microcontroller has to connect the meter to the access point and then read the VRMS (voltage), IRMS (current), AENERGY (active energy) and VAENERGY (apparent energy). With the information received from the meter, the conversions are made to electrical quantities by means of the calibration constants (WGAIN and VAGAIN), obtained according to the manufacturer's instructions in the datasheet.

The server waits for a message format to be received and, therefore, the microcontroller also needs to tailor the measurements to the expected format. After such adjustment is made, the data is sent to the Wi-Fi module, which then sends it to the ThingSpeak.

In order to make sure that the data has been received on the server, in the event of an error in the sending, the microcontroller reconnects to the access point and sends the data that was not sent to the server.

3.3 Accuracy tests

The energy meters have INMETRO (National Institute of Metrology, Quality and Technology) standards for their certification, and, in order to evaluate the accuracy of the developed device, a verification test was performed, based on the requirements of Administrative Rule 587 of November 5, 2012 of INMETRO. The energy meter AE-200 of Instruherm (Figure 11) was used as a reference meter for these tests.

In the absence of a device to generate controlled loads, fixed loads were used which, when associated, could emulate different types of load, with different power factors (PF). The loads were connected in parallel, with a voltage of 127V (single-phase) and by varying the load values, the values of active energy provided by the two meters (reference and developed prototype) were recorded.

Based on the data in Table 1, the meter can be classified in the INMETRO accuracy category A, with errors in the range of ± 2 %, with the mean absolute error being 0.89 %.

3.4 Energy consumption in the white tariff mode/policy

For billing in the white tariff mode, the energy consumption must be divided into periods with different tariffs, and for that, a plugin was created in ThingSpeak, as shown in Figure 12.

It is important to note that the start and end dates of the measurements are configurable, allowing the user flexibility to consult, since billing does not always occur on the first day of each month. In addition, daily, weekly, or any other desired period is possible.

In Figure 12, it is possible to visualize the electricity consumption in each tariff station, during the month of September 2016, in a residence in which the meter was installed.

3.5 Daily load curve

If the user wants a more detailed monitoring, this is possible through another plugin, where the accumulated electric energy consumption is presented at each hour of the

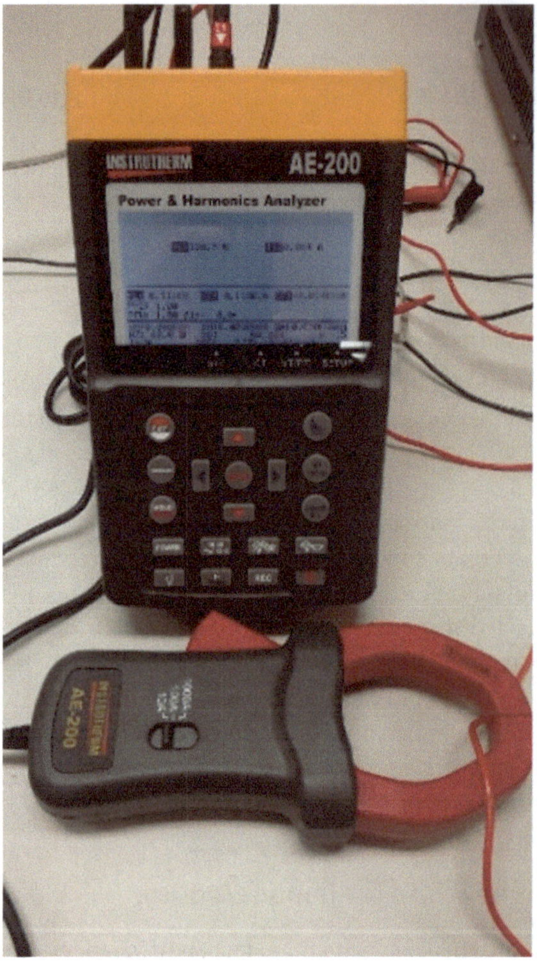

Figure 11: Energy meter AE-200.

Table 1: Active Energy (Reference Meter x Developed Prototype).

Current (A)	PF	AE200 (kWh)	Prototype (kWh)	Error (%)
0.88	1			1.4
0.97	0.48 ind			−0.06
1.0	0.46 cap			−1.7
5.2	1			0.8
5.2	0.54 ind			0.03
5.4	0.46 cap			0.7
10.2	1			0.35
10.0	0.64 ind			−1.7
9.8	0.66 cap			−1.3

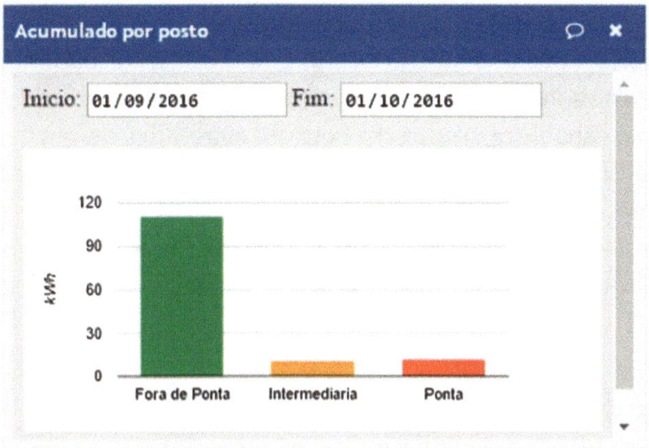

Figure 12: Energy consumption in each different period – peak (red), intermediate (yellow), off- peak (green).

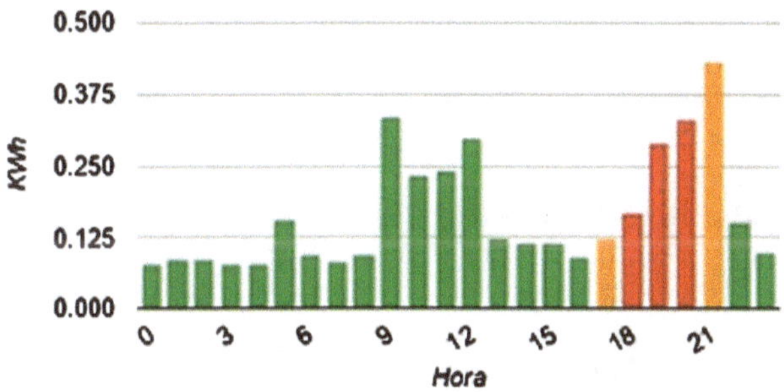

Figure 13: Daily load curve.

day, as shown in Figure 13 Through this plugin, in which it is possible to select the day to be observed, the user can keep up with the hourly consumption, and thus compare daily, if the change of habits can transfer consumption to off-peak hours (green).

4 Conclusion

Through open source platforms, it was possible to develop an energy meter, using a noninvasive method (current transformer) and, with web interface, so that the consumer can monitor their electricity consumption. The results obtained from tests carried out, comparing the prototype developed with a commercially available meter, attest that the developed project meets the tolerance required by INMETRO and ANEEL requirements for billing measurement in the white tariff.

In order to obtain a more efficient energy network, hourly rates of different forms have been used all over the world, in view of the prospect of gain with tariff modalities of this kind. However, it should be noted that not only incentive policies (hourly tariffs) and devices (energy meters) are capable of making the network more efficient, but the set of factors associated with the disposition of citizens who consume electricity in changing their habits consumption.

As future work, it is suggested to add intelligence to the device in a way that makes autonomous decisions, such as the disconnection of certain loads that do not need to be connected.

Bibliography

[1] Colak, I.: Introduction to Smart Grid. In: 3rd International Smart Grid Workshop And Certificate Program (ISGWCP), Istambul, Turkey, March 21–25. Institute of Electrical and Electronics Engineers (IEEE) (2016).

[2] Barbosa, P., Brito, A., Almeida, H.: A Technique to provide differential privacy for appliance usage in smart metering. Inf. Sci. 370, 355–367 (2016).

[3] Bhatt, J., Shah, V., Jani, O.: An instrumentation engineer's review on smart grid: Critical applications and parameters. Renew. Sustain. Energy Rev. 40, 1217–1239 (2014).

[4] Cunha, A. P.: Da "Conceptual Basis for Implementing Smart Electrical Distribution Grids". PhD Thesis, University of São Paulo, São Paulo (2011) (in portuguese).

[5] Faruqui, A., Harris, D., Hledik, R.: Unlocking the 53 billion savings from smart meters in the EU: How increasing the adoption of dynamic tariffs could make or break the EU's smart grid investment. Energy Policy 38, 6222–6231 (2010).

[6] ANEEL Normative Resolution n. 414, Sep. 2010.

[7] Leite, D. R. V.: Electronic Meters: Economic Viability Analysis in the Contexto of Smart Grids. University of Brasília, Brasília (2013) (in portuguese).

[8] ANEEL White Tariff [Online]. Available: http://aneel.gov.br/tarifa-branca. Accessed: March, 2018.

[9] Analog Devices [Online]. Available: http://analog.com. Accessed: March, 2018.

[10] YDHC 100A Split core current transformer SCT-013. [Online]. Available: http://yhdc.com/en/product/320/. Accessed: March, 2018.

[11] Microchip [Online]. Disponível em: http://www.microchip.com/wwwproducts/en/ATmega328. Accessed: March, 2018.

[12] Arduino. [Online]. Available: http://arduino.cc. Accessed: March, 2018.

[13] Espressif. [Online]. Available: https://espressif.com/en/products. Accessed: March, 2018.

[14] Sparkfun. [Online]. Available: http://sparkfun.com/products/ 13678. Accessed: March, 2018.

[15] D-LINK. [Online]. Available: https://dlink-manuals.org/dlink-di-524-user-manual/1/. Accessed: March, 2018.

[16] BRASIL. PNBL [Online]. Available: http://mc.gov.br/programa-nacional-de-banda-larga-pnbl. Accessed: March, 2018.

[17] THINGSPEAK [Online]. Available: https://thingspeak.com/channels/126055. Accessed: March, 2018.

Roberto Zanetti Freire, Bernhard Lenz,
Gerson Henrique dos Santos, Joseph Virgone, and
Abdelkrim Trabelsi

Distinct approaches to reproduce hygrothermal behavior of building materials based black-box models

Abstract: The presence of moisture in building envelopes caused by infiltration or condensation, especially in insulation layers, can have serious consequences in the whole-building energy performance and thermal comfort. Accurate prediction of moisture transport in buildings depends on properly understanding how water migrates across an interface, and it is usually performed by associating experimental analysis of different types of porous media or by numerical simulation. With the objective of reducing the energy consumption of buildings, computational tools are being used to simulate new and retrofitting buildings. In this type of application, it is common to find nonlinear behavior affecting temperature and relative humidity profiles in building structures, mainly due to modeling difficulty and highly moisture-dependent properties, increasing the difference between the results found by computational simulations and what happens inside building materials. Based on these concepts, this chapter presents two black-box approaches, with nonlinear identification focus, adopting Multivariate Adaptive Regression Splines (MARS) and Least Squares Support Vector Machines (LS-SVM). The first technique was considered to reproduce the behavior of highly hygroscopic building materials. Considering an experimental data set acquired using an experimental plant developed to study moisture effects on building

Acknowledgement: Funding for this research was provided by a grant from "La Region Rhone-Alpes" – France, and both Fundação Araucária (grant CP 17/2015) and Coordination for the Improvement of Higher Education Personnel (CAPES) – Brazil, and Brazilian National Council for Scientific and Technological Development (CNPq – grant 304783/2017-0).

Roberto Zanetti Freire, Pontifical Catholic University of Parana (PUCPR) – Polytechnic School (EP) – Industrial and Systems Engineering Graduate Program (PPGEPS), Rua Imaculada Conceição, 1555, 80215-901 Curitiba, Brazil; and Univ Lyon, CNRS, INSA-Lyon, Université Claude Bernard Lyon 1, CETHIL UMR5008, F-69621 Villeurbanne, France, e-mail: roberto.freire@pucpr.br
Bernhard Lenz, University of Applied Sciences Karlsruhe, Dept. of Architecture and Construction Engineering, Energy Optimisation and Building Science Group, 76133 Karlsruhe, Germany, e-mail: bernhard.lenz@hs-karlsruhe.de
Gerson Henrique dos Santos, Federal Technological University of Parana - UTFPR, Department of Mechanical Engineering, Av. Monteiro Lobato, Km 04, 84016-210 Ponta Grossa, Brazil, e-mail: gsantos@utfpr.edu.br
Joseph Virgone, Abdelkrim Trabelsi, Univ Lyon, CNRS, INSA-Lyon, Université Claude Bernard Lyon 1, CETHIL UMR5008, F-69621 Villeurbanne, France, e-mails: joseph.virgone@insa-lyon.fr, abdelkrim.trabelsi@univ-lyon1.fr

https://doi.org/10.1515/9783110584455-005

surfaces. MARS models were built in order to predict heat flux, mass flow rates, and both temperature and relative humidity profiles considering just indoor and outdoor surface temperature and relative humidity as inputs. In the second approach, by adopting multiples MISO (multiple-input, single-output) Nonlinear Auto-Regressive with eXogenous inputs (NARX) models, LS-SVM, a maximum margin model based on structural risk minimization, was used to predict vapor flux, sensible heat flux, latent heat flux and mould growth risk in roofs surfaces. In this second case study, outdoor weather conditions were considered as input for the models. To evaluate the proposed black-box regression and identification techniques, five performance coefficients were analyzed for both training and validation phases. Results of applying artificial intelligence based approaches in predicting the hygrothermal behavior of building materials showed consistent precision when compared to the results of both experimental and numerical model results.

Keywords: Multivariate adaptive regression splines, support vector machines, machine learning, black-box hygrothermal model, mould growth

1 Introduction

Building envelopes are exposed to different climate conditions over the year, where highly humid conditions severely affect the conduction loads through the envelope influencing on material properties. The dynamic behavior of building materials associated to both temperature and humidity variations has significant impact on thermal engineering design, especially when occupants' thermal comfort is taken into account. By considering that heat flux within building surfaces is affected by several parameters such as solar radiation, air temperature, rain, wind speed, season, shading, among others, models to properly represent the dynamic of building envelops subject to distinct contour conditions have been proposed since interest on energy efficiency of buildings has increased.

In the last decades, as alarming statistics have proved that buildings are responsible for a considerable amount of the total energy demand of emerging and developed countries, including Brazil, Europe and the United States of America [6, 15, 66], regulations and standards have been created by governments in order to reduce the energy consumption of new and retrofitting buildings.

In many cases, Building Performance Simulation (BPS) has been adopted as the main tool for certification when energy consumption is taken into account. The development of building simulation software have started with specific computational codes in order to describe the heat transfer thought a single building surface. Today, whole-building and energy simulation tools are available, where simulations of building structures were integrated to other systems, for example, heating, ventilation and air conditioning (HVAC), and lighting, in order to provide an accurate approximation of the building physics and a consistent analysis associated to energy usage [31, 44].

Moisture presence can considerably affect thermal gains or losses in buildings with direct relation to the energy consumption [52]. However, most of whole-building simulation tools normally do not take into account the moisture effects due to modeling difficulty, numerical divergence caused by nonlinear behavior, highly moisture-dependent properties and costly computational effort [16]. Moreover, water accumulation in building materials can contribute to mould and mildew growth on surfaces, affecting occupants' thermal comfort, and both durability and efficiency (in terms of insulation) of the building envelope [73]. As alternative, specific computational codes and experimental procedures were developed in order to analyze the particularities caused by moisture presence.

In a material level, considerable works dealt with thermal properties analysis of building materials, especially insulation materials, under distinct temperature situations [9, 21, 38, 51]. However, when moisture accumulation occurs, to determine thermal properties became a difficult task, and to simulate a given building envelope some uncertainties may occur due to the difficulty of modeling moisture transfer, as inaccurate hygrothermal properties and boundary moisture conductance [22]. In this way, a mixed approach involving experimental and numerical approach can be used to solve this problem.

Based on the problem involving the development of models considering moisture presence addressed above, this chapter presents two approaches to computationally reproduce the hygrothermal behavior of hygroscopic building materials. The first approach is based on an experimental procedure considering an insulation material; while the second one is based on numerical results, those obtained using a well-established computational code used for building hygrothermal simulation. The main idea is to present distinct approaches and strategies of using black-box models in both regression and identification approaches, where the final products are precise models capable to reproduce heat and moisture transfer.

The next section presents a literature review about alternatives to model the effects of moisture presence in building materials. The relation among moisture presence, energy efficiency, thermal comfort, and mould growth in building surfaces are also addressed, showing that one possibility to obtain consistent results in simulating heat transfer in building surfaces, when highly hygroscopic materials are taken into account, is assuming black-box models instead of traditional numerical analysis. In the sequence, Section 3 brings concepts about the two techniques for black-box modeling adopted in this study, MARS and LS-SVM, showing model structures and mathematical notation. Data acquisition procedures based on experimental and numerical approaches were presented in Section 4, where a plant capable to capture hygrothermal variations within building material was described. A dataset generate using this plant was assumed as the first case study of this chapter considering MARS technique. Additionally, a numerical model, which was used to generate data for the second case study of this chapter, was presented. The second case study consists in the analysis of heat and moisture transfer in concrete tiles, emphasizing the possibility of mould growth due to high humidity conditions. Simulation parameters and experimental configuration were also addressed in this section. Section 5 describes the results of both regression and identification approaches considering the two case studies. To conclude, remarks about this study and suggestions for future works in black-box system identification and regression applied to buildings were presented in Section 6.

2 Literature review

As an alternative to include moisture effects on numerical approaches to reproduce heat transfer in building materials, specific models were proposed in order to provide a realistic representation of the hygrothermal variations within building structures. According to [45], it was emphasized that moisture migrating through building envelopes can lead to poor indoor air quality as elevate ambient moisture level result in microbial growth. The authors also presented a nonlinear system of partial differential equations used to provide transient temperature and moisture distributions in building envelopes, as well as temperature and moisture content for building's indoor air subjected to outdoor weather conditions. Results were compared to experimental data in order to validate the proposed model. This information can be also verified in studies focusing in mould growth on building surfaces [34, 69, 70].

In the work presented in [11], a method was presented in order to reduce the discrepancies between real buildings and numerical approaches in order to calculate coupled heat and mass transfer in building materials. Authors showed that problems involving heat and moisture migration in porous building materials arise in a number of engineering interests, such as wall drying, the solar house designing, cooling load calculating of air conditioning, among others. This information is emphasized in [48], where it was reported that moisture presence does not affect just energy consumption of buildings, but also their service life, as highly moisture level can cause metal corrosion, wood decay and structure deterioration. In this research, authors presented a two-dimensional set of equations for evaluating the simultaneous heat and moisture migration in porous building materials.

In [43], significant differences in thermal loads of buildings when moisture effects are neglected were presented, showing that the total cooling, heating and the yearly load are usually underestimated in whole-building energy simulation when moisture transfer is neglected. Additionally, authors report that especially in hot-humid climate, building thermal performance is much influenced by the moisture transfer and storage within porous building materials.

Experimental procedures have also been conducted with the objective of understanding the hygrothermal behavior of building materials, especially in estimating thermal properties when distinct contour conditions are assumed. In the experimental procedures presented in [33], authors highlighted the important influence of temperature and relative humidity gradients on the hygrothermal properties of multilayer walls. This approach was extended to insulation layers, especially those applied in specific regions of the building envelop affected by solar radiation and/or moisture accumulation, for example, building horizontal surfaces. Another analysis considering specific regions in building structures was presented in [23], were thermal bridges located in intersections between building surfaces were analyzed in terms of mould growth considering a Fuzzy inference system associated to Kalman filter to predict moisture accumulation in these particular regions. The presented black-box approach showed consistent results in terms of precision when compared to a validated numerical model. Additionally, the proposed model was faster than the original method in terms of computational effort.

Building roofs are constantly affected by moisture presence. The model presented in [64] was adopted to analyze heat and moisture transfer in green roofs in China. By integrating both soil and roof in a unified model, this approach was validated using experimental results, where two energy balance equations were considered for both plant and soil. In the works presented in [20, 24], a coupled heat and moisture transfer model was adopted to predict moisture accumulation in the surface of roofs. In this case, effects of moisture adsorption and desorption on the thermal performance of concrete tiles due to an additional transport mechanism were shown, and a well-known model for predicting mould growth was applied illustrating a risk index.

One of the main issues involving coupled heat and moisture transfer modeling in building materials are the variances of thermal properties due to moisture presence, especially when highly hygroscopic materials are adopted. In these cases, a nonlinear behavior may occur affecting temperature profiles within building structures. This situation is constantly discussed in the literature as a difficult task due to modeling difficulty and highly moisture-dependent properties. In the work presented in [40], moisture induced degradation of the thermal conductivity in insulation slabs composed by silica aerogels was discussed, and an experimental analysis was performed to confirm this statement. This type of material is currently been used as thermal insulation due to their excellent thermal conductivity. However, experimental studies shown that thermal conductivity of the aerogel blankets can be increased approximately 20–40 % after wetting them, causing problems on thermal insulation. Additional studies developed to both determinate and evaluate thermal properties of building insulation materials can be found in [27, 30].

Taking into account the advances on experimental procedures in order to determinate how building materials are affected by moisture presence, the difficult in modeling transient conditions, and the presence of nonlinearities caused by coupled heat and moisture oscillation, artificial intelligence paradigms have been successfully applied in many applications in building energy simulation [14, 46]. Black-box mathematical models have proved to be a consistent approach for approximating computational simulation and reality. In [3], black-box models, those composed by different categories of Artificial Neural Networks (ANNs), were used to predict energy performance and occupants thermal comfort of an specific category of buildings considering low computational effort, and Multivariate Adaptive Regression Splines (MARS) models [25].

One important part of the building envelop is the roof. In warm climates, during clear sky conditions, up to about 1 KW/m^2 of solar radiation can directly reach the roof surface, where between 20 % and 95 % of this radiation is typically absorbed [60]. In cases of high radiation incidence, the use of proper insulating or higher roof solar reflectance can significantly reduce the solar energy absorbed by the roof, providing economy in the usage of air conditioning systems.Moreover, when a cold weather is taken into account, heat loses can be avoided if projects consider energy efficiency of this specific part of the building structure.

In this context, reflective roof and/or insulation coatings have been utilized for increasing the energy savings potential of building envelopes. Several research works can be found in the literature in this area such as [1, 4, 7, 12, 17, 50, 60]. Nevertheless, due to the low cost, concrete or ceramic tiles are widely used in roofs in Brazil. Those tiles enable the mould or algae to grow and normally no paint layer or impermeable films are provided. The mould growth decreases the tiles durability, degrades the aesthetic appearance of buildings and increases the solar absorptivity.

Taking into account that the cost of both heating and cooling in buildings is directly affected by roofs performances, this chapter also presents an approach combining a numerical computational code and an artificial intelligence method in order to predict the hygrothermal behavior of building roofs, focusing on the evaluation of mould growth risk. The main idea is to perform a nonlinear system identification by using data obtained from the results of the numerical model.

The SVM (Support Vector Machines) have already proven to be a promising approach in nonlinear identification. Based on statistical learning (details in [68]), it was originally created to solve classification problems. SVM is a kernel-based method, similar to artificial neural network (ANN) models, which constitute an approximate implementation of the structural risk minimization principle [59]. Considering structures called nucleus (kernels), SVM goes beyond the hyperplanes, and have been widely applied in classification, as it can be seen in [18, 26, 37], and nonlinear regression [10, 71, 72] areas. The objective of SVM is to map the input data in a space of characteristics of high dimensionality.

The variation of the SVM, known as LS-SVM (Least Squares Support Vector Machines), one of the strategies adopted in this chapter, was compared to the classical version of SVM in [28, 61] for regression/identification tasks, showing consistent improvements in performance. It involves the equality constraints only, where the solution can be obtained by solving a system of linear equations.

3 Black-box approaches for regression and identification

This section presents two approaches adopted in this chapter in order to reproduce the hygrothermal behavior of building materials. Two distinct data acquisition procedures were considered in order to generate data for both techniques. We will start with a promising regression technique called Multivariate Adaptive Regression Splines (MARS), followed by the Least Squares Support Vector Machines (LS-SVM), which was assumed for system identification.

The main difference between regression and identification is the way that data was applied in the development of the model. In regression procedures, the whole dataset can be considered to build a model, while in system identification only a set of actual and previous inputs/outputs can be considered.

In order to test each method, two distinct case studies were presented in the sequence of this chapter. The first assumes a dataset acquired from an experimental plant, while in the second one, the dataset was obtained using a well-established computational code.

3.1 Multivariate adaptive regression splines

This section describes the regression technique adopted to reproduce the hygrothermal behavior of high hygroscopic insulation materials, which is the first case study of this chapter. The Multivariate Adaptive Regression Splines (MARS) technique was proposed in [25] by Friedman. The model takes the form of an expansion in a product of spline basis functions, where the number of basis functions as well as the parameters associated to each one are automatically determined by data characteristics. The MARS technique can be classified as a nonparametric method, and it is capable to predict continuous dependent variables using a set of predictors.

Normally applied to high dimensional problems, MARS produces continuous models considering continuous derivatives. The relationship between inputs and the output is derived from a set of coefficients and basis functions that are driven from the data set without making any assumptions about any underlying functional relationship between the variables.

The process of building models using MARS is composed by two stages: (i) the forward phase, where basis functions are added in pairs to the model until the residual error criterion is touched, or the maximum predefined number of terms was reached, and (ii) the backward phase, where overfitted models are adjusted by pruning the least effective term in the model until the best submodel is found.

In terms of basis functions, MARS divides the data set into spline functions, those separate piecewise linear segments with different slope values. These splines are polynomials of order n connected at fixed points called knots. It means that, considering the lower (t_L) and the upper (t_U) limits of the data set, it is automatically defined ($L+1$) subintervals separated by L interior boundaries (ε_l) called knots. The final model consists of splines that are connected through knots while each piecewise curve is called basis function. In this case, both linear and nonlinear behavior are replicate according to [49, 55]:

$$h_1(x) = (t-x)_+ \begin{cases} t-x, & \text{if } t > x \\ 0, & \text{otherwise,} \end{cases} \tag{1}$$

$$h_2(x) = (t-x)_+ \begin{cases} x-t, & \text{if } x > t \\ 0, & \text{otherwise.} \end{cases} \tag{2}$$

The development of MARS models are similar to stepwise linear regression. In this case, functions in collection H and its multiple (interaction effects) are adopted as

$$H = \{(X_i - t)_+, (t - X_i)_+\}, \quad t \in \{x_{1j}, x_{2j}, \ldots, x_{nj}\}, \quad j = 1, \ldots, p. \tag{3}$$

If there are no repeated data in the data set, collection H includes 2np basic functions, considering the following structure:

$$f(x) = \beta_0 + \sum_{m=1}^{M} \beta_m h_m(X), \tag{4}$$

where each h_m (x) is a function from H, or multiple two, or several functions from H. The parameter M can be defined as the number of functions in the model identified after forward stage. By considering h_m functions, it is possible to estimate β_m through the minimum sum of squares errors.

In order to simplify the explanation of MARS technique, a single input variable is considered. In this case, the following data can be assumed: $((x_1, y_1), \ldots, (x_n, \ldots, y_n))$, where (x_i, y_i) is representing ith input–output observation. At first, the model only includes y-intercept term, which is:

$$\hat{f}_1(X) = \hat{\beta}_0. \tag{5}$$

Since $\hat{\beta}_0 = \bar{y}$, which can be described as model 1, it is possible to create the second model by selecting the model structure that includes the minimum sum of squares errors when compared to other models.

$$\begin{aligned} \hat{f}_{21}(X) &= \hat{\beta}_0 + \hat{\beta}_1(X - x_1)_+ + \hat{\beta}_2(x_1 - X)_+, \\ \hat{f}_{22}(X) &= \hat{\beta}_0 + \hat{\beta}_3(X - x_2)_+ + \hat{\beta}_4(x_2 - X)_+, \\ \hat{f}_{2n}(X) &= \hat{\beta}_{2n-1} + \hat{\beta}_3(X - x_2)_+ + \hat{\beta}_4(x_2 - X)_+. \end{aligned} \tag{6}$$

Assuming that the model $\hat{f}_{22}(X)$ the minimum square error, it can be selected and rewritten as presented in equations (7) and (8) given by:

$$\hat{f}_2(X) = \hat{\beta}_0 + \hat{\beta}_3(X - x_2)_+ + \hat{\beta}_4(x_2 - X)_+, \tag{7}$$

$$\hat{f}_2(X) = \hat{\beta}_0 + \hat{\beta}_3 h_3(X)_+ + \hat{\beta}_4 h_4(X). \tag{8}$$

The third model can be developed by including additional basic functions to the second one. If the new model reduces the sum of squares errors, it will be selected from the following models presented in equation (9) (model 3). When a new model is created, a new coefficient β is estimated. In this case,

$$\begin{aligned}\hat{f}_{31}(X) &= \hat{\beta}_0 + \hat{\beta}_1 * h_3(X) + \hat{\beta}_2 * h_4(X) + \hat{\beta}_3 * h_1(X) + \hat{\beta}_4 * h_2(X),\\ \hat{f}_{32}(X) &= \ldots + \hat{\beta}_3 * h_5(X) + \hat{\beta}_4 * h_6(X),\\ \hat{f}_{3(n-1)}(X) &= \ldots + \hat{\beta}_3 * h_{(2n-1)}(X) + \hat{\beta}_4 * h_{2n}(X).\end{aligned} \tag{9}$$

This procedure is repeated until the maximum number of basis functions, defined by the user, is evaluated. By choosing the model that provides the minimum sum of squares errors, an overestimation may occur. At this point, the backward phase starts, where basis functions will be removed from the model increasing the sum of squares error as little as possible. This procedure will run until all basis functions except y-intercept are removed from the model. This process is also called pruning.

In the sequence, $(2k-2)$ models are developed, where each one of them is a candidate for the final model. Note that k represents the number of basis functions. To define which one will be considered, the generalized cross-validation (GCV) criterion is assumed. The GCV, which is presented in equation (10), takes into account both the residual error and the model complexity:

$$\mathrm{GCV} = \frac{\sum_{i=1}^{N}(y_i - f(x_i))^2}{h}. \tag{10}$$

In the GCV calculation, $C = 1 + c.d$, where N is the number of data set samples, d is the effective degrees of freedom, which is equal to the number of independent basis functions. The parameter c is the penalty for adding a basis function. In the Friedman work [25], the recommended value is $c = 3$, but in [36], values in the range $[2, 4]$ are proposed, and the author suggests that a value of 0 penalizes only terms, not knots. The value of 0 was adopted in the present study, as it can be useful with a large data set and low noise. Experiments have shown that the best value for N can be found somewhere in the range $2 < d < 3$. Finally, the model that has the lowest GCV among $(2k-2)$ will be selected as the final model.

Two recent applications of MARS involving buildings were presented in [13, 55]. In [55], MARS and Extreme Learning Machine (ELM) were adopted for predicting both heating and cooling loads of residential buildings. The hybrid method provided better prediction results when compared to most disseminate techniques as linear regression, Gaussian processes and Radial Basis Function Neural Networks (RBF-NNs). In the second work proposed by Cheng and Cao [13], comparisons between MARS, Neural Network with Error Backpropagation Training (NNEBT), RBF-NN, Classification and Regression Tree (CART), and Support Vector Machine (SVM) showed promising results of MARS in terms of predicting building energy performance of twelve distinct geometries of buildings.

3.2 Least Squares Support Vector Machines (LS-SVM)

This section introduces the machine learning technique adopted to identify the hygrothermal dynamic of building roofs, which was second case study presented in this chapter. The Least Squares Support Vector Machines (LS-SVM) can be classified within a class of models that used for pattern recognition, those that use a set or subset of training data in the prediction stage based on kernel. These methods perform predictions from combinations of the outputs of functions centered on each of the points available. The functions used for weighting a given set of training data are called kernels.

At first, SVMs were used to train classifiers based on the concept of structural risk minimization [8]. Besides, the SVMs were developed using the method known as statistical learning. Statistical learning theory was developed for solving problems whose small amount of data and little prior knowledge about the system are available, which differs from the traditional methods.

The SVM technique is designed to adjust the vectors defined for supporting a hyperplane, which aims to separate the input data. The SVM estimated the relationship between output y_i and an input pattern x by the following equation:

$$y_i = w\varphi(x) + b, \tag{11}$$

where b is a bias term, w is a weighting vector and Θ is a nonlinear function that map the input pattern x into a higher-dimensional feature space. The coefficient vector w and bias term b are unknown, and can be obtained solving an optimization problem.

When LS-SVM is applied for system identification tasks, the following optimization problem linked with the minimization of the risk function J can be defined [59]:

$$\min J = \frac{1}{2}\|W\|^2 + \gamma\frac{1}{2}\sum_{i=1}^{n} \varepsilon[y_i, f(x_i)], \tag{12}$$

subject to

$$\varepsilon[y_i, f(x_i)] \begin{cases} 0, & |y_i, f(x_i)| \leqslant \xi \\ |y_i, f(x_i)| - \xi, & \text{otherwise.} \end{cases} \tag{13}$$

In equation (12), W is the vector of weights, and ε is a given real number and γ is a regularization parameter that provides balance between model complexity and training error. The first part of the objective function given by equation (12) is used to regulate the weights and penalize those with higher values. Due to regularization, weights tend to converge to smaller values. This is necessary because heavy loads cause excessive variance in the model dynamic, deteriorating the generalization ability of LS-SVM. The second part of equation (12) represents the regression error of training data. The equality constraint imposed by equation (13) provides the definition of the regression error.

In the case of nonlinearly separable patterns, the model needs to add variables to the problem, by introducing loss variables, ζ_i and ζ_i^*. In this case, it is possible to transform equation (12) in a primal objective function given by:

$$\min J = \frac{1}{2}\|W\|^2 + \gamma\frac{1}{2}\sum_{i=1}^{n}(\zeta_i^* + \zeta_i), \tag{14}$$

subject to

$$\begin{cases} y_i - W\varphi(x_i) - b \leqslant \varepsilon + \zeta_i \\ W\varphi(x_i) + b - y_i \leqslant \varepsilon + \zeta_i^*, \end{cases} \quad \text{where } i = 1, \ldots, N \text{ and } \zeta_i, \zeta_i^* \geqslant 0. \tag{15}$$

By introducing the Lagrange multipliers α_i and α_i^* (support vectors), the regression function given by equation (11) can be written as:

$$f(x, \alpha_i, \alpha_i^*)G(x_i, x_j) + b, \tag{16}$$

where $G(x_i, x_j)$ is the core function, and vectors α_i and α_i^* are obtained solving the linear system of equations, the following Karush–Kuhn–Tucker The vector $G(x_i, x_j)$ equals the inner product of two vectors x_i and x_j in the space of characteristics, $\varphi(x_i)$ and $\varphi(x_j)$, it says that $G(x_i, x_j) = \varphi(x_i)^T\varphi(x_i)$. The fact of adopting nucleuses to replace the calculation of $\theta(x_i)$ and $\theta(x_j)$ is complex, and can be made in a simpler way by means of an approximate function.

These nucleuses generate a mapping between the input space and a high dimensional space, called feature space. The SVM hyperplane generated by this space of characteristics, to be mapped back to input space, becomes a nonlinear surface. Finally, the separation hyperplane becomes no longer a linear function of the input vectors, but a linear function of the space vector of characteristics.

In this technique, a Radial Basis Function (RBF) kernel was adopted, which is given by

$$G(x_i, x_j) = \exp\left(\frac{\|x_i - x_j\|^2}{2\sigma^2}\right), \tag{17}$$

where σ is the spread of Gaussian kernel. In this application, to solve the linear programming training problem of LS-SVM, the SIMPLEX method have been adopted [56].

4 Data acquisition

High hygroscopic materials are constantly adopted for building envelops, as different types of wood and concrete based elements. Numerical models capable to reproduce the hygrothermal behavior of these materials were developed for building simulation mainly to attend building and energy certification programs [5, 40, 43, 47]. Under high moisture conditions, the thermal behavior of these materials can be significantly affected, and the quality of the numerical representation can be reduced.

In order to validate the available numerical approaches, distinct experimental procedures were proposed. In this way, the first case study presented in this section is based on an experimental plant developed to analyze building materials in high moisture conditions. This plant was considered to generate data based on an insulation material for the regression approach using MARS technique, which was described in Section 3.1.

The second case study described in this section was based in a validated numerical method. This method was assumed to simulate heat transfer in concrete tiles and the roof structure of a building. Results were obtained in order to generate a dataset for a system identification procedure based on the LS-SVM technique, where mould growth risk in concrete tiles can be predicted.

4.1 Case study 1: data acquisition based on an experimental plant

Based on an experimental plant introduced in [57, 58], a data set was collected and used in the regression procedure presented in this research assuming MARS technique. The main objective of the previous mentioned study was to present an experimental device capable to measure hygroscopic behavior of building materials, permitting the analysis of multilayered configurations that are commonly found in building envelops.

Some setups similar to the one adopted in this study were built to test variable boundary conditions [48], and have been performed in a scale analysis [67]. Additionally, some approaches are dedicated to evaluate hydric storage [35, 63]. However, in this research, a real scale setup is considered in order to provide the same conditions available on real building environments in terms of hygrothermal variations.

The equipment adopted here, which is presented in Figure 1, is composed by two chambers of 0.5 m × 1.0 m × 1.0 m, these representing the indoor and the outdoor environments. In this setup, 1 m^2 multilayer walls can be evaluated. The air temperature and relative humidity are controlled, and it is possible to consider both natural and forced convection effects.

By using temperature, humidity, heat flux and mass flow rate sensors, the equipment can improve the understanding of the hygroscopic phenomena involving build-

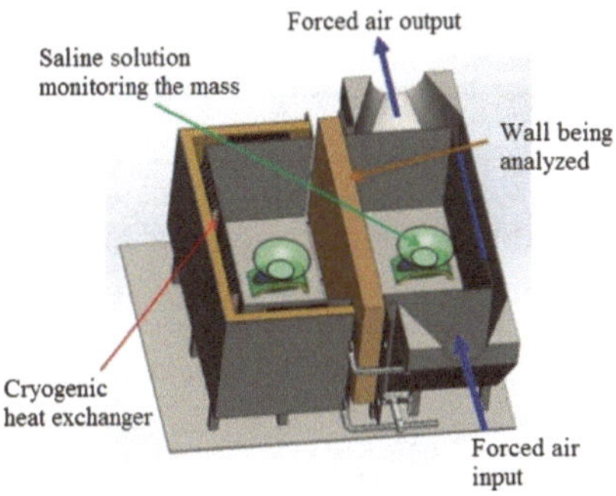

(a)

(b)

Figure 1: (a) Experimental plant scheme; (b) plant image: data acquisition being performed [57, 58].

ing materials. The humidity of the first compartment was regulated using a saturated salt solution (SSS), while its temperature was controlled by fluid-air exchangers linked to a circulation cryostat. The second compartment has two operating modes, natural and forced convection. The first mode works with humidity regulation using the SSS and temperature control with a climatic chamber in remote mode. The second mode uses a climatic chamber to pulse treated air parallel to the wall material being studied considering a controlled flowrate. Ventilators were installed to homogenize the temperature and humidity in each compartment.

Additionally, this equipment can also be used for the characterization of transport properties. Table 1 shows an overview about the data acquisition systems available on the experimental plant, and Table 2 presents sensors range and precision.

In Table 1, $T(x,t)$ and $RH(x,t)$ represent the evolution on temperature and relative humidity profiles at time t (in min) according to depth x (in cm).

4.1.1 Experimental procedures

This study adopts a data set obtained using the setup presented in the previous subsection considering a nonisothermal experiment. In order to test the potential of the regression technique described in Section 3, a wood fiber panel, which is frequently used for insulation in buildings, was adopted as case study. The idea is to consider a material from a particular group that can be classified by their highly hygroscopic properties, those that can increase the nonlinear behavior in terms of temperature variations when moisture presence is detected.

The thickness of the wood fiber panel is 8 cm. Temperature and relative humidity sensors were include on both surfaces, and also at 2,4, and 6 cm depths from internal to external surfaces. Heat flux and mass flow rate were measured on the internal surface, and just mass flow rate was monitored on the external surface. Additional information about the experimental procedures can be found in [57, 58].

Table 1: List of sensors and experiment configuration.

Location	location	Objective
Chambers 1 and 2	1 differential pressure sensor 2 thermocouple Pt100 4 capacitive hygrometers 10 thermocouples (internal surface) 2 mass scales	to control boundary conditions to verify both temperature and relative humidity homogeneity to measure vapor flowrate exchange between the air in the compartment and the wall
Material surface	4 thermocouples type K 6 capacitive hygrometers 1 heat flux sensor	to maintain boundary conditions and to verify both temperature and relative humidity uniformity to measure heat flux
Inside the analyzed material	3 capacitive hygrometers (distance of 2 cm from each other) (each one covers 8 cm) 3 capacitive hygrometers (distinct high from the previous)	to define both $T(x,t)$ and $RH(x,t)$ profiles to obtain the response of the wall to the loading and to provide second measurements
Input/Output Chamber 2	2 hot-wire anemometers 2 capacitive hygrometers	

Table 2: List of sensors and experiment configuration.

Sensor	Range	Accuracy
Temperature (SHT75)	[5–45] [°C]	0.17 [°C]
Relative Humidity (SHT75)	[40–90] %	2 %
Temperature (Pt100)	[5–45] [°C]	0.15 [°C]
Temperature (thermocouples)	[5–45] [°C]	0.16 [°C]
Mass	4200 g/3000 g	0.01 g
Liquid flow	–	0.04 g/h

Four variables were considered as inputs for the regression model, they are the internal and external surfaces temperature and relative humidity, two in each chamber. The objective of the MARS approach is to predict temperature and relative humidity at 2, 4, and 6 cm depth, internal heat flux, and internal and external mass flow rates (nine outputs). Figure 2 shows the input data obtained during the experimental procedure considering 1.04 min sample time. It represents almost 8.5 days of data acquisition and 11,255 samples of each variable.

Figures 3 and 4 present the model output. As it can be verified in Figure 3, mass flow rates measurements were obtained using a longer sample time. In this case, a linear interpolation was performed. As it can be seen in Figures 2–4, there are two particular events that happened in hours 41 and 147. These events modified the system dynamic for a nonisothermal case due to opening of the chambers doors for exper-

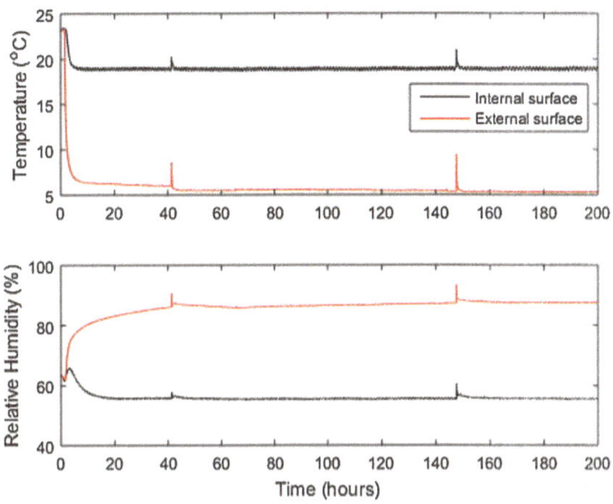

Figure 2: Inputs: internal and external surfaces – temperature and relative humidity.

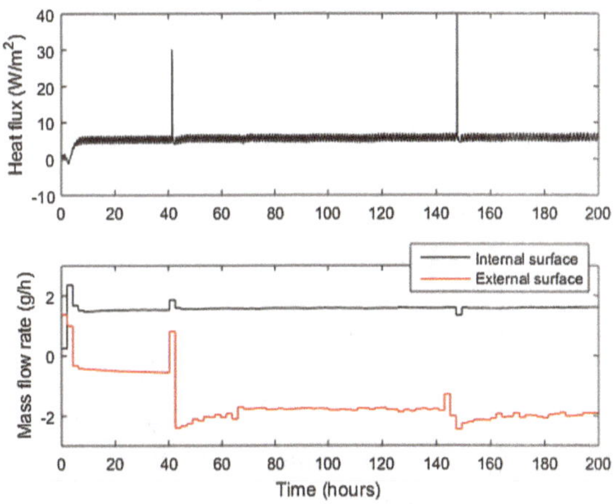

Figure 3: Outputs: internal heat flux and internal and external mass flow rates.

iment adjustments. This procedure included a perturbation on the system useful to verify the model adaptation capability.

4.1.2 Simulation parameters

For the nonlinear regression procedures, MARS technique was adopted in order to generate one model to each output considering four input signals. The data set was divided in two parts, where 80 % of the data set was adopted for the estimation (training) phase, and 20 % for the validation phase. More about this configuration will be commented in the sequence of this chapter.

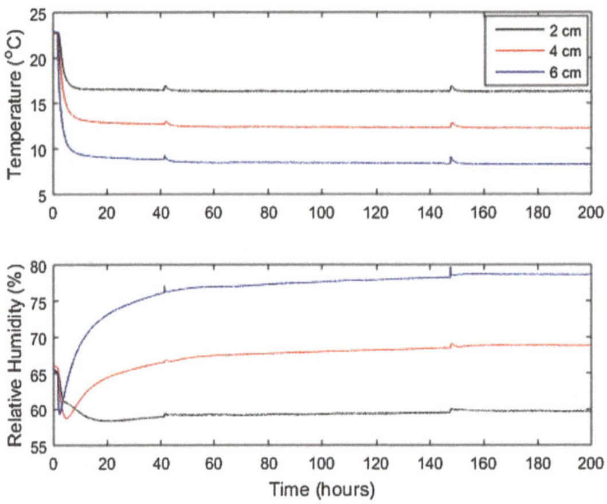

Figure 4: Outputs: temperature and relative humidity at 2, 4 and 6 cm depths.

In order to evaluate the model performance in predicting temperature and relative humidity at different depths, heat flux at the internal surface and mass flow rate at both internal and external surfaces, multiple performance indicators were adopted. The coefficient of determination (R^2), the mean square error (MSE), the root mean square error (RMSE), the relative root mean square error (RRMSE) and the mean absolute error (MAE) were selected to evaluate the model performance for both estimation and validation phases.

The R^2 measures how well considered independent variables account for the measured dependent variable. Higher values correlate with greater model predictive capability [55]. While MSE measures the average of the squares errors, RMSE computes the square error of the prediction compared to actual values, and also the square root of the summation value, representing the average distance of a data point from the fitted curve provide by the model along a vertical line. This performance index is efficient at assessing undesirably large differences. The RRMSE can be calculated by dividing RMSE with average value of measured data, and MAE is usually adopted to measure how close forecasts or predictions are to the eventual outcomes. All these performance indicators were described in equations (18)–(22):

$$R^2 = 1 - \frac{\sum_{t=1}^{N}[y(t) - \hat{y(t)}]^2}{\sum_{t=1}^{N}[y(t) - \hat{y})]^2}, \tag{18}$$

$$\mathrm{MAE} = 1 - \frac{\sum_{t=1}^{N}|y(t) - \hat{y(t)}|]}{N}, \tag{19}$$

$$\mathrm{MSE} = 1 - \frac{\sum_{t=1}^{N}[y(t) - \hat{y(t)}]^2}{N}, \tag{20}$$

$$\mathrm{RMSE} = \sqrt{\mathrm{MSE}}, \tag{21}$$

$$\mathrm{RRMSE} = \frac{\mathrm{RMSE}}{\frac{1}{n}\sum_{t=1}^{N} \hat{y}(t)}. \tag{22}$$

In equations (18)–(22), N represents the number of samples, $y(t)$ is the output of the real system (experimental data), $\hat{y}(t)$ is the output estimated by the model (MARS models) and $\bar{y}$ is the mean value of the system output set. In this case, t represents each sample time. All the performance indicators have been adopted for both estimation and validation phases.

The first test was performed in order to evaluate how efficient is MARS in predicting nine distinct outputs just considering temperature and relative humidity at the surfaces of the building material. This first analysis was used to define which output provides the most complex dynamics. Assuming the same parameters to create a model to each output, a maximum number of 15 basis functions was assumed.

In the sequence, the percentages of the data set for both estimation and validation phases were evaluated taking into account only the most complex dynamic (internal heat flux), considering a maximum of 30 basis functions.The percentage of the data set adopted for both training and validation phases were evaluated considering from 10 % to 90 % of the data set. Less than 80 % provided irrelevant approximation in terms of prediction, and more than 80 % overestimated models in the validation phase. In this case, 80 % of the data was adopted for the estimation, and consequently 20 % of the date set was used for the validation procedure.

The next procedure was to build a MARS model to each output in order to reach the best model prediction in terms of R^2, MSE, RMSE, RRMSE and MAE. The main idea was to obtain the best model performance for both estimation and validation phases considering less basis function as possible. In order to reach these design specifications, a maximum of 130 basis function was considered for the first output, which represents the internal heat flux, and 30 basis function for the other outputs.

Finally, the last procedure is the comparison between MARS models and a NARX multiple-input, multiple-output (MIMO) model, where a WNN was considered as nonlinear approximation tool.

4.2 Case study 2: data acquisition based on a numerical model

This case study presents a roof hygrothermal condition analysis. The model adopted for the porous media domain (Figure 5(a)) has been elaborated considering the differential governing equations for moisture, air and energy balances and was originally proposed in [53]. The transient terms of each governing equation have been written in terms of the driving potentials to take more advantage of the Multitridiagonal-Matrix Algorithm (MTDMA) solution algorithm [47]. This model provides detailed information about heat and mass transfer in building structures. Additionally, a mould growth risk model, based on the study presented in [70], was also included. An example of mould and algae growth on concrete tiles can be found in Figure 5(b).

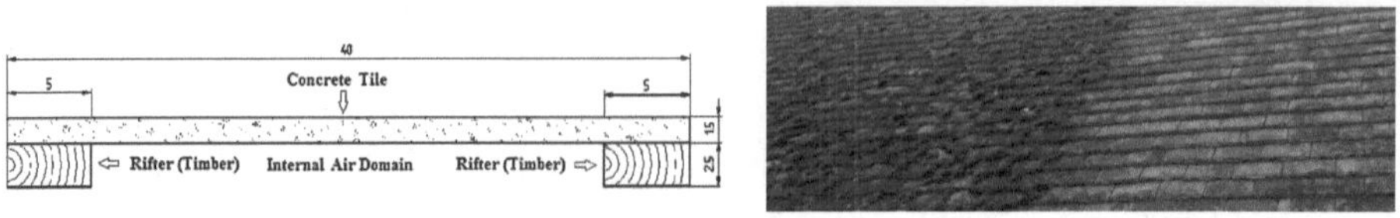

(a) (b)

Figure 5: (a) Physical domain and dimensions (cm) of the roof [24]; (b) Example of mould and algae growth in concrete tiles [65].

This model was addressed in details in the work presented in [24], and will not be discussed in details in this chapter as the main objective here is the system identification procedure.

4.2.1 Experimental procedures

The roof analyzed is composed of concrete tile (1.5 cm) and two rifters (2.5 cm) as shown in Figure 5. The hygrothermal properties have been obtained from [39] for concrete and from IEA Annex 14 Report [32] for timber (rifters). The internal emissivity was considered equal to 0.9 for both concrete tile and timber roofs and the solar absorptivity equal to 0.6 was adopted.

A regular 2-D mesh (2.5 mm^2) for the discretization using the finite-volume method and a 30 s constant time step have been applied for all simulations. The computational code was implemented in C programming language in order to enable dynamic memory allocation, and the sample time for data generation was set to 6 hours. Due to high computational time consumption of the computational code, the simulation using the hygrothermal model was performed for almost 1 year and 8 months, when the index M = 3 was reached for the mould growth (Figure 6). According to the model proposed in [70], $M = 3$ means some growth detected visually.

A temperature of $24^{\circ}C$ and a relative humidity of 50 % (conditioned environment) were considered for indoor conditions. The outdoor climate conditions were represented by the TRY (Test Reference Year) weather data for the city of Curitiba, Brazil, which can be found in [41], and are presented in Figure 7 for the first week of January (summer period), and in Figure 8 for the first week of July (winter period). Constant convective heat transfer coefficients of 3 and 10 W/m^2.K have been used at the internal and external surfaces. The external and internal convective water vapor transfer coefficients are calculated through Lewis's relation for each control volume. The other surfaces were considered adiabatic and impermeable. The sky temperature correlation presented in [62] was assumed. Gas (moist air) pressure has been considered constant at all surfaces.

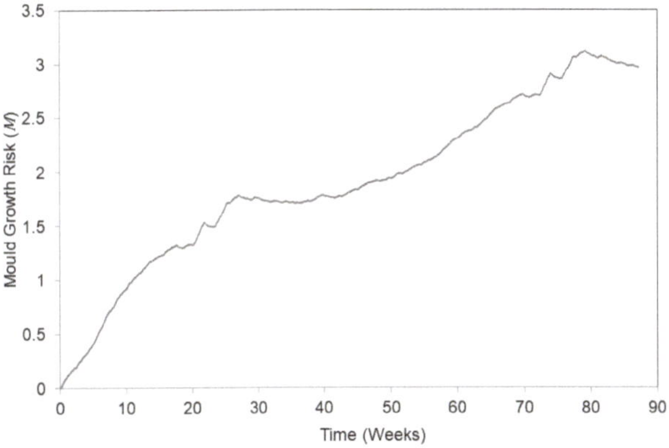

Figure 6: Mould index evolution in the concrete tile (adapted from [24]).

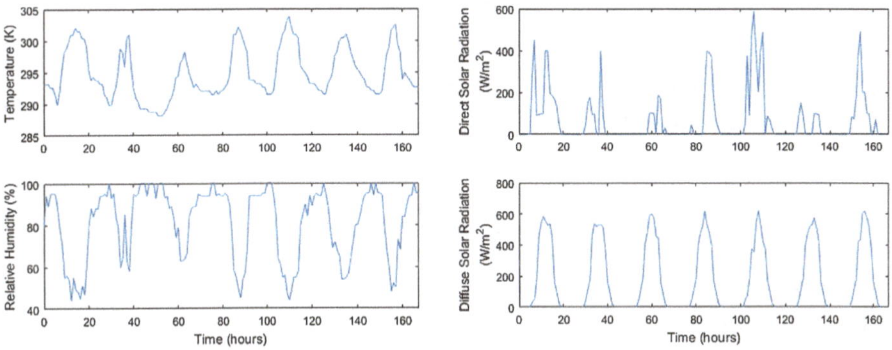

(a) (b)

Figure 7: Weather data for the first week of January (summer period) for Curitiba, Brazil: (a) temperature and relative humidity; (b) direct and diffuse solar radiation [24].

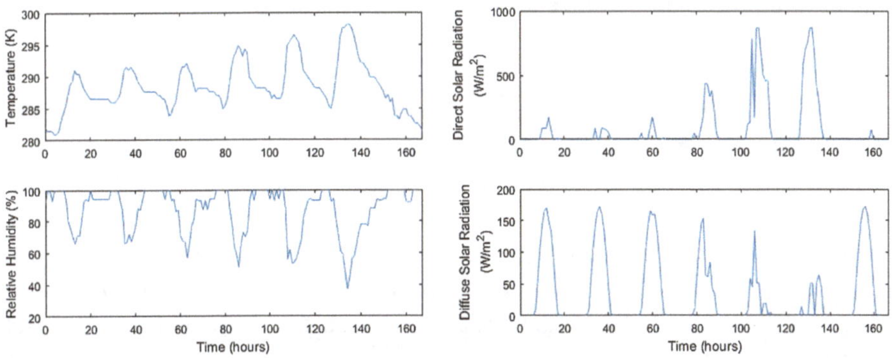

(a) (b)

Figure 8: Weather data for the first week of July (winter period) for Curitiba, Brazil: (a) temperature and relative humidity; (b) direct and diffuse solar radiation [24].

4.2.2 Simulation parameters

A Nonlinear AutoRegressive with eXogenous inputs (NARX) model structure was adopted in this case for the application of the LS-SVM technique. Four inputs were considered on the system identification procedures: external temperature (in K), external relative humidity (in %), direct solar radiation (W/m^2), and diffuse solar radiation (W/m^2). Four outputs have been dentified in a one-step ahead prediction considering a multiple-input, single-output (MISO) structure. The identified outputs are: sensible heat flux (W/m^2), latent heat flux (W/m^2), vapor flux ($kg/(m^2.s)$), and the mould growth risk index. A NARX model can be defined as product (equation (23)) to create a nonlinear form presented in equation (24):

$$[a_1, a_2, \ldots, a_{n_a}, b_1, b_2, \ldots b_{n_b}] * \begin{bmatrix} y(t-1), y(t-2), \ldots, y(t-n_a) \\ u(t-n_d-1), u(t-n_k-2), \ldots, u(t-n_d-n_b-1) \end{bmatrix}^T \tag{23}$$

$$\hat{y}(t) = f\begin{pmatrix} y(t-1), y(t-2), \ldots, y(t-n_a), u(t-n_d-1) \\ u(t-n_d-2), \ldots, u(t-n_d-n_b-1) \end{pmatrix} \tag{24}$$

where t represents the current time, and d the delayed sample.

The nonlinear function can be expressed in terms of the model regressors, and the nonlinear mapping can be performed using nonlinear estimators. In equations (17) and (18), $y(t)$ represents the current output of the model, $y(t-d)$ is a finite number of past outputs, $u(t-d)$ the inputs, $e(t)$ is a white-noise error that is introduced in the difference equation and $\hat{y}(t)$ the predicted output of the system. The model structure is entirely defined by three integers, where n_a represents the number of poles, $(n_b - 1)$ is the number of zeros and n_d is the time delay of the systems.

At the beginning of this analysis, 1 year and 6 months of data were collected (2444 samples of inputs and outputs) considering the physical domain presented on Figure 5(a).

The training set was divided into 10 subsets following the l-fold cross validation method [2], in order to train the classifier 10 times, each time leaving out one of the subsets from training, but using the omitted subset to compute the classification errors using the Mean Absolute Error (MAE) presented in equation (19) as minimization criterion.

In terms of the model structure, n_a and n_b were set equal 2 according to previous analysis of the number of regressors using the neighborhood component analysis for regression [29]. Additionally, no time delay was considered between the inputs and the outputs ($n_k = 0$).

In order to define the percentage of data used for training, an analysis using 10 % to 90 % of dada for training was performed. In this case, besides MAE, the coefficient of determination (R^2 in equation (14)) and the Mean Square Error (MSE in equation

(20)) were also adopted. In this case, 50 % of the data set was considered for training, while the other 50 % was considered for the validation procedure.

5 Results

Based on the previous case studies, this section presents the results of both MARS and LS-SVM methods. MARS technique was applied to create regression models with the objective of predicting temperature and relative humidity at 2, 4 and 6 cm depth, internal heat flux, and internal and external mass flow rates (nine outputs). Four variables were considered as inputs. They are the internal and external surfaces temperature and relative humidity, two in each chamber.

In the sequence, the results of LS-SVM method are presented in Section 5.2, where a system identification approach considering a NARX model structure is assumed. LS-SVM method was used to predict four outputs based on the data set provided by a well established numerical model. These outputs are: sensible heat flux, latent heat flux, vapor flux and Mould growth risk. As inputs for the LS-SVM, we adopted the external temperature, external relative humidity, direct solar radiation and diffuse solar radiation.

5.1 MARS results for case study 1: experimental plant dataset

This section presents the simulation results using MARS to compute a model to predict temperature and relative humidity profiles (at 2, 4 and 6 cm), heat flow and mass flux rates in a wood fiber board based on the data acquisition procedures presented in Section 4.1.

The first approach to predict hygrothermal behavior of building materials on a nonisothermal case, as mentioned before, was the definition of the more complex dynamic. A specific analysis of this procedure will not be provided as it can be verified using the complementary results addressed in this section, where 109 basis functions should be adopted to reproduce properly the internal heat flux behavior, while 30 basis functions were enough to represent the other outputs. More information will be presented in the sequence of this section.

Figures 9–11 show the results obtained using the models created by MARS technique. Both estimation and validation phases were include in each figure, where the vertical dashed blue line indicates the transition between the two phases representing 80 % of the data set.

Figure 9 shows the results for the internal heat flux prediction, where the maximum absolute error in the validation phase is lower than 2 W/m^2. Note that the model could even predict the dynamic when the chamber door was open for adjusts during

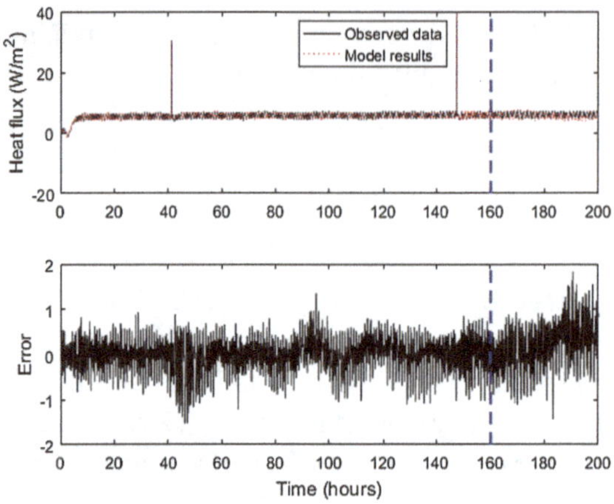

Figure 9: Inputs: Model results: Internal heat flux prediction.

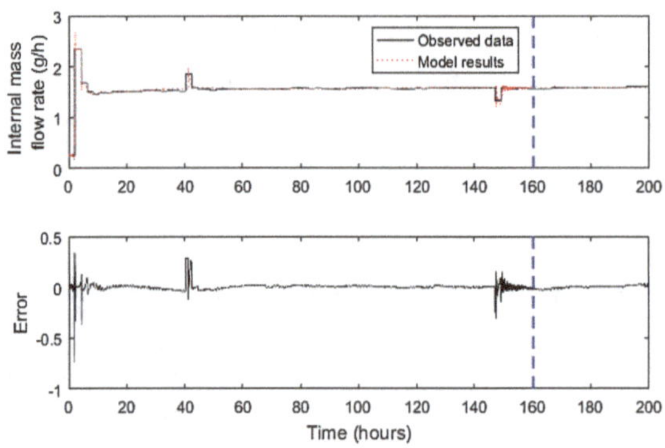

Figure 10: Model results: Internal mass flow rate prediction.

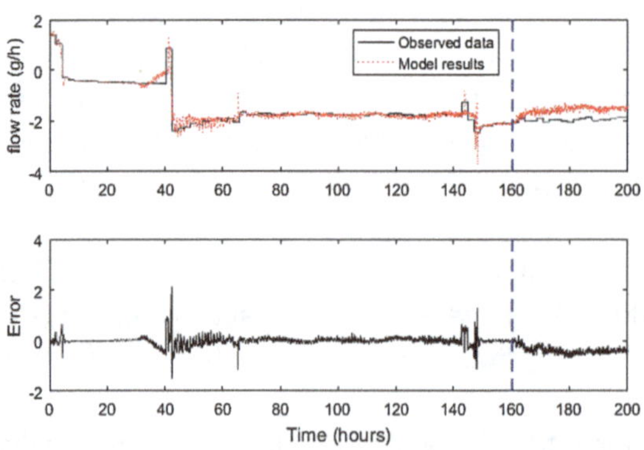

Figure 11: Model results: External mass flow rate prediction.

the data acquisition. In the same figure, an error analysis is presented for both estimation and validation phases.

Figures 10 and 11 present the model response when compared to the experimental results in terms of internal and external mass flow rates. As it can be verified, the maximum error occurs when the chambers were open. This situation does not represent a normal behavior of strongly hygroscopic and highly capacity materials submitted to normal weather conditions, where boundary conditions will variate in a smoother way. However, this situation provided important information about the consistency of the MARS model presented in this case.

Figures 12–14 illustrate MARS models behavior in predicting temperature and relative humidity at 2, 4 and 6 cm depths, respectively. These figures illustrate the best results found by the MARS model when compared to the previous outputs. In terms of error analysis, the worst result for temperature prediction was found during the estimation phase at 2 cm depth, where a difference of 0.24 °C between experimental and model results was reported. In terms of relative humidity, the maximum difference between MARS model and the data set was also found at 4 cm depth, were 1 % difference was reported. In terms of model precision, MARS models provided promising results for predicting temperature and relative humidity at different depths.

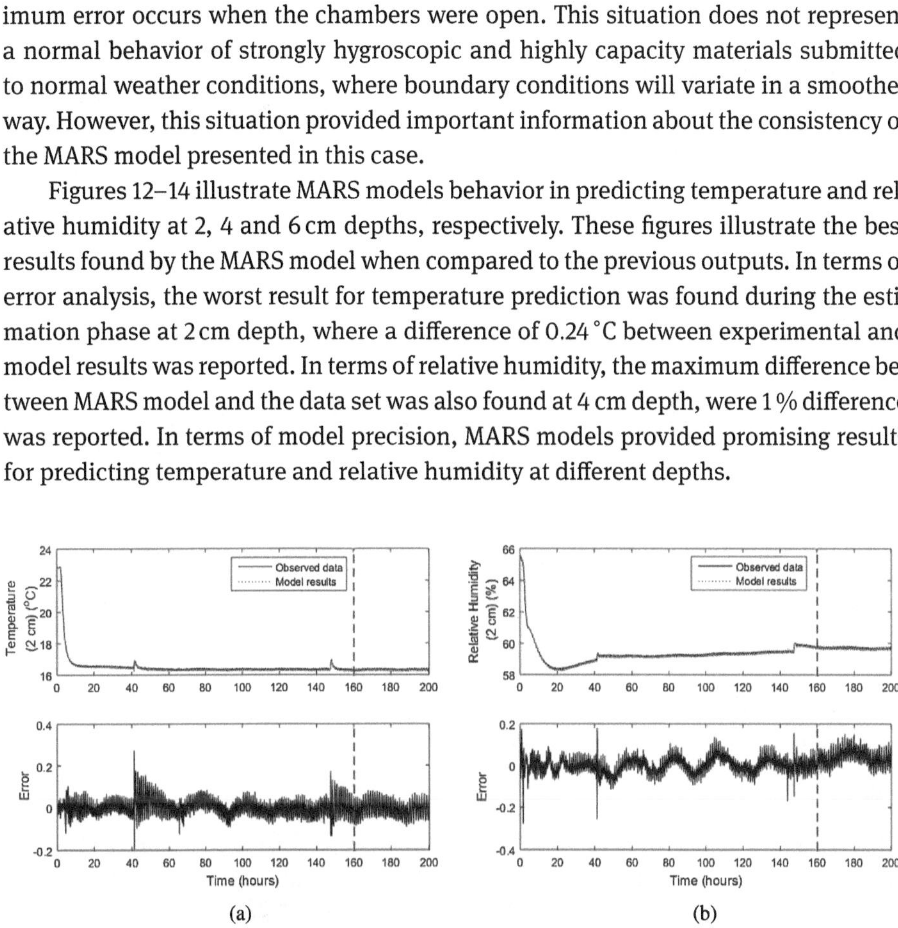

Figure 12: Model results: (a) temperature at 2 cm depth; (b) relative humidity at 2 cm depth.

Table 3 presents the MARS models performance in taking into account the performance indicators R^2, MSE, RMSE, RRMSE and MAE. Additionally, it provides the number of basis functions adopted to build each model.

According to [19, 42], model accuracy is considered excellent when RRMSE < 10 %, good if 10 % < RRMSE < 20 %, fair if 20 % < RRMSE < 30 % and poor if RRMSE > 30 %. However, R^2 values between 0.9 and 1.0 are considered reasonable approximations for many applications in identification, control and forecasting fields [54]. It is important to emphasize that models that produce reasonable predictions

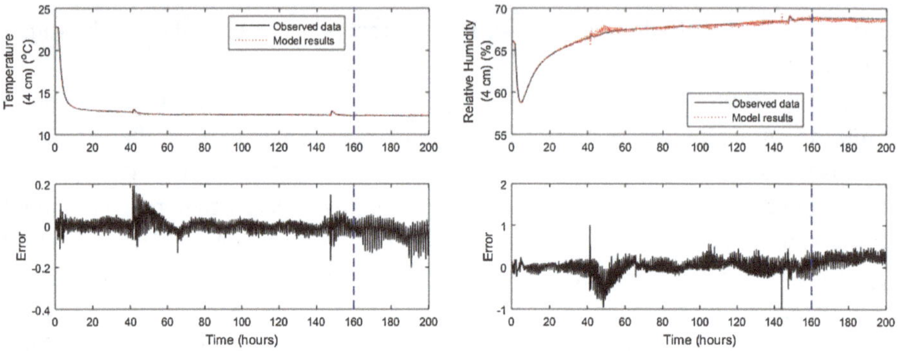

(a) (b)

Figure 13: Model results: (a) temperature at 4 cm depth; (b) relative humidity at 4 cm depth.

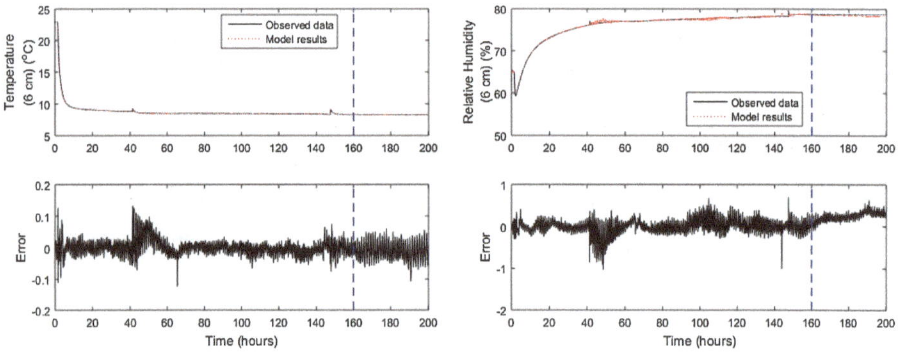

(a) (b)

Figure 14: Model results: (a) temperature at 6 cm depth; (b) relative humidity at 6 cm depth.

considering less basis functions can be obtained, e. g., for the first output 83 basis functions can provided R^2 values for both estimation and validation phases higher than 0.9. However, in this case, RRMSE higher than 30 % was obtained for the validation phase.

To conclude this analysis, MARS does not consider any delayed inputs or outputs in the model, which means that MARS approach is much more suitable for building simulation purposes as it reduces the accumulated error and the number of parameters adopted during prediction.

5.2 LS-SVM results for case study 2: numerical model dataset

This section presents the results of the LS-SVM method when applied for system identification of case study 2.

Table 3: MARS results in terms of R^2, MSE, RMSE, RRMSE and MAE.

Output	#Basis Functions	Phase	R^2	MSE	RMSE	RRMSE	MAE
1	109	Estimation	0.938248	5.935047E-05	0.007704	0.248499	0.005800
		Validation	0.912673	7.497438E-05	0.008659	0.005800	0.006474
2	27	Estimation	0.966497	2.567005E-04	0.016022	0.183038	0.006709
		Validation	0.965463	2.135416E-04	0.014613	0.185842	0.006356
3	26	Estimation	0.956099	1.906878E-03	0.043668	0.209525	0.023419
		Validation	0.950510	1.871827E-03	0.043265	0.222464	0.025255
4	30	Estimation	0.998012	3.460761E-05	0.005883	0.044586	0.004599
		Validation	0.997560	3.449758E-05	0.005873	0.049398	0.004681
5	27	Estimation	0.998382	2.273013E-05	0.004768	0.040226	0.003708
		Validation	0.997543	2.854575E-05	0.005343	0.049573	0.004158
6	28	Estimation	0.999495	8.412195E-06	0.002900	0.022474	0.022474
		Validation	0.999161	1.145737E-05	0.003385	0.028963	0.028963
7	28	Estimation	0.994363	2.441491E-04	0.015625	0.075079	0.010770
		Validation	0.992699	2.945681E-04	0.017163	0.085446	0.012661
8	28	Estimation	0.999834	2.233225E-06	0.001494	0.012878	0.001064
		Validation	0.999770	2.532967E-06	0.001592	0.015157	0.001180
9	30	Estimation	0.997822	6.491749E-05	0.008057	0.046668	0.005691
		Validation	0.996729	8.649780E-05	0.009300	0.057196	0.007005

Figures 15 and 16 present the comparison between LS-SVM and the numerical method in terms of prediction. Figure 15 presents the validation phases for outputs 1 (vapor flow) and 2 (sensible heat flow), while Figure 16 shows the results in terms of validation phase for outputs 3 (latent heat flow) and 4 (mould growth risk). Training results were not presented in the figures due to the amount of data available and the similarity to the validation phase graphics. The error is also presented in these figures. Additionally, Table 4 reports the values of the coefficient of determination for both training and validation phases.

As it can be observed in Figures 15 and 16, the highest absolute error values can be found in the mould growth index approximation, which can be justified by the different on the dynamic of the mould growth model. As the M index provides different growth behavior in distinct stages of growth, those defined by distinct equations, all these stages should be used on the LS-SVM training stage. As the LS-SVM training data set presented in this chapter considered just the behavior between $\Theta \leqslant M < 2$, where higher values of M were just considered in the validation phase, the model was still capable to reproduce mould growth with a considerable precision.

As it can be viewed in Table 4, the model presented consistent approximation for all the four outputs. Higher values of R^2 for the mould growth risk can be justified by its behavior prior to exceed the $M = 2$ limit, as the coefficient of determination is a cumulative measure.

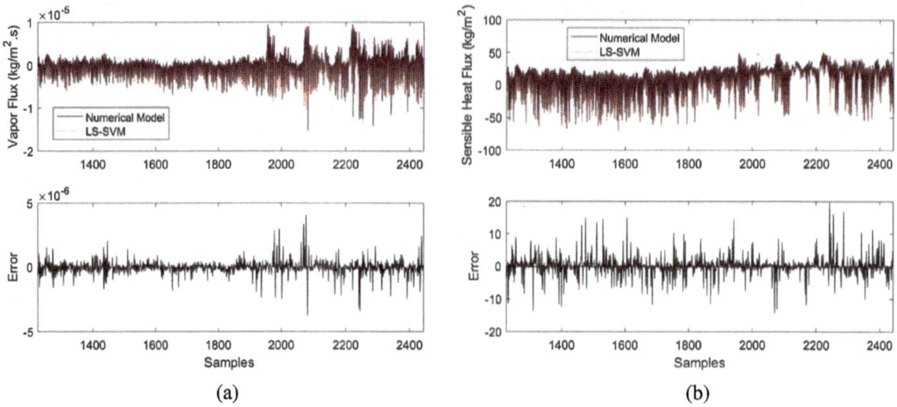

Figure 15: Comparison between the numerical and the LS-SVM models – validation phase (a) output 1: vapor flux; (b) output 2: sensible heat flux.

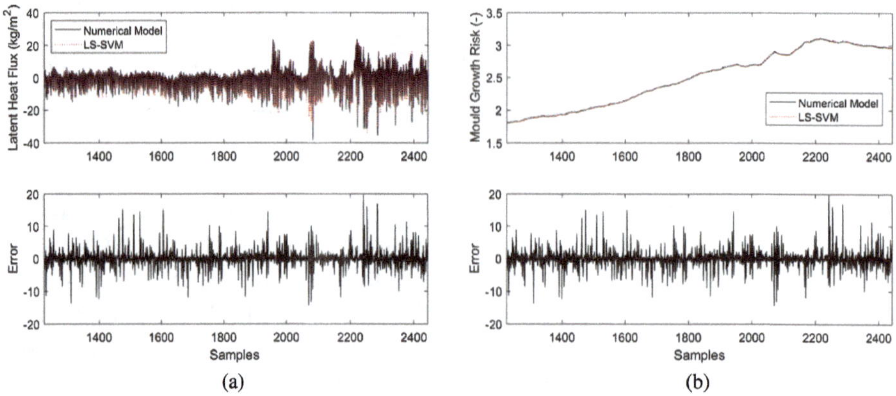

Figure 16: Comparison between the numerical and the LS-SVM models – validation phase (a) output 3: latent heat flux; (b) output 4: mould growth risk.

Table 4: MARS results in terms of R^2, MSE, RMSE, RRMSE and MAE.

Sensor	Range	Accuracy
Temperature (SHT75)	[5–45] [°C]	0.17 [°C]
Relative Humidity (SHT75)	[40–90] %	2 %
Temperature (Pt100)	[5–45] [°C]	0.15 [°C]
Temperature (thermocouples)	[5–45] [°C]	0.16 [°C]
Mass	4200 g/3000 g	0.01 g
Liquid flow	–	0.04 g/h

In terms of computational effort, the whole simulation considering both training and test phased does not take more than 30 s, while the traditional method took about 120 hours.

6 Conclusion

This chapter presented two black-box approaches to estimate hygrothermal variations within building materials. Two case studies were considered to generate data and to validate both regression and system identification methods.

In order to predict internal heat flux, internal and external mass flow rates, and both temperature and relative humidity profiles of highly hygroscopic building materials, a wood fiber board material was adopted, which is commonly used in building envelops as insulation. This material was monitored under controlled conditions considering an experimental plant developed in France. Base on data generated during this experimentation, a Multivariate Adaptive Regression Splines (MARS) technique was used to develop regression models considering nine outputs and just four inputs. The previous mentioned case became the first case study of this chapter.

In order to build less complex models as possible, procedures to determine the complexity of the model outputs, the percentage of data used for both estimation and validation phases, and the minimum number of basis functions that allows satisfactory performance were considered. The model built to each output took the form of an expansion in product spline basis function, where the number of basis functions and the parameters associated to each of them are automatically calculated using the dataset. Five different performance criteria were adopted in order to evaluate the proposed models, showing promising results of MARS in predicting hygrothermal behavior of saturated materials considering five performance coefficients: R^2, MSE, RMSE, RRMSE and MAE.

Finally, it is important to emphasize that MARS considered a small number of inputs and has no necessity of delayed outputs in its model structure. Since the present work, MARS regression technique has not been applied in predicting coupled heat and mass transfer in building materials, an area dominated by both numerical methods based on physical modeling and system identification techniques based on concepts of artificial intelligence.

In a second case study, an approach to predict vapor flux, sensible heat flux, latent heat flux and mould growth risk in concrete tiles based on the external weather conditions was proposed. By creating four MISO (multiple-input, single-output) models, external temperature, relative humidity and direct and diffuse solar radiation were considered as input of a Nonlinear AutoRegressive with eXogenous inputs (NARX) model structure. In order to determine the parameters of these models, Least Squares Support Vector Machines (LS-SVM) technique was considered. The proposed approach presented promising results considering just two regressors (delayed inputs/outputs), reaching values for the coefficient of determination with smaller value over 0.97 for the validation phase.

For future works, in terms of the first case study, it can be suggested the comparison of MARS models to numerical models and distinct black-box system identification techniques in terms of computational effort and performance related to the experimental data approximation. Models evaluation for an isothermal case is also proposed. To conclude, a novel model approach considering less input parameters, e. g., internal and external air temperature and relative humidity, will be considered, as well as four additional outputs in order to predict the surfaces temperature and relative humidity.

In terms of the second case study, an idea is to include the changes on roof solar absorptivity in the presented model, so that mould growth and the effects on the hygrothermal performance of the building could be verified more precisely.

Bibliography

[1] Al-Sanea, S. A.: Thermal performance of building roof elements. Build. Environ. 37(7), 665–675 (2002). https://doi.org/10.1016/S0360-1323(01)00077-4.

[2] An, S., Liu, W., Venkatesh, S.: Fast cross-validation algorithms for least squares support vector machine and kernel ridge regression. Pattern Recognit. 40, 2154–2162 (2007). https://doi.org/10.1016/j.patcog.2006.12.015.

[3] Ascione, F., Bianco, N., De Stasio, C., Mauro, G. M., Vanoli, G. P.: Artificial neural networks to predict energy performance and retrofit scenarios for any member of a building category: A novel approach. Energy 118, 999–1017 (2016). https://doi.org/10.1016/j.energy.2016.10.126.

[4] Belusko, M., Bruno, F., Saman, W.: Investigation of the thermal resistance of timber attic spaces with reflective foil and bulk insulation, heat flow up. Appl. Energy 88(1), 127–137 (2011). https://doi.org/10.1016/j.apenergy.2010.07.017.

[5] Berger, J., Mazuroski, W., Mendes, N., Guernouti, S., Woloszyn, M.: 2D whole-building hygrothermal simulation analysis based on a PGD reduced order model. Energy Build. 112, 49–61 (2016). https://doi.org/10.1016/j.enbuild.2015.11.023.

[6] Brazilian Energy Research Company (EPE). Brazilian Energy Balance (2016). Retrieved from www.epe.gov.br.

[7] Buonomano, A.: Code-to-Code Validation and Application of a Performance Analysis. Energies 9(4), 301 (2016). https://doi.org/10.3390/en9040301.

[8] Burges, C. J. C.: A Tutorial on Support Vector Machines for Pattern Recognition. Data Min. Knowl. Discov. 2(2), 121–167 (1998). https://doi.org/10.1023/A:1009715923555.

[9] Busser, T., Berger, J., Piot, A., Pailha, M., Woloszyn, M.: Dynamic experimental method for identification of hygric parameters of a hygroscopic material. Build. Environ. 131, 197–209 (2018). https://doi.org/10.1016/j.buildenv.2018.01.002.

[10] Chen, Y., Xu, P., Chu, Y., Li, W., Wu, Y., Ni, L., Wang, K.: Short-term electrical load forecasting using the Support Vector Regression (SVR) model to calculate the demand response baseline for office buildings. Appl. Energy 195, 659–670 (2017). https://doi.org/10.1016/j.apenergy.2017.03.034.

[11] Chen, Z. Q., Shi, M. H.: Study of heat and moisture migration properties in porous building materials. Appl. Therm. Eng. 25(1), 61–71 (2005). https://doi.org/10.1016/j.applthermaleng.2004.05.001.
[12] Cheng, H., Wang, X., Zhou, M.: Optimized Design and Feasibility of a Heating System with Energy Storage by Pebble Bed in a Solar Attic. Energies 10(3), 328 (2017). https://doi.org/10.3390/en10030328.
[13] Cheng, M.-Y., Cao, M.-T.: Accurately predicting building energy performance using evolutionary multivariate adaptive regression splines. Appl. Soft Comput. 22, 178–188 (2014). https://doi.org/10.1016/j.asoc.2014.05.015.
[14] Chou, J. S., Bui, D. K.: Modeling heating and cooling loads by artificial intelligence for energy-efficient building design. Energy Build. 82, 437–446 (2014). https://doi.org/10.1016/j.enbuild.2014.07.036.
[15] Ciulla, G., Lo Brano, V., D'Amico, A.: Modelling relationship among energy demand, climate and office building features: A cluster analysis at European level. Appl. Energy 183, 1021–1034 (2016). https://doi.org/10.1016/j.apenergy.2016.09.046.
[16] Crawley, D. B., Hand, J. W., Kummert, M., Griffith, B. T.: Contrasting the capabilities of building energy performance simulation programs. Build. Environ. 43(4), 661–673 (2008). https://doi.org/10.1016/j.buildenv.2006.10.027.
[17] D'Orazio, M., Di Perna, C., Di Giuseppe, E., Morodo, M.: Thermal performance of an insulated roof with reflective insulation: Field tests under hot climatic conditions. J. Build. Phys. 36(3), 229–246 (2013). https://doi.org/10.1177/1744259112448181.
[18] Dai, C., Zhang, H., Arens, E., Lian, Z.: Machine learning approaches to predict thermal demands using skin temperatures: Steady-state conditions. Build. Environ. 114, 1–10 (2017). https://doi.org/10.1016/j.buildenv.2016.12.005.
[19] Despotovic, M., Nedic, V., Despotovic, D., Cvetanovic, S.: Evaluation of empirical models for predicting monthly mean horizontal diffuse solar radiation. Renew. Sustain. Energy Rev. 56, 246–260 (2016). https://doi.org/10.1016/j.rser.2015.11.058.
[20] Dos Santos, G. H., Mendes, N.: Numerical analysis of hygrothermal performance of reflective insulated roof coatings. Appl. Therm. Eng. 81, 66–73 (2015). https://doi.org/10.1016/j.applthermaleng.2015.02.017.
[21] Fabiani, C., Pisello, A. L.: Coupling controlled environmental forcing and Transient Plane Source method: an innovative thermal characterization procedure for building insulation materials. Appl. Therm. Eng. 130, 254–263 (2018). https://doi.org/10.1016/j.applthermaleng.2017.10.155.
[22] Fang, X., Athienitis, A. K., Fazio, P. P.: Methodologies for shortening test period of coupled heat-moisture transfer in building envelopes. Appl. Therm. Eng. 29(4), 787–792 (2009). https://doi.org/10.1016/j.applthermaleng.2008.04.006.
[23] Freire, R. Z., Coelho, L., dos Santos, G. H., Mariani, V. C.: Predicting building's corners hygrothermal behavior by using a Fuzzy inference system combined with clustering and Kalman filter. Int. Commun. Heat Mass Transf. 71, 225–233 (2016). https://doi.org/10.1016/j.icheatmasstransfer.2015.12.015.
[24] Freire, R. Z., Santos, G. H., Coelho, L. S.: Hygrothermal dynamic and mould growth risk predictions for concrete tiles by using Least Squares Support Vector Machines. Energies 10(8) (2017). https://doi.org/10.3390/en10081093.
[25] Friedman, J. H.: Multivariate Adaptive Regression Splines. Ann. Stat. 19(1), 1–67 (1991). https://doi.org/10.1214/aos/1176347963.
[26] Friedman, N., Kumar, P., Zander, E., Matthies, H. G.: Bayesian calibration of model coefficients for a simulation of flow over porous material involving SVM classification. Proc. Appl. Math. Mech. 16(1), 669–670 (2016). https://doi.org/10.1002/pamm.201610323.
[27] Gallegos-Ortega, R., Magaña-Guzmán, T., Reyes-López, J. A., Romero-Hernández, M. S.: Thermal behavior of a straw bale building from data obtained in situ. A case in Northwestern México. Build. Environ. 124, 336–341 (2017). https://doi.org/10.1016/j.buildenv.2017.08.015.

[28] Gestel, T. V., Suykens, J. A. K., Baesens, B., Viaene, S., Vanthienen, J., Dedene, G., Vandewalle, J.: Benchmarking Least Squares Support Vector. Mach. Learn. 54(1), 5–32 (2004).
[29] Goldberger, J., Roweis, S., Hinton, G., Salakhutdinov, R.: Neighbourhood Components Analysis. Science 290(5500), 2323–2326 (2000).
[30] Haba, B., Agoudjil, B., Boudenne, A., Benzarti, K.: Hygric properties and thermal conductivity of a new insulation material for building based on date palm concrete. Constr. Build. Mater. 154, 963–971 (2017). https://doi.org/10.1016/j.conbuildmat.2017.08.025.
[31] Hamdy, M., Nguyen, A.-T., Hensen, J. L. M.: A performance comparison of multi-objective optimization algorithms for solving nearly-zero-energy-building design problems. Energy Build. 121, 57–71 (2016). https://doi.org/10.1016/j.enbuild.2016.03.035.
[32] Hens, H.: IEA Annex 14: Condensation and Energy. J. Build. Phys. (1992). https://doi.org/10.1177/109719639201500307.
[33] Hou, S., Liu, F., Wang, S., Bian, H.: Coupled heat and moisture transfer in hollow concrete block wall filled with compressed straw bricks. Energy Build. 135, 74–84 (2017). https://doi.org/10.1016/j.enbuild.2016.11.026.
[34] Hukka, A., Viitanen, H. A.: A mathematical model of mould growth on wooden material. Wood Sci. Technol. 33(6), 475–485 (1999).
[35] James, C., Simonson, C. J., Talukdar, P., Roels, S.: Numerical and experimental data set for benchmarking hygroscopic buffering models. Int. J. Heat Mass Transf. 53(19–20), 3638–3654 (2010). https://doi.org/10.1016/j.ijheatmasstransfer.2010.03.039.
[36] Jekabsons, G.: ARESLab: Adaptive Regression Splines toolbox for Matlab/Octave (2011). Retrieved May 3, 2017, from http://www.cs.rtu.lv/jekabsons/.
[37] Khan, U., Schmidt-Thieme, L., Nanopoulos, A.: Collaborative SVM classification in scale-free peer-to-peer networks. Expert Syst. Appl. 69, 74–86 (2017). https://doi.org/10.1016/j.eswa.2016.10.008.
[38] Khoukhi, M., Fezzioui, N., Draoui, B., Salah, L.: The impact of changes in thermal conductivity of polystyrene insulation material under different operating temperatures on the heat transfer through the building envelope. Appl. Therm. Eng. 105, 669–674 (2016). https://doi.org/10.1016/j.applthermaleng.2016.03.065.
[39] Künzel, H. M., Holm, A. H., Krus, M.: Hygrothermal Properties and Behaviour of Concrete. In: WTA-Almanach – Wissenschaftlich-Technische Arbeitsgemeinschaft für Bauwerkserhaltung und Denkmalpflege, pp. 161—181 (2008).
[40] Lakatos, Á.: Investigation of the moisture induced degradation of the thermal properties of aerogel blankets: Measurements, calculations, simulations. Energy Build. 139, 506–516 (2017). https://doi.org/10.1016/j.enbuild.2017.01.054.
[41] Lamberts, R.: Wheather files in format TRY, SWERA, CSV and BIN (2017) (in Portuguese). Retrieved June 28, 2017, from http://www.labeee.ufsc.br/downloads/arquivos-climaticos/formato-try-swera-csv-bin.
[42] Li, M. F., Tang, X. P., Wu, W., Liu, H. B.: General models for estimating daily global solar radiation for different solar radiation zones in mainland China. Energy Convers. Manag. 70, 139–148 (2013). https://doi.org/10.1016/j.enconman.2013.03.004.
[43] Liu, X., Chen, Y., Ge, H., Fazio, P., Chen, G.: Numerical investigation for thermal performance of exterior walls of residential buildings with moisture transfer in hot summer and cold winter zone of China. Energy Build. 93, 259–268 (2015). https://doi.org/10.1016/j.enbuild.2015.02.016.
[44] Loonen, R. C. G. M., Favoino, F., Hensen, J. L. M., Overend, M., Attia, S.: Building Performance Simulation Tools: Selection Criteria and User Survey. J. Build. Perform. Simul. 1493, 1–19 (2010). https://doi.org/10.1080/19401493.2016.1152303.
[45] Lü, X., Modelling of heat and moisture transfer in buildings: I. Model program. Energy Build. 34, 1033–1043 (2002). https://doi.org/10.1016/S0378-7788(02)00021-X.

[46] Mat Daut, M. A., Hassan, M. Y., Abdullah, H., Rahman, H. A., Abdullah, M. P., Hussin, F.: Building electrical energy consumption forecasting analysis using conventional and artificial intelligence methods: A review. Renew. Sustain. Energy Rev. 70(December, 1108–1118 (2017). https://doi.org/10.1016/j.rser.2016.12.015.
[47] Mendes, N., Philippi, P. C., Lamberts, R.: A new mathematical method to solve highly coupled equations of heat and mass transfer in porous media. Int. J. Heat Mass Transf. 45(3), 509–518 (2002). https://doi.org/10.1016/S0017-9310(01)00172-7.
[48] Qin, M., Aït-Mokhtar, A., Belarbi, R.: Two-dimensional hygrothermal transfer in porous building materials. Appl. Therm. Eng. 30(16), 2555–2562 (2010). https://doi.org/10.1016/j.applthermaleng.2010.07.006.
[49] Rounaghi, M. M., Abbaszadeh, M. R., Arashi, M.: Stock price forecasting for companies listed on Tehran stock exchange using multivariate adaptive regression splines model and semi-parametric splines technique. Physica A 438, 625–633 (2015). https://doi.org/10.1016/j.physa.2015.07.021.
[50] Saber, H. H., Maref, W., Sherrer, G., Swinton, M. C.: Numerical modeling and experimental investigations of thermal performance of reflective insulations. J. Build. Phys. (2012). https://doi.org/10.1177/1744259112444021.
[51] Salvalai, G., Imperadori, M., Scaccabarozzi, D., Pusceddu, C.: Thermal performance measurement and application of a multilayer insulator for emergency architecture. Appl. Therm. Eng. 82, 110–119 (2015). https://doi.org/10.1016/j.applthermaleng.2015.02.062.
[52] Santos, G. H., Mendes, N.: Corner Effects on the hygrothermal performance of buildings. In: 13th Conference of International Building Performance Simulation Association (BS2013), pp. 3546—3553. Chambéry, France (2013).
[53] dos Santos, G. H., Mendes, N.: Analysis of Hygrothermal Performance and Mould Growth on Concrete Tiles. In: 23rd ABCM International Congress of Mechanical Engineering (COBEM 2015), Rio de Janeiro, Brazil, pp. 1—8 (2015).
[54] Schaible, B., Xie, H., Lee, Y.-C.: Fuzzy logic models for ranking process effects. IEEE Trans. Fuzzy Syst. 5(4), 545—556 (1997).
[55] Sekhar Roy, S., Roy, R., Balas, V. E.: Estimating heating load in buildings using multivariate adaptive regression splines, extreme learning machine, a hybrid model of MARS and ELM. Renew. Sustain. Energy Rev. (2017). https://doi.org/10.1016/j.rser.2017.05.249.
[56] Sentelle, C., Georgiopoulos, M.: A Fast Revised Simplex Method for SVM Training. In: 19th IEEE International Conference on Pattern Recognition (ICPR2008), pp. 3—6. IEEE, Tampa, FL, USA (2008). https://doi.org/10.1109/ICPR.2008.4761810.
[57] Slimani, Z., Trabelsi, A., Virgone, J.: Design of an experimental device for the study of hygrothermal behavior for very hygroscopic walls. In: International Conference on Materials and Energy (ICOME'16). La Rochele, France (2016).
[58] Slimani, Z., Trabelsi, A., Virgone, J.: Study of hygrothermal behavior of very hygroscopic insulation. In: International RILEM Conference Materials Systems and Structures in Civil Engineering 2016 (MSSCE 2016) on Moisture in Materials and Structures. Lyngby, Denmark (2016).
[59] Souto, G., Guilherme, L., Luvizotto, J., Cocco, V., Coelho, S.: Least squares support vector machines with tuning based on chaotic differential evolution approach applied to the identification of a thermal process. Expert Syst. Appl. 39(5), 4805–4812 (2012). https://doi.org/10.1016/j.eswa.2011.09.137.
[60] Suehrcke, H., Peterson, E. L., Selby, N.: Effect of roof solar reflectance on the building heat gain in a hot climate. Energy Build. 40(12), 2224–2235 (2008). https://doi.org/10.1016/j.enbuild.2008.06.015.
[61] Suykens, J. A. K., Gestel, T. V., Brabanter, J. D., De Moor, B. D., Vandewalle, J.: Least Squares Support Vector Machines. World Scientific Pub. Co., Singapore (2002). https://doi.org/10.1109/IJCNN.1999.831072.

[62] Swinbank, W. C.: Long wave radiation from clear skies. Q. J. R. Meteorol. Soc. 89(381), 339–348 (1963). https://doi.org/10.1002/qj.49708938105.

[63] Talukdar, P., Olutmayin, S. O., Osanyintola, O. F., Simonson, C. J.: An experimental data set for benchmarking 1-D, transient heat and moisture transfer models of hygroscopic building materials. Part I: Experimental facility and material property data. Int. J. Heat Mass Transf. 50(23–24), 4527–4539 (2007). https://doi.org/10.1016/j.ijheatmasstransfer.2007.03.026.

[64] Tian, Y., Bai, X., Qi, B., Sun, L.: Study on heat fluxes of green roofs based on an improved heat and mass transfer model. Energy Build. 152, 175–184 (2017). https://doi.org/10.1016/j.enbuild.2017.07.021.

[65] Trondheim, R.: Et Naturlig Valg For Renhold I Trondheim (2018). Retrieved from http://renholdtrondheim.org/wp-content/uploads/2017/05/takvask-21.jpg.

[66] U. S. Energy Information: Annual Energy Outlook 2013. Office of Integrated and International Energy Analysis. 1, 1–244 (2013). https://doi.org/https://doi.org/DOE/EIA-0383(2013).

[67] Van Belleghem, M., Steeman, M., Willockx, A., Janssens, A., De Paepe, M.: Benchmark experiments for moisture transfer modelling in air and porous materials. Build. Environ. 46(4), 884–898 (2011). https://doi.org/10.1016/j.buildenv.2010.10.018.

[68] Vapnik, V. N.: Statistical Learning Theory. Interpreting (1998). https://doi.org/10.2307/1271368.

[69] Vereecken, E., Roels, S.: Review of mould prediction models and their influence on mould risk evaluation. Build. Environ. 51, 296–310 (2012). https://doi.org/10.1016/j.buildenv.2011.11.003.

[70] Viitanen, H., Ojanen, T., Peuhkuri, R.: Mould growth modelling to evaluate durability of materials. In: Proceedings of the 12DBMC – International Conference on Durability of Building Materials and Components, Porto, Portugal, pp. 1–8 (2011). Retrieved from http://www.irbnet.de/daten/iconda/CIB22377.pdf.

[71] Yildiz, B., Bilbao, J. I., Sproul, A. B.: A review and analysis of regression and machine learning models on commercial building electricity load forecasting. Renew. Sustain. Energy Rev. 73, 1104–1122 (2017). https://doi.org/10.1016/j.rser.2017.02.023.

[72] Zhang, F., Deb, C., Lee, S. E., Yang, J., Shah, K. W.: Time series forecasting for building energy consumption using weighted Support Vector Regression with differential evolution optimization technique. Energy Build. 126, 94–103 (2016). https://doi.org/10.1016/j.enbuild.2016.05.028.

[73] Zhang, M., Qin, M., Rode, C., Chen, Z.: Moisture buffering phenomenon and its impact on building energy consumption. Appl. Therm. Eng. 124, 337–345 (2017). https://doi.org/10.1016/j.applthermaleng.2017.05.173.

Part II: **Modern urban agriculture**

Sérgio Manuel Serra da Cruz, Anja Steglich, Pedro Vieira Cruz, and
Ana Claudia Macedo Vieira

Investigating the challenges and opportunities of urban agriculture in global north and global south countries

Abstract: The size of the world's largest cities is increasing; the urbanization process is complicated and different in developed and developing countries. However, if well managed, urban spaces may offer valuable opportunities for economic and social development. This chapter investigates the current challenges and opportunities in Urban Agriculture (UA) and discusses if the adoption of Urban Computing (UC) and information and communications technologies (ICT) can aid urban dwellers, farmers and planners to progress UA in global north or global south countries like Germany and Brazil. Besides, the chapter seeks to explore new joint-research possibilities addressed in the 8th Brazil-Germany Frontiers of Science and Technology Symposium (BRAGFOST).

Keywords: Food security, urban agriculture, big data, urban computing, urban farmer, urban land use

1 Introduction

Cities around the world are growing on an unprecedented scale. At the beginning of the century, an estimated 40 % of the developing world's population (about 2 billion people) lived in urban spaces. By 2015–2020, over half of the world's population will be living in urban areas, the majority of them in developing countries [10, 14].

The size of the world's largest cities is increasing; the urbanization process is complex and quite different in global north developed countries and global south developing countries [15, 26, 30, 38]. The latter presents high levels of poverty, unemployment and food insecurity. The urban poor spend most of their income just to feed them-

Acknowledgement: This work has been partially supported by the Brazilian Agencies CNPq, CAPES, FNDE, the German A. V Humboldt Foundation and the Ibero-American Program for Science and Technology for Development (CYTED2014-515RT0489 – BigDSSAgro) and German Federal Ministry of Education and Research. Likewise, the authors want to thank Professors Gabriela Celani and Olfa Kanoun for the kind invitation to submit the manuscript and the reviewers of this chapter.

Sérgio Manuel Serra da Cruz, UFRRJ, Rio de Janeiro, Brazil, e-mail: serra@pet-si.ufrrj.br
Anja Steglich, TU-Berlin, Berlin, Germany
Pedro Vieira Cruz, UFRRJ, Rio de Janeiro, Brazil
Ana Claudia Macedo Vieira, UFRRJ, Rio de Janeiro, Brazil

https://doi.org/10.1515/9783110584455-006

selves, and their children suffer levels of malnutrition that are usually smaller than those found in rural areas. Thus, to survive, millions of favela (informal settlements or slums) dwellers have resorted to growing their food on marginalized pieces of land: in backyards, along rivers, roads, railways and under power lines [1].

If well managed, urban spaces may offer valuable opportunities for economic and social development [3]. Urban agriculture (UA) is an interdisciplinary research topic, which is gaining traction in many cities across the world. The movement is generating the highest amount of excitement and interest in many countries [7]. UA is developed under different conditions in the global north and the global south due to different economic and spatial contexts, and for historical reasons [36].

Developing agricultural capacity within or close to urban spaces either in developing and or developed countries such as Brazil and Germany have the potential to reduce social costs and environmental impacts, provide economic development opportunities and increase access to healthful food. Regardless of these potential advantages, there are several challenges to establishing the feasibility of urban production as compared to conventional agricultural practices, including space availability, production scalability, reuse of waters, labor costs, demographics [4, 8, 34].

Urban agriculture presents formidable opportunities and challenges in many cities around the world. For instance, in the case of the city of Rio de Janeiro (Brazil), even while living in a tropical country blessed with water, plenty of solar energy and lush vegetation, most of the inhabitants cannot take advantage of the natural resources and unconventional food plants around them. Many usually buy packaged goods, sometimes contaminated with agrochemicals.

Organic products and nonconventional food plants are still inaccessible from a logistical and financial standpoint for a large part of the population, especially urban favelas residents. Furthermore, the city has few projects related to urban farming (such as Hortas Cariocas [32] and PMPANC [37]). On the other hand, in the last decade, the city of Berlin (Germany) has become a hot spot and the international "capital" of UA. It is very active when it comes to fostering a broad variety urban agriculture and gardening projects within the reunited city. Many of the endeavors are very light on the land, creating vegetable gardens that may be moved to accommodate the changes in urban spaces that characterize a developed city that is rapidly growing (such as Prinzessinnengarten and ROOF WATER-FARM [35]).

Tackling these challenges seemed almost impossible years ago given the complexity of the cities and the different maturity level of information and communication technology (ICT), internet of thing (IoT) [29] and urban computing (UC) [39]. Nowadays, urban sensors, mobile technologies, autonomous vehicles, social networks, smart cities applications and large-scale computing infrastructures have produced massive amounts of unstructured data (big data) in urban spaces of these countries. This big data implies rich knowledge about urban activities which may help to enhance UA when appropriately used.

This chapter discusses the challenges and opportunities of urban agriculture in an interdisciplinary way. Besides, it seeks to explore new joint-research possibilities addressed in the 8th Brazil-Germany Frontiers of Science and Technology Symposium (BRAGFOST). Hence, the central thesis of this chapter is to investigate the current challenges and opportunities in UA and discuss if the adoption of UC can aid urban dwellers to improve UA either in Global North or Global South countries like German and Brazil.

The rest of the chapter is organized as follows. In Section 2, we introduce the main dimensions and concepts of UC and ICT that can be applied to UA and discuss how they can be connected. In Section 3, we address the challenges and opportunities of UA regarded to use of urban data in global north and global south countries. In Section 4, we describe two leading projects in UA, one in Berlin and one in Rio de Janeiro, showing their key characteristics and discussing how they can take advantage of the digitalization of UA and data-centric technologies. Finally, in Section 5, we conclude the chapter and point out a few future directions of this interdisciplinary research theme.

2 Background

2.1 Dimensions of urban computing

The term urban computing (UC) [19, 20], it is still an imprecise concept with many open research questions [39]. UC is an interdisciplinary concept fusing the computing science with traditional fields like engineering, architecture, ecology, economy and sociology in the context of urban spaces. UC connects ICT, IoT, data science and, advanced management of large volumes of data and data-centric methods to propose efficient solutions to problems faced by cities.

Urban computing is the process of acquisition, integration and analysis of large amounts of heterogeneous and unstructured data generated by diverse sources in urban spaces to tackle the major issues that cities face [39]. UC seeks to understand the nature of urban and social phenomena to better plan the future of cities, improve the urban environment and increase the quality of life of its inhabitants. According to Foth et al. [12] and Zheng et al. [39], UC is situated at the intersection of three dimensions: urban spaces, human resources and technology (Figure 1).

Figure 1 describes the flows of data within the dimensions and the knowledge generation. Each dimension generates massive amounts of unstructured data that are consumed by the "technology" dimension. Such dimension is composed of several computational technologies (such as web, computer-supported cooperative work (CSCW), cloud computing, IoT, big data, human-computer-interaction (HCI), mobile

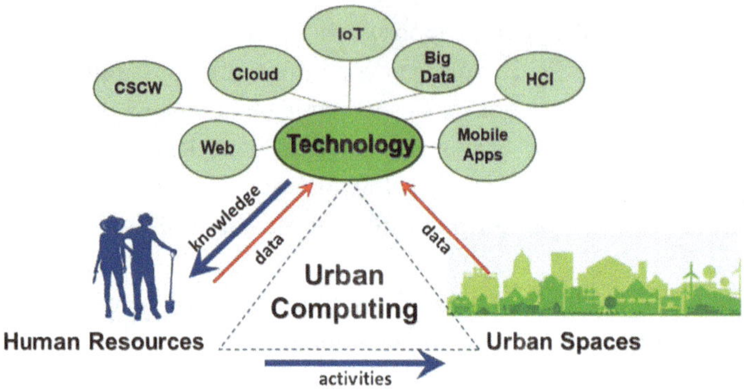

Figure 1: Dimensions of the urban computing and the flows of data, knowledge and activities.

applications) that can compute the data and produce explicit knowledge used at the "human resources" dimension. The human resources dimension is composed of people that may perform different roles (such as urban dwellers, urban farmers, policy-makers, urban planners), besides they execute actions or activities that can change the urban spaces to fit their needs.

One of the most significant contributions of investigating the relationship between UC and UA is that it may reframe the relationship between human resources and the urban spaces. A better understanding of the role computational technologies in UA may contribute to enhance the wealth of urban dwellers, the sustainability of cities, the smart reuse of urban assets and also promote the development of innovative projects promoted by single individuals, community organizations, universities, charities, cooperatives and social enterprises.

2.2 Urban agriculture

Urban agriculture is the process of growing plants, raising of animals and distributing food products, using resources and local materials from the urban spaces where the action takes place [5].

It can be developed independently or collectively by people for self-consumption or commercialization purposes. UA is splited into two categories: "intraurban", which describes vegetable gardens found within urban spaces and "periurban", which are gardens that occur on the periphery of metropolitan centers. Intraurban and periurban agriculture provides food products from different types of crops (grains, root crops, mushrooms, fruits); animals (poultry, rabbits, goats, cattle, pigs, guinea pigs, fish, etc.) as well as non-food products (aromatic and medicinal herbs, ornamental plants, tree products) [11].

Urban agriculture is performed in small areas like backyards, terraces, rooftops, patios, along rivers, roads and railways, or under power lines with the purpose of subsistence or small-scale sale in local markets. There are, however, more ambitious urban farming initiatives in community lots in gentrified urban spaces in Asia, Europe and North America [1]. UA offers the potential of producing high-quality food at an affordable cost, especially for populations with low access to these resources. The practice also promotes socioeconomic development and food security, as well as improving environmental resilience by conserving biodiversity. UA offers the potential to ameliorate urban environmental problems by increasing vegetation cover and, therefore, contributing to a decrease the urban heat island intensity and increase the reuse of waters, improving the livability of cities.

The benefits of UA are many but vary in developing and developed countries. Although several cities and countries have begun to loosen restrictions on UA, and even to encourage it with financial incentives, it has remained an open question how urban farmers and urban planners can take advantage of UC as a potential force for sustainability. From a computational perspective, we foresee that UA can be related with UC techniques, sharing the three basic parameters of any big data problem, they are expressed as the "3 V's": velocity, volume and variety (Table 1).

Table 1: The "3 V's": velocity, volume, variety.

Parameter	Definition
Velocity	It is the speed at which data are streaming in from sensors, social media and mobiles applications, and the rate at which it needs to be fused and interpreted to generate knowledge to the urban dwellers and planners
Volume	It is the amount of data being collected from farmers or growing platforms about the plants, climates, crops and livestock conditions. The volume can be enormous regarding the amount of data to be analyzed
Variety	It is a measure of the different types of data, such as environmental, plant health, labor-related metrics, or exogenous variables such as climate and marketplaces

To the best of our knowledge, few works discuss the combination of the dimensions of UC and UA and the use of big data to extract knowledge aiming to respond to the challenges of UA in the cities of global north and global south.

2.3 Dimensions of urban agriculture

According to Mougeot [27, 28], most authors define UA in general terms; their studies rarely use their findings to refine the UA concept or to clarify how UA relates to computational technologies (e. g., UC, ICT, CSCW and big data). Mougeot l. c. pointed out that UA is grounded in five dimensions: economy, society, environment, health, and technology (Figure 2).

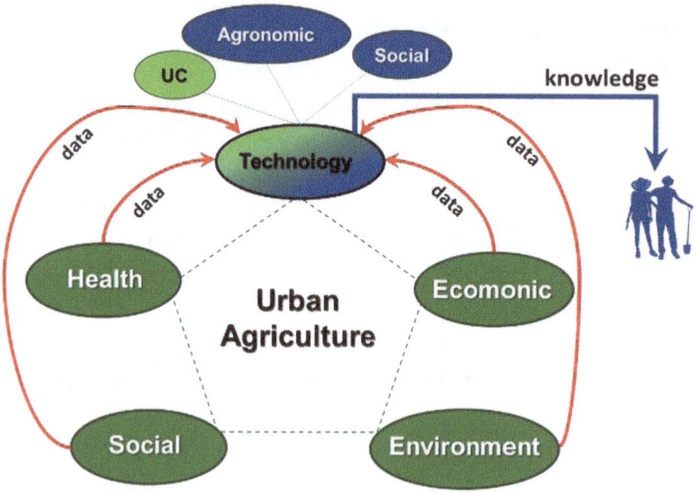

Figure 2: Dimensions of UA and the flows of data and knowledge.

As far as we are concerned, on a conceptual level, each dimension produces lots of data sets that can be explored at the "technology" dimension, addressing several challenges of UA. We stress that the "technology" is an extensive and evolving dimension because it may encompass several related topics like agronomic techniques (such as aquaponics, aeroponics, vertical farming, water reuse), social technologies and computational technologies, just to name a few. Furthermore, the "technology" dimension operates as a platform upon which knowledge generation and social interaction occur.

The data sets in UA are composed of unstructured data which is either machine or human generated (Figure 2). Unstructured data do not have a pre-defined model or is not organized in a pre-defined manner, it is typically text but may contain information such as dates, numbers, images, multimedia and facts as well [23]. Here, we summarize the data produced by each dimension (Table 2).

3 Challenges and opportunities in urban agriculture

In light of the merits of UA, urban farms and gardens are popping up across global north and the global south countries. However, the benefits and limitations that urban planners, dwellers, and growers face must be fully understood and addressed if urban farms are to become widespread, profitable and even sustainable.

Table 3 indicates the main opportunities and challenges associated with the five dimensions discussed in Section 2.

Table 2: Summary of data produced by UA dimensions.

Dimension	Description
Social	Consists of data about the social and urban spaces, such as youth development and education, food security, sociality integrated aging, gender participation, gentrification of depressed urban areas. When used aggregately with demographic data, these data sets can aid the visualization and mapping of city assets or understand urban anomalies
Environment	Consists of meteorological data (humidity, temperature, pressure, wind speed and weather conditions); air quality data (concentration of CO_2, NO_2 and SO_2); ecological data (awareness of food system ecology, stewardship, storm and waste waters management, soil improvement). When used aggregately with sensors and satellite data, these data sets be used to identify a city's issues (such as polluted and drought/flooding areas, heat islands)
Economic	Consists of economic data representing a city's economic dynamics. For example, local economic stimulation, job growth, job readiness, food affordability, carbon emissions, stock prices, transportation bottlenecks, housing prices and people's incomes. When used aggregately, these data sets can capture the economic rhythm of a city, therefore, predicting the future of the economy
Health & Educational	There are already abundant educational and health care and disease data generated by schools, hospitals and clinics. When used aggregately, these data sets can show the impact of education and food quality/eating change on people's health

4 Projects of urban agriculture projects in Germany and Brazil

This subsection discusses some examples of leading projects in UA in developed countries like Germany and developing countries like Brazil, showing their key characteristics and discussing how they can take advantage of the digitalization of UA.

4.1 Urban farming in Germany

According to the German Museum of Small Gardens in Leipzig, allotment gardening in Germany has its roots in 1814 in Kappeln [2]. Since that time, different forms of UA have gained increased interest and participation in Germany. From small allotment gardens over community gardens to semientrepreneurial self-harvest farms and fully commercialized agriculture (Urban Farming), Germans are very engaged in UA [16].

In the Bonn-Rhein-Sieg region, there are approximately 27 different urban gardening projects where the predominant type is the community garden [16]. A community garden is a piece of small land planted with fruits, vegetables and herbs in joint voluntary work. The gardening activities are determined by a set of rules. A large proportion

Table 3: Opportunities and challenges associated with the five dimensions.

Opportunities	Challenges
Social	
– Increase social interaction, strengthening social ties and facilitating new social connections and intergenerational relationships – New meeting places for community members to cooperate, particularly important in cities where open green spaces are rare – Increase the perception of safety/reduction in crimes, and consequent strengthening of dweller's pride of place – Facilities to the neighborhood of diverse backgrounds to interact who otherwise would not have such an impetus – Facilities for immigrants to develop ties with hosts and other ethnic communities, expand culturally, gain a sense of belonging and maintenance of cultural heritage – Establish new policies to repurpose unused urban land/properties	– Instigate different organizational structures and decision makers to support and the development of urban farming – Support initiatives led by lower-income communities, they usually experience disparities in access to land, political support, and government funding compared to UA efforts led and middle-class groups
Environment	
– Increase the biodiversity of the neighborhood. Urban food production also means that healthy, fresh produce is readily available to urban dwellers – Microclimate regulation (e. g., reduction in the urban heat island) through transpiration processes and reduction of emissions of greenhouse gas associated with food transportation – Carbon sequestration by crops and vegetation through filtration of particulates by plants – Increase rainwater drainage and reuse, reducing the risks of drought and floods – Increase the recycling of organic wastes and urban soils	– Soil management and fertilizer use practices by UA growers may not be ecologically sound – Find reliable and safe water sources can be tricky. Technologies such as irrigation deliver water where and when it is desired can help conserve it – Reusing wastewater and rainwater may provide additional water, but those sources must be monitored for contaminants, and perhaps treated
Economic	
– Increase employment and household, particularly for low-income and socially excluded populations – Increased property values surrounding urban gardens, particularly in gentrified neighborhoods – Entrepreneurial UA may attract venture capital and make profitable business opportunities, particularly in repurposed urban spaces	– UA projects may offer job opportunities that require additional knowledge beyond technical farming skills, which may need more staff or higher labor costs – UA projects may require financial and political support; several projects cannot survive on profits from produce, mainly if incorporating other social missions

Table 3 (continued)

Opportunities	Challenges
Health and Education	
– Engage, activate and train youth and educators in schools; youth can play an essential role in increasing knowledge and understanding about healthy eating and gaining access to fresh food – Learn about the provenance of food, agricultural processes, nutrition, and sustainability – Promote youth development as an alternative to the exposure to drug and crime economies, including early wage-earning opportunities – Source of physical activity and mental health benefits, including stress reduction, providing focused activities, cognitive stimulation, creating a sense of pride, and provisioning a connection to nature	– Develop UA projects which provide comprehensive education beyond technical farming skills require additional expertise, which may require more staff, time and elevated labor costs – UA projects may not be supplying enough food to communities in which they are located – Local food may not be physically or economically accessible to residents – Health risks to growers, consumers, and local community from water and soil contaminants if adequate preventative measures to reduce exposures were not taken

of beds are cultivated collectively or individually. The garden is open to the public for an appointed time. There is no membership restriction and no use of digitalization techniques of UA.

On the other hand, the self-harvest garden is starting to appear the same region, requiring some degree of digitalization and data management policies. The self-harvest garden is a cooperation of consumers with farmers (commercial and sometimes organic). The farmers plant a wide variety of vegetables in long rows on arable land which is within easy reach of a city. The farming area is divided up into strips so that the whole range of vegetables is grown on each strip. Usually, for a small fee, a strip can be leased. Community projects with self-harvest garden are usually a good extra income for farmers [16].

According to Gauder et al. [13], the primary incentive for the participants of the self-harvest gardens seems to be their engagement in the production of local and healthy food. The participants characterized themselves as having a middle or high income, a sustainable lifestyle, high level of education and high nutritional awareness.

In Berlin, there is a large variety of different types of UA projects. A collection of data carried out in 2012 by more than 100 community gardens, excluding the additional presence of allotment gardens, which often are organized rather individually. These numbers increase every year, showing the considerable interest in activating new community gardens in the urban spaces.

For example, in Berlin, a variety of crops are being grown on a renovated site in Kreuzberg. For example, the 6,000-square-meter urban farm named Prinzessinnengärten aims to raise awareness about the issues associated with global-industrialized farming, seed-distribution monopolies and a global decline in biological diversity. Today, the urban space is producing fresh and healthy food and hopes to decrease CO_2 emissions associated with farming activities, protect the local climate and even increase Berlin's biological diversity. The philosophy behind the project is to be more than just an urban garden, Prinzessinnengärten is an effort to fulfill the need for social learning and cultural change. TonSteineGärten is another community garden running on the same site. It started its first season in summer 2009, and its size is approximately 1500 square meters and does not have a legal structure or organizational form. This means some cooperation with charities or NGOs once it comes to funding purposes.

Rosa Rose is another community garden now located on a public green area at Jessnerstraße (Berlin-Friedrichshain) where the gardeners grow fruits, vegetables and ornamental plants. The use of the garden is free of charge, it is assured by a contract with the borough office. The garden tries to facilitate urban gardening independently from financial background or ownership structures. Every urban dweller who wants to take part is invited. Since its beginning, the garden had to move from different urban spaces (from Kinzigstraße in 2004 to Jessnerstraße in April 2010) due to economic interests. Through these places, parts of the original group and the plants have also been renewed. According to Zezza and Tasciotti [38], what remains intact is the central idea of a communal garden, a garden for everybody.

Through these stages, Rosa Rose's history has repeated a trend, which is entirely typical for the international UA movement in global north countries. Many gardens begin as "guerrilla gardens" in local neighborhoods, and many of them are devastated sooner or later by the city's economic dynamics. Despite this unfortunate history, a strong neighborhood cohesion has built up over time for Rosa Rose. It became a frequent object of academic studies [38].

In many European cities, existing water infrastructures of supply and disposal are at stake. The process of maintaining and retrofitting them is a big challenge [22]. This process is linked to dynamics of urban transformation, i. e., growth and shrinking of urban settlements and is accompanied by a number of (german) research initiatives. The aim of this initiatives is to explore, in theory and built practice, the structural transformations of linear centralized infrastructures toward decentralized approaches, integrated urban water management, and design [6, 17].

A common aim is to promote the efficient usage of water resources to meet the challenges of demographic developments and climate change, in order to enhance water security both in supply and disposal. Interdisciplinary and transdisciplinary cooperation is needed to meet the challenge [24]. When other challenges are taken into account, such as the linkage of water and food supply, this inter and transdisciplinary cooperation becomes really necessary. Water and soil are limited factors within the urban context, but they are needed for food production. The development of innovative concepts and technologies – facing the linkage of water and food supply – is especially sought in the urban context. The result could be multifunctional infrastructural technology that links urban water management and food production via, i. e., water-based farming including an innovative data management for crop production.

The Roof Water-Farm project (http://www.roofwaterfarm.com/en/) investigates the production of fresh food in urban settings via hydroponics (water-based plant cultivation) and aquaponics (combined fish and plant culture), while being linked to water treatment technologies for greywater and blackwater, as well as management of rainwater. It is an on-site approach, combining urban water management with urban farming. The building-integrated combination of water treatment with the production of fish and plants is a spatial- and resource-efficient urban design strategy that uses the spatial potential of buildings' roofs within the city context. A central aspect of the research is also to explore tools and methods to incorporate citizen participation and education to foster a sustainable implementation of the Roof Water-Farm [25].

Water-based farming strategies like the ROOF WATER-FARM concept show the potential for the integration of urban computing and urban farming. As argued before, there are a lot of driving forces: Soil and space are limiting factors for UA within the urban shape. Treated water (i. e., rainwater, grey water, black water) is an urban resource, which could be reused and recycled via farming strategies. Especially building-integrated, water-based farming strategies like the ROOF WATER-FARM concept need data sets and computing technology to unfold their full potential of productivity and diversity within the urban context.

4.2 Urban farming in Brazil

While in the global North UA is drawing attention from the main European cities, it remained a pillar of food systems in the global south [38]. The Brazilian UA projects are entirely different from German ones; the social impact of UA is predominant in small or marginalized communities, and no digitalization initiatives are perceived. Usually, the goals of the projects are related to urgent needs of the inhabitants, such as food security issues, increase household income and the lack of job opportunities in the traditional urban spaces. Besides, the project occurs especially in favelas or in other marginalized urban spaces of a city.

The use of smartphones, tablets, mobile computing and social networks are growing in the Brazilian cities, opening new opportunities to use data science to help urban farmers to make more informed decisions about their farming practices, potentially leading to higher yields and improved profits. However, at the time of the writing of this chapter, no Brazilian projects are related to the intensive use of UC, ICT or big data in UA. For instance, there are several small UA projects in Rio de Janeiro (such as "Favela Organica" [32], "Hortas Cariocas" [31] and "PMPANC" projects (Medicinal Plants and Unconventional Food Plants) [37]). These projects, like the German ones, do not consider the disruptive role of big data, UC and ICT in UA.

The price of food in many Brazilian cities continues to rise steadily since 2008, including in Rio de Janeiro. Programs of UA with socioeconomic potentials, such as organic agriculture and urban gardens, are particularly important for creating the enabling conditions to promote the wealth of marginalized communities. Nowadays, almost 12 million people live in the larger metropolitan area of Rio de Janeiro. More than 6 million people live in the municipality. One in five (about 1.4 million) live in favelas, and the numbers are increasing [18].

The "Favela Organica" project encourages the creation of community gardens and generates incomes from the reuse of the food produced in them [9]. The project became a success and is now being replicated in several cities throughout Brazil. In Rio (in 2015), it employed women all residents of Babilônia–Chapéu Mangueira favelas to work part-time and to train favela residents how to prepare meals by using the peels, rinds, seeds and stems of food items that are typically discarded. The Favela Organica project offers more than 450 recipes based on the reuse of foods avoiding wastes [32].

The "Hortas Cariocas" project was created by the Rio de Janeiro city hall in 2012. The city hall invests in UA and administers over 30 organic gardens implemented in public urban spaces (17 of which are in public schools) in various poor areas and favelas across the city [31]. The rationale of the production is quite simple; the gardeners distribute half of their crops to schools kitchens and also to at-risk families previously identified by the residents' association. The other half of the production is sold by them, and the profits divided among the workers or reinvested back into the gardens.

The workers of the "Hortas Cariocas" are former prison inmates or member of the community (e. g., Maguinhos and Boreo favelas). To stimulate the urban farms, the city of Rio provides uniforms, seeds, basic farming equipment, individual protective gear and organic fertilizers. The "Hortas Cariocas" gardeners are trained in agroecology and paid a stipend of approximately US$120 per month to work full-time in the gardens or administrative tasks. The project is one of the few municipal-led, social development programs that aim to bring poverty alleviation to the impoverished people of favelas [33].

Another example of urban farming project without the support of UC in Brazil is the "PMPANC" project (http://www.farmacia.ufrj.br/labfbot/). The project combines UA and pharmaceutical assistance to the inhabitants and small farmers of periurban areas. It is being developed in the city of Magé, which is located at Baixada Fluminense, a region in the fringes of the metropolitan area of Rio de Janeiro state.

The "PMPANC" project started in 2012 and is being developed in collaboration with the Educational Tutorial Program (PET-SI/UFRRJ) (http://r1.ufrrj.br/petsi/mageagrofamiliar/), the laboratory of Pharmacobotanics (LabFBot) Faculty of Pharmacy of Federal University of Rio de Janeiro (UFRJ), EMATER-Rio and the municipality of Magé. The project work with residents of an environmental protected area named "APA Guapimirim" to establish a community space with small farmers and families to produce organic medicinal plants and unconventional food plants (UFP) ("Plantas Alimentícias Não-Convencionais" – PANC, in Portuguese). UFP is the term refers to plants that have uncommon processing methods and usually do not have market value or are only commercialized on small scales [21].

A handful of volunteer residents and small farmers cultivate and maintain long organic UFP garden beds. They obtain financial remuneration for their efforts when they sell their goods at street fairs. Moreover, they receive specialized training given by LabFBot to identify and how to cultivate and prepare natural medicines from medicinal herbs and cultivate UFP. They define their working hours, making all decisions about what is grown and how to distribute and sell what is produced. The farmers help each other maintain the plots and resolve their interpersonal conflicts. They appreciate the spaces as a productive and therapeutic social space.

A range of traditional fruits, vegetables, medicinal herbs and seasonings are grown year-round. The production has increased the income of the small farmers, and by extension, the nutrition intake consumed by its beneficiaries. The gardens fulfill an essential role in diversifying food habits and providing nutrients from new sources, especially vitamins and mineral salts from plants.

Brazil is one of the most biodiverse countries in the world, with a significant number of species. However, this diversity is rarely exploited in UA projects. Differently, from related works, the "PMPANC" project stimulates the small farmers to cultivate native UFP. The production and consumption of UFP may be an alternative for a better food diversification and to increase the primary income of small farmers. Because UFP are present in the region where there is still exertion of influence of traditional foods, but they are still unknown for a significant part of the population.

Though UA plays only a minor role in economic development in the city of Magé, it is critical to the sustainable development landscape. In this case, UA directly impacts on the success and quality of urbanization because all food is organic, and thus is free from agrochemical and pesticides. Furthermore, the PMPANC project uses the same medicinal plants prescribed by the Brazilian national healthcare system, preserve the traditional populations' knowledge of medicinal plants, promote environmental conservation, and strengthen the Brazilian pharmaceutical research sector.

5 Final considerations and future work

Historically, urban agriculture is being developed under different conditions in the global north and the global south due to different economic and spatial contexts. However, both have a window of opportunities; they generate large amounts of data but do not integrate various information and ICT solutions securely to manage their assets and aid the urban dwellers, farmers and planners. As far as we are concerned, UA should be evaluated for the multidimensional nature of its outcomes, and not merely for its potential benefits regarding food production or economic development.

In this chapter, we evaluated the dimensions of UA and UC and discussed the challenges and opportunities of UA. Besides, we discuss leading UA projects in Germany and Brazil. The projects we have reported in the previous sections has changed the face of several communities. They increased biodiversity, created jobs, increased access to fresh, nutritional food and (re)created new urban spaces. However, they still lack to embrace the emerging UC and ICT trends such as the rise of social networking applications, big data, data science, Web 2.0 services, the increasing ubiquity of mobile technology and real-time sensor networks to name a few.

This chapter was written in the hope that sharing the underlying thinking and expectations as well as hopes and aspirations of a group of interdisciplinary researchers will enable a new level of constructive study that contributes to pushing the UA agenda forward. Although a large variety of social and technical disciplines influences UA, we believe that is critical to bring about UC and ICT in this domain. In this sense, this chapter is also an invitation to join this growing interdisciplinary community.

Bibliography

[1] Badami, M. G., Ramankutty, N.: Urban agriculture and food security: a critique based on an assessment of urban land constraints. Global Food Secur. 4, 8–15 (2015). https://doi.org/10.1016/j.gfs.2014.10.003.

[2] Bell, S., Fox-Kämper, R., Keshavaraz, N., Benson, M., Caputo, S., Noori , S., Voigt, A. (eds.) Urban Allotment gardens in Europe. Routledge, New York, USA (2016).

[3] Clinton, N., Stumacher, M., Miles, A., Aragon, N. U., Wagner, M., Geogescu, M., Herwig, C., Gong, P.: A Global Geospatial Ecosystem Services Estimate of Urban Agriculture. Earth Future 6(1), 40–60 (2018). https://doi.org/10.1002/2017EF000536.

[4] Costanza, R., de Groot, R., Sutton, P., van der Ploeg, S., Anderson, S. J., Kubiszewski, I., Turner, R. K.: Changes in the global value of ecosystem services. Global Environ. Change 26, 152–158 (2014). https://doi.org/10.1016/j.gloenvcha.2014.04.002.

[5] Cruz, S. M. S.: Is data the new soil in urban agriculture? Challenges and opportunities. In: BRAGFOST/17 Agenda Book, Potsdam, Germany (2017).

[6] DIFU Deutsches Institut für Urbanistik: Intelligente und multifunktionelle Infrastruktursysteme für eine zukunftsfähige Wasserversorgung und Abwasserentsorgung Vorstellung der Verbundprojekte: Ziele, Projektpartner und geplantes Vorgehen (2013). Bonn: BMBF 2013 has been replaced by 2nd ed. 2014.

[7] d'Amour, C. B., Reitsma, F., Baiocchi, G., Barthel, S., Güneralp, B., Erb, K. H., Seto, K. C.: Future urban land expansion and implications for global croplands. Proc. Natl. Acad. Sci. USA 114(34), 8939–8944 (2017). https://doi.org/10.1073/pnas.1606036114.

[8] Erickson, D. L., Lovell, S. T., Méndez, V. E.: Identifying, quantifying and classifying agricultural opportunities for land use planning. J. Landsc. Urb. Plan. 118, 29–39 (2013). https://doi.org/10.1016/j.landurbplan.2013.05.004.

[9] Favela Organica: Favela Orgânica – The project (2018). Retrieved from http://www.favelaorganica.com/english/.

[10] FAO – Food and Agriculture Organization of the United Nations: Profitability and Sustainability of Urban and Peri-Urban Agriculture (2007). Retrieved from http://www.fao.org/tempref/docrep/fao/010/a1471e/a1471e00.pdf.

[11] FAO – Food and Agriculture Organization of the United Nations (2018). Urban agriculture. Retrieved from http://www.fao.org/urban-agriculture/en/.
[12] Foth, M., Choi, J. H-J., Stachell, C.: Urban Informatics. In: Proceedings of the Conference on Computer Supported Cooperative Work (CSCW). ACM Press, Hangzhou, China (2011).
[13] Gauder, M., Hagel, H., Golimann, N., Stängle, J.: Motivation and background of participants and providers of self-harvest gardens in Germany. Ren. Agric. Food Syst. (2018). https://doi.org/10.1017/S174217051800008XP.
[14] Godfray, H. C. J., Beddington, J. R., Crute, I. R., Haddad, L., Lawrence, D., Muir, J. F., Toulmin, C.: Food security: The challenge of feeding 9 billion people. Science 327(5967), 812–818 (2010). https://doi.org/10.1126/science.1185383.
[15] Hamilton, A. J., Burry, K., Mok, H.-F., Barker, S. F., Grove, J. R., Williamson, V. G.: Give peas a chance? Urban agriculture in developing countries. A review. Agron. Sustain. Dev. 34, 45–73 (2013). https://doi.org/10.1007/s13593-013-0155-8.
[16] Hirsch, D., Meyer, C. H., Klement, J., Hamer, M., Terlau, W.: Urban AgriCulture and Food Systems Dynamics: Urban Gardening and Urban Farming of the Bonn-Rhein-Sieg region, Germany. In: Proceedings in System Dynamics and Innovation in Food Networks (2016). https://doi.org/10.18461/pfsd.2016.1646.
[17] Hoyer, J., Dickhaut, W., Kronawitter, L., Weber, B.: Water Sensitive Urban Design: Principles and Inspiration for Sustainable Stormwater Management in the City of the Future. Jovis, Berlin (2011).
[18] IBGE – Brazilian Institute of Geography and Statistics. Census (2010).
[19] Kindberg, T., Chalmers, M., Paulos, E.: Guest editors' introduction: Urban computing. Pervasive Comput. 6(3), 18–20 (2007). https://doi.org/10.1109/MPRV.2007.57.
[20] Kostakos, V., O'Neill, E.: Cityware: Urban computing to bridge online and real-world social networks. In: Handbook of Research on Urban Informatics. Information Science Reference, Hershey, PA (2008). https://doi.org/10.4018/978-1-60566-984-7.ch059.
[21] Kinupp, V. F., Lorenzi, H.: Plantas Alimentícias Não Convencionais (PANC) no Brasil: guia de identificação, aspectos nutricionais e receitas ilustradas. Instituto Plantarum de Estudos da Flora, São Paulo, Brasil (2014).
[22] Libbe, J., Beckmann, K.: Infrastruktur und Stadtentwicklung: Technische und soziale Infrastrukturen – Herausforderungen und Handlungsoptionen für Infrastruktur- und Stadtplanung. Deutsches Institut für Urbanistik, Berlin (2010).
[23] Liu, L., Ozsu, T.: Encyclopedia of Database Systems. Springer (2009).
[24] Löwe, H., Schade, J., Höckele, V.: In: Funding Priority "Sustainable Water Management" (NaWaM) Within the Framework Programme on "Research for Sustainable Development". BMBF (Bundesministerium für Bildung und Forschung), Bonn, Germany (2012). http://www.fona.de/mediathek/pdf/BMBF-Nawam-ENG_barrierefrei.pdf.
[25] Million, A., Bürgow, G., Steglich, A.: Urban Waters for Urban Agriculture – ROOF WATER-FARMs as Participatory and Multifunctional Infrastructure. In: Roggema, R. (ed.) Sustainable Urban Agriculture and Food Planning, pp. 142–165. Routledge, New York (2016). ISBN 978-1-138-18308-7. Book Chapter 9.
[26] Mok, H.-F., Williamson, V. G., Grove, J. R., Burry, K., Barker, S. F., Hamilton, A. J.: Strawberry fields forever? Urban agriculture in developed countries: A review. Agron. Sustain. Dev. 34, 21–43 (2013). https://doi.org/10.1007/s13593-013-0156-7.
[27] Mougeot, L. J. A.: Urban Agriculture: Definition, Presence, Potentials and Risks, and Policy Challenges (2000).
[28] Mougeot, L. J. A.: AGROPOLIS – The Social, Political and Environmental Dimensions of Urban Agriculture. Retrieved from https://idl-bnc-idrc.dspacedirect.org/bitstream/handle/10625/28341/IDL-28341.pdf.

[29] Ng, I. C. L., Wakenshaw, S. Y. L.: The Internet-of-Things: Review and research directions. Int. J. Res. Mark. 34(1), 3–21 (2017). https://doi.org/10.1016/j.ijresmar.2016.11.003.
[30] Orsini, F., Kahane, R., Womdim, R. N., Gianquinto, G.: Urban agriculture in the developing world: A review. Agron. Sustain. Dev. 33(4), 695–720 (2013). https://doi.org/10.1007/s13593-013-0143-z.
[31] Prefeitura do Rio de Janeiro: Secretaria Municipal de Meio Ambiente (2018). Retrieved from www.rio.rj.gov.br/web/smac/hortas-cariocas.
[32] Rekow, L.: Fighting insecurity: experiments in urban agriculture in the favelas of Rio de Janeiro. Field Actions Sci. Rep. 8 (2015).
[33] Rekow, L.: Urban Agriculture in the Manguinhos Favela of Rio de Janeiro: Laying the Groundwork for a Greener Future. Sustain. Econ. Dev. 155–185 (2017). https://doi.org/10.1007/978-3-319-45081-0_10.
[34] Santo, R., Palmer, A., Kim, B.: Vacant lots to Vibrant plots: a review of the benefits and limitations of urban agriculture. Retrieved from https://www.jhsph.edu/research/centers-and-institutes/johns-hopkins-center-for-a-livable-future/_pdf/research/clf_reports/urban-ag-literature-review.pdf.
[35] Steglich, A.: Roof Water Farm – Urban waters for urban agriculture. In: BRAGFOST/17 Agenda Book. Postdam, Germany (2017).
[36] Tornaghi, C.: Critical geography of urban agriculture. Prog. Hum. Geogr. 38, 551–567 (2014). https://doi.org/10.1177/0309132513512542.
[37] Vieira, A. C. M.: Conhecendo, conservando e comendo Plantas Alimentícias Não Convencionais (PANC) nos municípios de Magé e Guapimirim (RJ). 7° Congresso Brasileiro de Extensão Universitária. Brazil (2016). ISBN 978-85-93426-00-2.
[38] Zezza, A., Tasciotti, L.: Urban agriculture, poverty, and food security: Empirical evidence from a sample of developing countries. Food Policy 35, 265–273 (2010). https://doi.org/10.1016/j.foodpol.2010.04.007.
[39] Zheng, Y., Capra, L., Wolfson, O., Yang, H.: Urban Computing: Concepts, Methodologies, and Applications. J. ACM Trans. Intell. Syst. Technol. 5(3), 38 (2014). https://doi.org/10.1145/2629592.

Milena Serafim

Social technology and urban agriculture in Brazil: the social technology network and the social technology DataBank project

Abstract: This paper explores the synergy between urban farming project and movement for social technology, and the similarities and common challenges of some urban farming projects promoted by Banco do Brasil Foundation (FBB/BTS) and Social Technology Network (RTS). Both organizations were fundamental in boosting, financially and politically, social technology and urban farming projects in Brazil. Besides that, it can be observed the role of the networks as communication and divulgation platforms of experiences between policy makers, citizens, scientists, farmers and also as tools for public policy and several projects.

Keywords: Urban farming, social technology, communication and divulgation platform, public policy

1 Introduction

Brazil has a huge problem with social exclusion (from different perspectives: economic, cultural, lack of housing and collective urban space, food insecurity, mobility, etc.). Almost 45 million people (or 9.3 million families) were socially vulnerable and suffered from food insecurity and malnutrition. According to the Human Development Index, Brazil is still a quite unequal country, with an HDI of 0.754. In 2003, the then newly elected federal government made inequality and poverty some of the core themes in its agenda. They launched a program called "Zero Hunger Strategy" to ensure the human right to nutritional and alimentary security of vulnerable people. Until 2016, more than US$60 billion were invested in this strategy, which is composed of four hubs and more than 20 programs, promoted by six main ministries: Ministries of Social Development, of Science, Technology and Innovation, of Labour and Employment, of Education, of Agrarian Development and of Agriculture. One of these programs is the National Urban Farming Program. Urban farming experiences in Brazil were strengthened by federal programs to promote the reduction of social exclusion problems. The emergence of spaces for urban farming, including those supported by public policies, is a trend that has been observed throughout the world. More than an alternative for production, urban agriculture may be understood as a way of redefining the relationship between individuals and the space they live in

Milena Serafim, University of Campinas, Campinas, Brazil, e-mail: milenaserafim@gmail.com

https://doi.org/10.1515/9783110584455-007

[6]. Mougeot [11] claims that urban agriculture is different from conventional farming in terms of its connection to the urban economic and ecological system. Behind the development of these experiences, there is not only the need to increase farming output or to rationalize the means of distributing food. It is through its constitutive processes that other meanings, values and interest are shaped. Experiences of urban farming projects have shown huge synergy between with and the movement for social technology (or technology for social inclusion), which has been promoted by social movements, the government (Ministries of Science, Technology and Innovation and Social Development), national agencies (Banco do Brasil Foundation and Social Technology Network), international agencies and the research community. The debate related to social technology, understood as products, techniques of reapplicable technologies and methodologies developed in the interaction with communities and leading to effective solutions for social problems, is similar to undergoing initiatives in countries such as India and China, in which they are called grassroots innovations [8]. The main point is that social technology and urban farming share, as a constitutive element, the idea of empowerment and the participation of users in the design and management of technologies and methodologies capable of improving their living conditions. In this sense, the purpose of this paper is to explore the similarities and common challenges of six urban farming projects, especially about their objectives, production techniques, composting processes and actors involved. These experiences –- Urban agriculture and the "Little buckets" revolution; Urban Agroecology and Food Security; Community garden – social and productive inclusion; Agroecological food production in urban area; Productive gardens: cities cultivating the future; and Organic fruit farms – were chosen because they are funded by Banco do Brasil Foundation, Social Technology Network and federal ministries. These actors were responsible for promoting the synergy between urban farming projects and the Social Technology Movement, which is concerned with the development of technologies for social inclusion.

2 Urban farming as a social technology: action and social network strategies

Urban farming has different meanings, values and forms through which it may transform local realities. Urban farms can be understood as spaces for alternative food production and rescue of food culture, or as inclusive and sustainable urban spaces, or also as spaces for food security in poor neighborhoods. In Brazil, different meanings, values and forms come together in the same experiences, but most of urban farming projects serve to reduce social exclusion and food insecurity. It is because of this that the urban farming projects are integrated within the social technology movement. Social technology comprises products, techniques or methods and seeks to promote social transformation through low cost innovation. Technology and knowledge play a crucial role in this process. But also the main idea of ST is the participation of users in the design and management of technologies capable of improving their living conditions. One of the main concerns is keeping the projects alive for a long term. For this, we need users to understand the sense of taking part in the preparation of gardens and farms, for example. Their involvement, in the design and management of the projects, is one strategy to achieve this. In this sense, the Banco do Brasil Foundation and the Social Technology Network have an important role as the basis for network actions and strategies involved with social technologies. In Brazil, public policies for social technology were inserted since the early 2000s in the strategies of various governmental agencies (such as the Ministry of Science, Technology and Innovation, the Ministry of Social Development and the Ministry of Labour and Employment) and non-governmental organizations, such as the Banco do Brasil Foundation (FBB), with the Social Technology DataBank (BTS) and the Social Technology Network (RTS). Banco do Brasil Foundation has acted as a supporter of social and research projects in the field of Science and Technology since 1985. In 2001, FBB created the Social Technology DataBank program, investing in mapping, fostering and diffunding reapplicable technologies effective in solving social problems. The FBB, through the Social Technology Award, recognizes and certifies experiences that are included in the DataBank, available on the Foundation's website, in order to foster the dissemination of experiences in social technology. The Social Technology Network was an issue network, constituted in 2005, from NGO and social movement representatives, government bodies and the research community. The network began operating with the participation of 640 institutions (of which nearly 60 % were NGOs, with the rest being either teaching and research institutions or government bodies) and eight supporting partners, including Banco do Brasil Bank, Caixa Bank, Ministries of Science and Technology of Social Development, Petrobras and Finep, which funded resources of about R$224 million (more than US$100 million) in the reapplication of eight large national programs [5]. RTS sought to promote new governance objective (network-building and collaborative practices) and in promoting the sustainable development, through the application of social technologies. Unfortunately, this network has been suspended in 2013 due to political differences, but their members continued funding social technologies. In this sense, Social Technology DataBank, (funded by Banco do Brasil Foundation) and the Social Technology Network (funded by other actors) have promoted network-building and practices of collaboration for and through Urban Farming. Currently, this database comprises a set of 1,000 certified Social Technology experiences in different areas (food, education, energy, housing, income, health, water resources and environment). The reapplication of some of these technologies became national policies with the support of the FBB and other organizations. Relevant examples of such experiences are PAIS-Integrated and Sustainable Agroecological Production; the National Program for Urban and Peri-urban Agriculture; the Agricultural Food Acquisition Program (PAA) and P1+2 -- the One Land, Two Waters Program.

3 Methodology

The research was based upon concepts and methods from different approaches that shape the field of science and technology studies, like sociotechnical analysis, critical analysis of technology, social technology, and of policy analysis. Thus, we propose an integrative sociotechnical perspective as a means of better understanding the numerous questions that will surely appear during the research process. The methodology of the research was based on two main steps. The first was the construction of a theoretical approach, supported by ideas and concepts of authors, such politics of artifacts [16]; social relevant groups and interpretative flexibility [3, 12]; technological systems [1, 9]; social construction of functioning, sociotechnical dynamics and trajectories [13, 14]; coconstruction of science, technology and society [2, 10]; tacit knowledge and knowledge transfer [4]; the ambivalence of technology [7]; user-producers relationships [15]. After that, it was structure the sets for project analysis, based on field research: (1) Description and contextualization; (2) Nature and sociotechnical environment; (3) Political, ecological and environmental sustainability; (4) Institutional and linkage arrangement. This research is part of another research project – "Social Technology (ST) in Latin America," funded by IDRC, Banco do Brasil Foundation and Polis Institute.

4 Results

The urban agriculture projects analyzed were those, highlighted by the Social Technology DataBank and reapplied in different Brazilian cities. The first experience, "Urban agriculture and the 'Little buckets' revolution," was reapplied in 25 farms and in 4 community gardens in the city of Florianopolis and integrated more than 600 people. This experience arose from the concern for waste treatment and wasteland. In this sense, the actors involved organized actions to deal with it, especially by transforming waste into organic compost. The technique used was organic aerobic composting and linear production technique. Actors involved were: local universities, city government and federal ministries. The second one is "Urban Agroecology and Food Security," which is responsible by structuring of 13 vegetable gardens, next to healthy unity, in the city of São Paulo, involving more than 800 people. They sought to enable food production (mainly vegetables) in urban areas while promoting social interaction. Organic aerobic composting and linear production techniques were employed. Actors involved were local universities, city government and ministries. The main focus was on food security and social interactions. The third experience was dubbed "Community gardens – social and productive inclusion," responsible for the reapplication of 22 gardens and involving 630 families in the city of Maring. To provide healthy food (vegetables and medicinal plants) and work/income to vulnerable people. Organic aerobic composting and linear production technique. Actors involved were local universities, city government and federal ministries. The focus was on food security and the improvement of health and well-being. The fourth experience was named "Agroecological food production in urban areas" (6 vegetable gardens reaching 30 families in São Paulo). It aimed toward stimulating agroecological practices and bioconstruction techniques in small urban spaces. Planting in suspended beds, rainwater harvesting and microsprinkler irrigation were some of the employed techniques. The actors involved were local universities, the city government and federal ministries.

The fifth project was "Productive gardens: cities cultivating the future." It involved the re-application of 12 gardens in the city of Belo Horizonte, which arose from the concern with work and income for vulnerable people. The project served to generate income for the community through the sale of products to local public schools and restaurants, leading to additional income for the community. Organic aerobic composting and a circular rotation production technique were employed. The actors involved were local universities, the city government and the RUAF Foundation (Resource Centre on Urban Agriculture and Food Security). The last experience is the "Organic fruit farms Program," which was reapplied in 201 areas and reached 100 families. It was aimed toward the rescue of traditional food crops with nutritional and functional properties (fruits, vegetables and grains).[1]

Aerobic organic composting and circular production techniques were used. The actors involved were the city government, EMBRAPA (a public Brazilian agricultural research company) and federal ministries. In Figure 1, we present a brief synthesis of the experiences and their main elements.

The urban farming experiences, their similarities and common challenges were analyzed. Based on field research, we could observe that the projects have a particular objective and focus. Each place has a specific situation, specific problems to be faced, the projects are designed to reflect these conditions. Additionally, three elements are noteworthy. The first one is about the similarity of production technique. Most experiences have organic aerobic composting with a linear production technique. Just two experiences used a circular rotation production technique and another project uses suspended bed production (because it occurs over a landfill site). The second element is related to the support and assistance provided by actors such as local universities. All the experiences (except one), besides the common actors (city administration and federal ministries), have the research community as supportive actors. The university is responsible for the systematic monitoring process and provides technical assistance. The third element is about the role of the networks. Looking at the results,

1 The RUAF Foundation is an international network of seven regional centers and one global resource center on Urban Agriculture and Food Security. See more in: www.ruaf.org

Figure 1: Overview about the urban agriculture projects.

we can recognize the social networks on the interaction between projects and actors. There are three types of actors: (1) Actors responsible for the funding and reapplication of technologies, such as the Social Technology Network, the Banco do Brasil Foundation and the federal ministries. The networks act as communication and reapplication platforms among stakeholders (policy makers, citizens, academics and farmers) and projects (same thematic projects or different projects). In this sense, we can observe a synergistic effect, driven by networks. These organizations acted as policy networks, which were instrumental in the construction of the social technology movement and as tools for public policy and several related projects. (2) Actors responsible for the implementation process, like users, city government and communities, neighborhood associations, public schools; (3) Actors involved with support and assistance process, provided by local universities (or by EMBRAPA, in one of the cases). It is because of the plurality of actors that the experiences are sustained even in adverse times, such as the economic crisis Brazil is presently undergoing. Looking at the similarities and common challenges of the experiences, we are able to identify four main similarities and common challenges. Regarding similarities, first, all of the projects had some kind of impact on health, eating and physical activities, and especially an impact on community self-esteem. It is impressive how communities, over time, incorporate improvements in their lives. It was possible to verify that some neighbors of the experience painted and arranged their houses, because they happened to receive more visits because of the garden. The second similarity is about the involvement of different participants in a support network, which are essential to foster these initiatives (financial resources and cognitive and political resources). The beginning of the experiments depended, in most cases, on the drive of government through public policies, but also they had the participation of local universities or research institute. The integration of existing networks around the projects is extremely important for the sustainability (environmental, economic, political, etc.) of urban agriculture projects. Although there is a centrality of government in the network concerning experience, actors such as the local research community are extremely important in the legitimacy and continuity of experiences at adverse times. The third similarity refers to the projects use organic or agroecological techniques in the promotion knowledge exchange. This element was expected, in view of the selection of the experiments to have been based on the integration between urban agriculture and the social technology movement. The use of pesticides would conflict with the model that urban agriculture intends to print in the logic of the city and in the logic of good living. However, the exchange of knowledge among users was a new factor. Finally, the fourth common point is the projects have a cooperative management. The logic visualized in the experiments was of self-management, where all were equally responsible for production and decision. At the same time, some common challenges were noted. The first is the nonparticipation of some users at the moment of the project formulation led to adjustments at the moment of project implementation, like a transformation in land design, the layout of the plants, the type of seeds, seedlings, etc. This nonparticipation meant that during the course of experiments there were important changes in the way of production and distribution of land. The second one is the lack of land proprietorship. The continuity of some experiments is fragile in view of the transitory assignment of the lands until there is another use for it, especially when the experience is dependent from city governments. Because of this, the projects have difficulty in accessing funding programs. This concern by participants is the third point. The last one is on the political sustainability of experiences depends, in the medium term, on broadening alliances between social actors in the support network. We observed that the success of these projects depends on the degree and continuity of community engagement and the performance of the social network of actors. Part of this community engagement has as a driving force the involvement of the government with the structuring of public policies. Cities in which local government support is low, the difficulties are even greater, such as the possibility of reappropriation of land use by the local government, causing great demotivation to users over time.

5 Closing remarks

Urban farming seeks to promote the transformation of our cities. This change consists of two intrinsic elements: social inclusion and technology. The relationship between technology and the processes of social inclusion, involving different aspects (cultural, economic, political, gender, age, etc.) is a topic of interest for academics, policy makers and social movements, and it is of particular importance to the social technology movement. This movement understands that social transformation also depends on technological aspects, because societies are shaped by technology just as technologies are shaped by society. Thus, social exclusion (and inclusion) are materialized by the technologies we decide to use. And some actors are key in terms of being able to provide the knowledge to build inclusive cities There is a lot of knowledge potential in social movements and communities, though this is often ignored by mainstream science practices and policy making. The success of the aforementioned projects depends heavily on community engagement, not only on technical and financial viability, and the performance of the social network of actors. Networks, as communication and reapplication platforms between policy makers, citizens, academics and producers, play a pivotal role in this urban farming process. They ensure that the role of the networks, as communication and diffusion platforms between policy makers, citizens, scientists and farmers, is noteworthy. Also, these organizations act/acted as policy networks which were instrumental in the construction of the social technology movement and as tools for public policy and several related projects. The main aspect of these networks is that they are relevant tools for public policy and several projects. The initial centrality of actors such as FBB and RTS was fundamental to boost the strengthening of TS, urban agriculture and their own experiences, both in sharing and reapplying of experiences (from the Social Technology DataBank), and in the decision-making agenda of the Ministries. They are a platform for the exchange of knowledge and experiences between local governments, citizens, scientists, farmers, etc. This is where the main potential of these experiences may be found. But it is not enough. There should always be a commitment to encourage more exchange of knowledge, of techniques of replicable technologies and methodologies developed in the interaction with communities and leading to effective solutions for social problems.

Bibliography

[1] Bijker, W.: Of bicycles, bakelites, and bulbs: toward a theory of sociotechnical change. The MIT Press, Cambridge (1995).

[2] Callon, M.: Society in the making: the study of technology as a tool for sociological analysis. In: Bijker, W., Hughes, T., Pinch, T. (eds.) The social construction of technological systems: new directions in the sociology and history of technology, pp. 83–105. The MIT Press, Cambridge (1987).

[3] Collins, H.: Changing order: replication and induction in scientific practice. Sage Publications, Londres (1985).

[4] Collins, H.: Tacit knowledge, trust and the Q of sapphire. Soc. Stud. Sci. 31(1), 71–85 (2001).

[5] Dias, R.: Sessenta anos de política científica e tecnológica no Brasil. Ed. Unicamp, Campinas, SP (2012). 256 pp.

[6] Dias, R. B., Serafim, M. P., Oliveira, J. M.: From the grassroots or by the state? Strategies for fostering urban agriculture in Brazil. In: Anderson, C., Buchanan, C., Chang, M., Sanchez, J., Wakeford, T. (eds.) Everyday Experts: How people's knowledge can transform the food system, 1st ed., pp. 427–439. Coventry University, Conventry (2017).

[7] Feenberg, A.: Transforming technology: A critical theory revisited. Oxford University Press, Oxford, UK (2002).

[8] Gupta, A. K., Sinha, R., Koradia, D., et al.: Mobilizing grassroots' technological innovations and traditional knowledge, values and institutions: Articulating social and ethical capital. Futures 35(9), 975–987 (2003).

[9] Hughes, T. P.: The Seamless Web: Technology. Science, Etcetera, Etcetera. Social Studies of Science, pp. 281–292 (1986).

[10] Law, J.: On Sociology and STS. Sociol. Rev. 56(4) (2008).

[11] Mougeot, L.: Agricultura urbana: conceito e definição. Rev. Agric. Urbana 1, 5–12 (2000).

[12] Pinch, T. J., Bijker, W. E.: The social construction of facts and artifacts: Or how the sociology of science and the sociology of technology might benefit each other. In: Bijker, W., Hughes, T., Pinch, T. (eds.) The social construction of technological systems: new directions in the sociology and history of technology, pp. 1–6. The MIT Press, Cambridge (1987).

[13] Thomas, H.: Estilos socio-técnicos de innovación periférica: la dinámica del SNI argentino, 1970–2000. In: IX Seminario Latino-Iberoamericano de Gestión Tecnológica: Innovación Tecnológica en la Economía del Conocimiento, San José de Costa Rica (2001).

[14] Thomas, H.: Estructuras cerradas vs. procesos dinámicos: trayectorias y estilos de innovación y cambio tecnológico. In: Thomas, H., Buch, A. (eds.) Actos, actores y artefactos. Sociología de la Tecnología, vol. 217. Universidad Nacional de Quilmes, Bernal (2008).

[15] Von Hippel, E.: The dominant role of users in the scientific instrument innovation process. Res. Policy 5(3), 212–239 (1976).

[16] Winner, L.: The whale and the reactor: A search for limits in an age of high technology. University of Chicago Press (2010).

Alessandra Pavesi

Orchards from the forest: Urban agriculture as a lab for multiple learning

Abstract: The destruction of Cerrado (Brazilian savannah), the second largest Brazil's biome after Amazonia, has become the main concern of urban collectives whose activity consists in reintroducing elements of this important ecosystem in the landscape of the cities and in the imagery of their inhabitants. This paper focuses on a conceptual and methodological approach aimed at analyzing and fostering the practice of Cerrado cultivation in urban settings, with an emphasis on the collaborative learning processes it entails. This approach is based on the cultural historical activity theory and it reframes urban farming for Cerrado regeneration from a systemic, developmental and participative perspective, in order to analyze structure and dynamics of the activity and to assist practitioners in identifying and handling the evolving tensions and contradictions internal and external to the activity system, by rethinking its object and designing new actions and artifacts needed to materialize it.

Keywords: Urban agriculture, Cerrado Orchard

1 Urban agriculture for Cerrado regeneration and conservation

In Brazil, the conversion of native habitats to large scale-farming has caused, among other environmental impacts, deforestation and fragmentation of natural systems along with a considerable loss in biodiversity [9, 17, 22].

If, on the one hand, the rate of deforestation of Amazonia has declined since 2005, thanks to control programs sponsored by Brazilian government, on the other hand insufficient efforts were made to restrain the destruction of the second largest Brazil's biome after Amazonia, the Cerrado. On the list of the world's 35 hotspots, it is estimated that in just 20 years, the Cerrado area has been reduced by 26 million hectares – or 260,000 km^2, equivalent to twice the area of England. Today Cerrado's native vegetation occupies 47 % of its original area and the area preserved by conservation units only corresponds to 8 % [1] – 6,6 %, if we consider reserves which afford higher levels of protection.

In urban settings, the Cerrado biome is no longer safe. Even if cities account for only 0.25 % of the total area of the country [16] – thus having a smaller direct impact

Alessandra Pavesi, Educator and researcher of the Arca do Cerrado, outreach activity at Universidade Federal de São Carlos (UFSCar), São Carlos, SP, Brazil, e-mail: sandrapavesi@gmail.com

https://doi.org/10.1515/9783110584455-008

on natural landscapes compared to agriculture – and despite they are considered a space-efficient solution to meet the needs of human societies [24], urban population is expected to reach 90 % of the total population by 2030. Such increasing concentration will exert a great deal of pressure on natural habitat, if not in terms of space [2] surely in ecological footprint [20].

It is true that Brazilian law provides for nature protection, but recently it has been loosened by federal and local administrations in order to meet the pretensions of urban developers. As a consequence, Cerrado remnants have been pushed to urban fringes. Abandoned or used for dumping garbage and residues of all kinds, and stigmatized as inhospitable and no-go areas, they constitute in most of Brazilian cities mere reserve of land for urban sprawl, waiting for formal or informal human settlements, road networks and industrial buildings, which will change irreversibly the original landscape, with effects on soil fertility and productivity, local and global climates (e. g., urban heat islands and emission of greenhouse gases), water resources and finally habitat and biodiversity [26].

Given the prospect of extinction of one of the richest biomes on the planet and the consequences that it would entail for the quality of human life and wildlife, local and international NGOs and scientific/research/academic institutions, among other actors, have been undertaking initiatives aimed at Cerrado conservation and regeneration. With access to very diverse resources and tools, they compose a heterogeneous and comprehensive movement with different motivations, specific contents and local expressions.

The production of seedlings and the cultivation of arboreal native species for commercial purposes is one of the activities the movement gave rise. The seedlings grown in commercial nurseries are intended for reforestation of the so-called "legal reserve areas" (the law, in fact, provides that part of the area of the agricultural property is destined to preservation or regeneration, when applicable, of native forests).

The movement includes also indigenous people who know in depth Cerrado's ecology. They cultivate since time immemorial its biodiversity, an essential resource for subsistence, mobilizing very sophisticated knowledge acquired in practice and handed down from generation to generation. In fact, Brazil is a country whose socioecological traits and dynamics reflect its great biocultural diversity and in which societies driven by globalized leading technologies coexist (sometimes side by side, as in the case of the metropolis of São Paulo), though not necessarily peacefully, with societies whose existence is inextricably linked to the features and resources of local landscapes and indigenous practices, knowledge and skills developed in the management of soils and biodiversity of fields and forests [8, 19].

On the contrary, in urban settings, where people tend to value domesticated, orderly and even aseptic sceneries, biodiversity is generally seen as an inconvenience. As a consequence, not even people who live near the remnants of native forests are aware and acknowledge ecological services they afford, e. g., plants and fruits, which are edible and provide therapeutic properties [4, 12]. According to McKinney [15], such a lack of ecological culture in highly urbanized societies hinders the many conservation opportunities commonly created by an informed and proactive public.

In this paper, urban farming and ecological restoration are considered as intermingled practices aimed precisely at creating a culture which promotes biodiversity by arousing sensitivities and attitudes of care and belonging to a community which comprehend all living beings and the landscape we shape together [10, 11, 13, 25]. Ecological restoration focuses mainly on revitalizing natural ecosystems, but local perspectives eventually seek to realign it beyond its initial rewilding purposes toward other interests like food and herbs production [14]. On the other hand, urban agriculture for restoration cultivates resilience by enhancing biological diversity and ecosystem services, such as pollination, soil enrichment and natural weed and pest control.

Such a complex and challenging task both demands and generates knowledge and learning. From this point of view, urban agriculture compares to a lab for multiple learning not only in the field of sciences and technologies for sustainability, but also in environmental education for ecosystems and biodiversity conservation. At the same time, it is articulated with the idea of community of practice, since participants – volunteers and collaborators (technicians, scientists, etc.) learn in practice, that is, by doing and interacting, and through the collaborative and transdisciplinary mobilization of information and resources, in order to overcome problems and contradictions inherent to activity and its object.

The following sections outline a methodological approach aimed at understanding and organizing learning processes which have the activity of urban agriculture as a system and unit of analysis and management. Such methodology is based on cultural historical activity theory [5, 7] and it will be discussed with empirical reference to the emergent activity of collectives of people who strive to bring back Cerrado to the landscapes of Brazilian cities and to the consciousness of their inhabitants.

2 Cultivating Cerrado in urban settings as wildfire activity

Of all the organizations for Cerrado conservation mapped in 2017 by the Institute for Population and Nature Society (ISPN) in partnership with Conservation International (CI), few of them have the purpose of producing native species seedlings for the recovery of degraded areas in rural environment. None of them operates in urban settings.

However, the ISPN report does not do justice to a more recent phenomenon that acquired some visibility thanks to social networks, consisting in interventions with the purpose of repopulating urban public areas with native vegetation from Cerrado. Individuals and groups engaged in this activity are not numerous, but strongly motivated, being their performance characterized by tenacity in the face of all sorts of constrains. After all, public space, especially on the periphery of capitalism, is characterized as a place of conflict among contradictory interests, uses and practices [27].

As a matter of fact, the activity of these urban groups has a contestatory character, of resistance against the predominant model of urbanization, heavily predatory of local biodiversity and natural resources, as well as against the way of conceiving and designing green areas, based on arbitrary aesthetics, which do not contemplate the ecological, ornamental and, ultimately, propaedeutic importance of utilizing Cerrado vegetation, a prerequisite for its conservation [23].

In the book, Guia de campo dos Campos de Piratininga (Field guide to Piratininga Fields) – where Piratininga Fields used to be the "landscape prior to colonization, having been extinguished after São Paulo urban development," [3, p. 13] visual artist, Daniel Caballero, describes a collective experience of recomposing a Cerrado landscape in a public square in São Paulo, by "looking for wild landscapes in the ditch and collecting memories of a discarded and residual nature, of no value" (p. 31), in order to compose a "collage of varied territories represented by plants and the harvested soil itself... as a practice of subversive relational art, mobilizing people with the intention of creating a decolonizing territory within the city" (p. 61).

Another active group in Cerrado restoration (in the city of Brasília) is headed by landscape architect, Mariana Siqueira. Her office develops projects and experiences in partnership with the Chico Mendes Institute for Biodiversity Conservation (ICMBio / MMA) and the University of Brasilia (UnB), whose researchers subsidize through technical knowledge the creation of a conceptual and methodological framework for "a landscape that expresses the Brazilian savannah." The purpose of this collective is to remedy a gap in the theory, teaching and practice of landscape architecture, by utilizing Cerrado species in gardens aimed at reconnecting users to territory, from an affective point of view, and also to recreate habitat for wildlife.

Finally, the collective Pomar do Cerrado (Cerrado Orchard) was created as a branch of a movement for the conservation of a natural area within the campus of the Federal University of São Carlos (SP) threatened by the ruling master plan, which has transformed the previous landscape into a predictable mosaic of buildings, parking lots and vast lawns. The immediate purpose of the orchard's mentors was to make beauty and worthiness of native species apparent to local people, especially the academic community. But the deeper motivation of their activity consists in reintroducing Brazilian Savannah into collective consciousness through a process of ecological place meaning [21], and ultimately in integrating a nature conservation program into the priorities of the institution.

The three collectives above are born from the passion and obstinateness of their founders. Over time, actors and collaborators joined the activity oriented toward a common object – which consists in keeping the Cerrado alive in urban and mental landscapes – with enough drawing power and motivational force to stimulate the search for sustainability and expansion in spite of a number of adversities and constraints, such as little monetary rewards or institutional support and excessive expenditures of time and energy. Engeström [6] conceptualizes these social production activities as wildfire activities, since they follow a pattern of development characterized by expansive swarming, sideways transitions and boundary-crossing.

This type of activity differs from traditional craft activities and from mass production (although they may seek symbioses with the vertical and linear structures of the latter) in part because they are use-value orientated and resistant to thorough commercialization or assimilation by institutional dynamics. They also differ from peer production, mainly because they develop outside of the sphere of digital virtuality. This does not mean that actors renounce to adoption and use of information and communication technologies, but with little emphasis and dependency on them. For example, their presence in social networks responds to the need to give visibility and attractiveness to an activity that is seen as a solution to the problem of erosion of Cerrado's biodiversity, but a time line or a blog are only tools among many others and not the object of the activity. Similarly, the creation of information systems such as electronic herbarium catalogs, in addition to making the private collection more accessible to professionals and other people, serves as a record of the results and learning achieved through a vast repertoire of actions, such as the collection of seeds and the planting of seedlings in specific contexts: a public square, a school, a botanical garden or a university campus.

Wildfire activities are also characterized by high mobility. For example, one of the main actions of participants is "plant hunting," that is, the practice of walking through the city or preserved natural areas to collect specimens, fruits and seeds. But the physical movement of the actors is just one dimension of mobility, since both the material terrain where the garden or orchard is located and the virtual terrain of the activity are permanently crossed by flows of things and information and their subsequent entanglements. As Pink [18] puts it, although the garden as a materiality is visible as a locality, the garden project is not a bounded entity, its edge being opened to plants, humans and other living beings, services (i. e., water), material inputs and tools that are moved between the borders, as well as to the weather. On the other hand, especially gardens and orchards for Cerrado restoration are meant to provide awareness and inspiration, experiences of sensory aesthetic, new socialities and relationships to nurture, besides learning opportunities about those gardens in particular and the making of them. In their turn, all these experienced aspects are constituted in relation to discourses, intentionalities, agencies and agendas, which also transcend the garden as a locality, thus emphasizing its character of "a site where agency can be exercised in the face of global culture" [18, p. 89].

By moving around in an unexplored territory, that is both material and experiential, people make, therefore, cognitive trails which lead to a progressively more stable conceptualization of that territory and of the way of moving in it. Thus, for instance, when walking through a natural or even degraded area of Cerrado in search of fruits and seeds, or just of evidences of its vegetation, that which may appear to the beginner as an indistinct green mass reveals to a more attentive look its diversity of shapes and textures, its seasonality and mutability: “As I learn to know each type of plant, my sight opens to this type of vegetation and I distance myself from the everyday landscape. What used to be a uniform green mass gains countless textures and volumes, as if I were cured of a type of myopia”.

Our perception – which corresponds to active engagement with the things that matter to us through our sensing and sensed carnal bodies – is a privileged source of awareness and knowledge of the landscape, its elements and transformations. In the sphere of activity which has as its general object the “scratching” of a tract of urban landscape in order to transform it into something that evokes (and invokes) the native landscape, the making of a Cerrado garden or orchard can be understood as a process of sensorial and embodied engagements (collecting and saving seeds, producing seedlings, digging planting holes, gardening and also observing the transformations the garden goes through...) as well as other imaginative and practical actions (planning, applying for funding or support, recruiting volunteers, researching, recording and publicizing the results...) “designed to change the way that the garden might be experienced/known” [18, p. 94].

Through this repertoire of actions and operations that configure the routine of the collectives, trails both across the territory and as cognitive objects leave marks in experience and in the environment – a garden, an orchard, a particular scenery, but also their representations, such a book, an electronic catalogue, an exhibition, among other narratives. Particularly the marks in the environment tend to persist and allow the enhancement of the ability to navigate through certain feature-domain [6] as well as to fit each one’s purpose in the activity.

Indeed, people gain their membership by virtue of contributing something to the endeavor of the collective, and once engaged in the activity, they work symbiotically, on the basis of a spontaneous, indirect coordination between agents or actions – this is another feature of wildfire activities which also suits the specific practice of planting and cultivating Cerrado in urban settings. Indeed, this activity is carried on by heterogeneous and floating collectives composed by agents coming from different cultural backgrounds (academics, artists, landscape architects, environmentalists, etc.) who join the groups drawn by particular interests, but still involved in the transformation or redefinition of a shared challenging object. “Encounters” (as Engeström define interactions between actors in the effort to construct a temporary yet effectively collaborative knot) include also agents pertaining to institutions vested with power, as it is the case of the project Jardins do Cerrado (Cerrado Gardens) of the architect Mariana Siqueira, a partnership with the Ministry of the Environment, and the project Cerrado Orchard. In this case, at a certain moment of its existence the collective sought opportunities for dialogue and the support of the university institution, necessary to guarantee the permanence of the material infrastructure and the continuity and expansion of the activity on the territory of the university.

Encounters between agents in their various trajectories generate questions, deliberations, negotiations and decisions that reflect a sort of balance between understandings, intentions and valuations that are often contradictory. This nonconflict-free process shapes the activity in its organization and dynamics. It affects, for example, the adoption of technologies and tools or the drafting of procedures and rules etc. At the same time, encounters multiply learning opportunities, by opening new terrain to be dwelled in and explored, and by constructing collective concepts that serve as platforms for expansive restructuring of the activity.

In the next section, it will be briefly described how the collective Cerrado Orchard has been constructing its own conceptual platform and infrastructure, through a learning process, which bridges fields of knowledge and sectorial practices (in this specific case, activism and social learning, academic research and landscape design). By the way, one of the main forces of learning in the context of wildfire activities is the crossing of boundaries and the tying of knots between actors operating in fractured and often poorly charted terrains [6], as it is the case of planning and designing for the enhancement and conservation of Cerrado's biocultural diversity.

The author of this paper has been part of Cerrado Orchard collective for the last 3 years and after she joined the group, she proposed that our subsequent actions were designed and evaluated based on the rationale of the cultural historical activity theory (CHAT) [5, 7], since it provides a systemic and prospective approach to the practice, particularly useful to collectively construct the meaningful object of our activity, identify and address inherent contradictions and build consensus on goals, principles and procedures.

In the 1990s, based on CHAT (initiated by Lev Vygotsky between the 1920s and 1930s, and further improved by his disciple, Alexei Leont'ev), Engeström and colleagues developed a model of societal activity. According to this model, the constituents of human activities are depicted within a triangular structure, known as activity system, designed in order to grasp the systemic whole and the multiple elements and relations which participate in activities (see Figure 1).

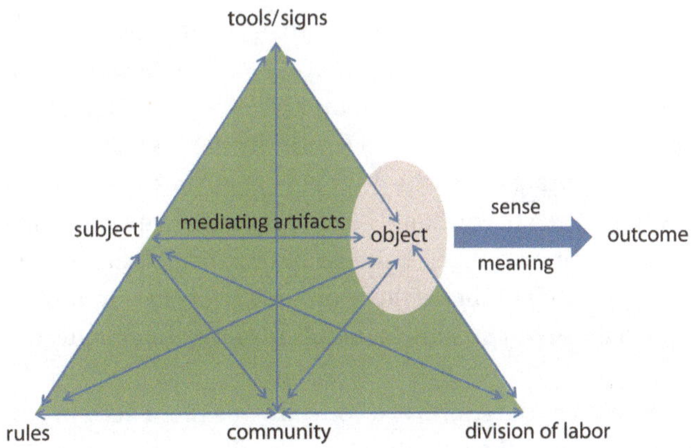

Figure 1: The structure of a human activity system [6].

The activity system, mediated by (physical and symbolic) artifacts and oriented to an object – that consists in the "raw material" or "problem space" at which the activity is directed and that is molded and transformed into outcomes with the help of those artifacts – is taken as the smallest and simplest unit of analysis. This is the first of five principles that help to summarize activity theory. The second principle is precisely the multivoicedness of activity systems, since they comprise all the subjects who participate in the activity, their traditions and backgrounds, values, interests and discourses. The third principle is that problems and potentials of activity systems can only be understood against the background of their history. The fourth principle is the central role of contradictions – understood as historically accumulating structural tensions within and between activity systems – as the driving force of change and development. According to activity theory, as contradictions intensify, questioning and innovations can lead to expansive transformations, to the end of which the object and motive of the activity are deliberately and collectively reconceptualized and a wider horizon of possibilities is embraced. The fifth principle refers to such potential inherent to activity systems, which corresponds to the capacity of subjects to engage in processes of organizational learning. In this new reconceptualization, CHAT has been applied in learning facilitation and monitoring in the most diverse fields of cultural production.

3 The Cerrado Orchard activity system

Cerrado Orchard collective owes its existence to a 10-year dispute over the use of a natural area of 50 Hectares on the campus of the Federal University of São Carlos (UFSCar): where the campus administrators see an undifferentiated piece of land intended to expand infrastructure, the collective that strives for the preservation of the biotic community which inhabits that fragment of Cerrado sees a tract of an ecological corridor (whose disruption could cause an irreversible loss of local biodiversity), but also an opportunity of coexistence with other living beings in urban environment.

Along such local dispute – which typify the deeper contradiction between conflicting exchange and use values resulting from the commodification of urban land – it became evident the neglect of native vegetation remarkably by the planning office, unwilling and unprepared to incorporate it into the designing of green areas on campus. So in a first moment, Cerrado Orchard presents itself as an initiative of guerrilla gardening – which consists of the unauthorized cultivation of native plants on one of the grass surfaces that occupy much of the territory of the university – in protest against invisibility of Cerrado to those responsible for administrating the campus territory and infrastructure.

Although still weakly coordinated, the efforts of a collective inexperienced in cultivation practices, which is dedicated to the activity in the free time, sought to mark a trail aimed at linking the Cerrado fragment threatened by the campus master plan to the daily life of the university community and campus users in general. Participants concentrated on performing basic operations such as preparing the soil, planting seedlings, keeping them alive during the dry season, protecting them from wind and invasive plants… (see Figure 2).

Figure 2: Cerrado Orchard collective at work.

Volunteers ignored the scientific denomination of most of plants, known through fancy names assigned ad hoc to evoke morphological or sensorial characteristics (such as perfume and texture), which would allow to associate the collected seeds to the mother plants in the forest. However, this system of identification limited the search for information in botanical collections or scientific publications that would allow the collective to insert the plants of the orchard in the realm of the ecological relations, of which they participate in the local context, thus valuing not only the plants of the orchard itself, but also the habitat from which they come (see Figure 3).

Awareness of the contradiction concerning insufficiency and inadequacy of tools and techniques led the group to develop the project of a virtual catalog of the orchard's plants and a database of scientific publications about Cerrado on UFSCar campus. Another contradiction identified by the collective stems from the current configuration of the orchard, far below expectations of stimulating the perceptive and aesthetic fruition of the orchard through its design (see Figure 4).

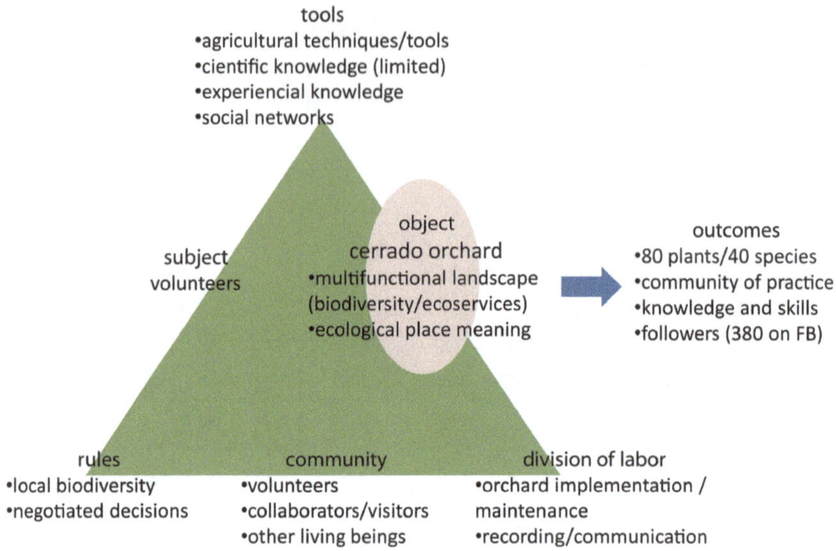

Figure 3: Cerrado Orchard activity system.

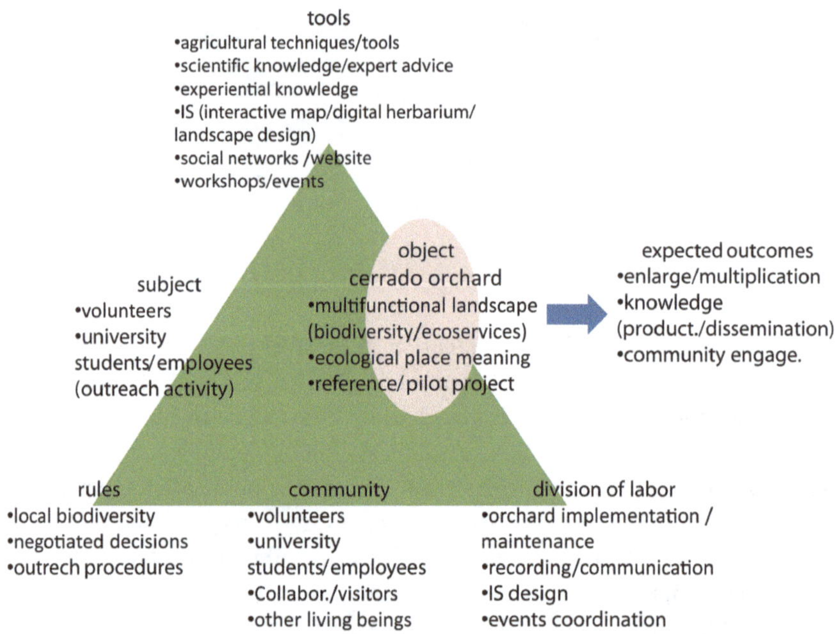

Figure 4: Cerrado Orchard's development.

The most important effect of publicizing the collective's results in social networks and of seeking a dialog with people which share the same interests was aggregation of collaborators, specialized knowledge, tools and techniques that contributed to reflect on and reformulate the object of the activity, as well as the actions needed to materialize it. On the other hand, more recent needs (i. e., of a catalog and a database) drove the collective closer to scientific community (see Figure 5).

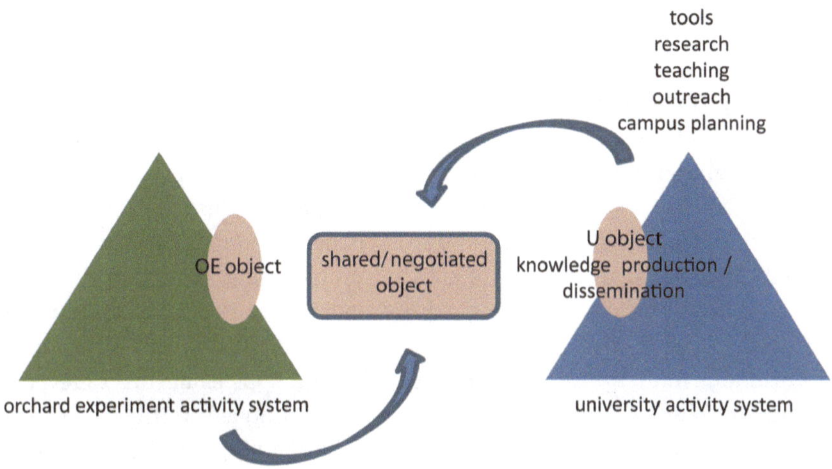

Figure 5: Collaboration between activity systems.

Indeed, the volunteers who created and care for the orchard have been working on a base of trial and error, using their own resources and with restricted access to academic and technical knowledge. This has clearly limited the results of their efforts, both in substantive terms (e. g., orchard's extent and configuration) and in terms of learning opportunities, which could be provided by research together with hands-on activities. The need to overcome such limitations, related to contradictions between the "orchard activity system" and the "university activity system," caused the practice to move along a path of expansive learning and development. As a result, an outreach program was created through which the collective has sought the collaboration and resources of the academic institution with the purpose of consolidating its own infrastructure and social impact.

4 Final considerations

In a recent article in The New York Times (March 12, 2016), titled The Global Solution to Extinction, Edward O. Wilson defends that "The only way to save upward of 90 percent of the rest of life is to vastly increase the area of refuges, from their current 15 percent of the land and 3 percent of the sea to half of the land and half of the sea." According to him, such amount "can be put together from large and small fragments around the world to remain relatively natural, without removing people living there or changing property rights" and he describes our sustained coexistence with the rest of life both a practical challenge and a moral decision.

This paper presents aspects related to the structure and development of the activity of urban collectives, which took very seriously the challenge launched by Wilson. Indeed, they engaged in learning processes aimed at reintroducing Cerrado's endangered biotic community not only in urban landscapes, but also in the collective imagery, through a process of production and ecological significance of places.

This practice, with local variations, has the characteristics of a wildfire activity [6], a model of human and organizational activity that pursues innovation and expansion along with efficiency and sustainability according to a pattern of development, which take multiple learning directions and crosses the boundaries of academic disciplines, fields of knowledge and ways of knowing and learning. Particularly, the activity addressed in this article has ecological sustainability as its central object; in a historic context in which government organizations as well as research institutions have shown serious limitations in fostering attitudes and policies needed to reverse the destruction of natural ecosystems, the collectives mentioned in this paper took on the task of criticizing and provoking the transformation of current cultural practices that place at risk not only the survival of wildlife, but the very basis of natural resources on which all human societies depend, regardless of their socioeconomic formations.

In the opening of physical and conceptual trails with the purpose of consolidating their own infrastructure, the collectives intersected other historical trails; in fact, as Engeström reminds us, the physical, cultural and symbolic landscape on which the collectives learn to move and leave the marks of their agency, "never is an empty space to begin with; it has dominant trails and boundaries made by others, often with heavy histories and power invested in them. When new dwellers enter the zone, they eventually have critical encounters with existing trails" (p. 14). Thus, the Cerrado planters' journey is unlikely to be free of obstacles and contradictions imposed by dominant cultural practices of urban land use. But precisely the overcoming of these obstacles and contradictions constitutes the main motivation and motor of development and self-renewal of wildfire activities, as they trigger processes of "transformative learning...built into the very operating principles and everyday social textures of these activities" (p. 5).

In the article, we seek to illustrate how CHAT can serve as a conceptual framework both to understand the structure and dynamics of the activity that consists in cultivating the Cerrado in urban settings and to facilitate organizational learning processes in a way that its participants gain awareness and control on the activity, greater efficiency in the use of resources and opportunities, sustainability and social impact.

Bibliography

[1] Beuchle, R., Grecchi, R. C., Shimabukuro, Y. E., Seliger, R., Eva, H. D., Sano, E., Achard, F.: Land cover changes in the Brazilian Cerrado and Caatinga biomes from 1990 to 2010 based on a systematic remote sensing sampling approach. Appl. Geogr. 58, 116–127 (2015). https://doi.org/10.1016/j.apgeog.2015.01.017. Accessed February 10, 2018.

[2] Bogaert, J., Vranken, I., Andre, M.: Anthropogenic effects in landscapes: historical context and spatial pattern. In: Hong, S.-K., Bogaert, J., Qingwen, M. (eds.): Biocultural landscapes: diversity, functions and values, pp. 89–112. Springer, Dordrecht (2014).

[3] Caballero, D.: Guia de campo dos Campos de Piratininga ou o que sobrou do cerrado paulistano ou como fazer seu próprio Cerrado Infinito. La Luz del Fuego, São Paulo (2016).

[4] Dias, J. E., Laureano, L. C.: Farmacopéia Popular do Cerrado. Goiás: Articulação Pacari (Associação Pacari), Goiás (2009).

[5] Engeström, Y.: Expansive learning at work: toward an activity theoretical reconceptualization. J. Educ. Work 14, 133–156 (2001). https://doi.org/10.1080/13639080020028747 Accessed February 10, 2018.

[6] Engeström, Y.: Wildfire activities: new patterns of mobility and learning. Int. J. Mobile Blended Learn. 1, 1–18 (2009).

[7] Engeström, Y.: Learning by expanding: an activity-theoretical approach to developmental research. Cambridge University Press, New York (2015).

[8] Hecht, S.: Kayapó savanna management: fire, soils, and forest islands in a threatened biome. In: Woods, W. I., Teixeira, W. G., Lehmann, J., Steiner, C., WinklerPrins, A., Rebellato, L. (eds.) Amazonian Dark Earths: Wim Sombroek's vision. Springer, Dordrecht (2009).

[9] Klink, C. A., Machado, R. B.: Conservation of the Brazilian Cerrado. Conserv. Biol. 19, 707–713 (2005). https://doi.org/10.1111/j.1523-1739.2005.00702.x. Accessed February 10, 2018.

[10] Krasny, M. E., Lundholm, C., Lee, E., Shava, S., Kobori, H.: Urban landscapes as learning arenas for sustainable management of biodiversity and ecosystem services. In: Elmqvist, T., Fragkias, M., Goodness, J., Güneralp, B., Marcotullio, P. J., McDonald, R. I., Parnell, S., Sendstad, M., Schewenius, M., Seto, K. C., Wilkinson, C. (eds.) Urbanization, Biodiversity and Ecosystem Services: Challenges and Opportunities. Springer, New York (2013).

[11] Kudryavtsev, A., Krasny, M. E., Stedman, R. C.: The impact of environmental education on sense of place among urban youth. Ecosphere 3, 29 (2012). https://doi.org/10.1890/ES11-00318.1.

[12] Maroni, B. C., Di Stasi, L. C., Machado, S. R.: Plantas medicinais do Cerrado de Botucatu: Guia ilustrado. Editora UNESP, São Paulo (2006).

[13] McCann, E.: Restoration-based education: teach the children well. In: Egan, D., Hjerpe, E. E., Abrams, J. (eds.) Human dimensions of ecological restoration: integrating science, nature, and culture, pp. 315–334. Island Press, Washington (2011).

[14] McCann, E. P., Schusler, T. M.: Ecological restoration. In: Russ, A., Krasny, M. E. (eds.): Urban Environmental Education Review, pp. 251–-260. Cornell University Press, Ithaca, N.Y. (2016).

[15] McKinney, M. L.: Urbanization, biodiversity, and conservation: the impacts of urbanization on native species are poorly studied, but educating a highly urbanized human population about these impacts can greatly improve species conservation in all ecosystems. Bioscience 52, 883–890 (2002). https://doi.org/10.1641/0006-3568(2002)052[0883:UBAC]2.0.CO;2. Accessed February 10, 2018.

[16] Miranda, E. E., Gomes, E. G., Guimarães, M.: Mapeamento e estimativa da área urbanizada do Brasil com base em imagens orbitais e modelos estatísticos. In: Anais XII Simpósio Brasileiro de Sensoriamento Remoto, pp. 3813–3820. INPE, Goiania, Brazil (2005).

[17] Pearce, F.: The Cerrado: Brazil's other biodiverse region loses ground. Yale Environment 360, Yale School of Forestry and Environmental Studies. http://e360.yale.edu/features/the_cerrado_brazils_other_biodiversity_hotspot_loses_ground. Accessed February 10, 2018.

[18] Pink, S.: Situating everyday life: practices and places. Sage Publications Ltd, London (2012).
[19] Posey, D. A.: Indigenous management of tropical forest ecosystems: the case of the Kayapo indians of the Brazilian Amazon. Agrofor. Syst. 3, 139–158 (1985).
[20] Rees, W.: Revisiting carrying capacity: area based indicators of sustainability. Popul. Environ. 17, 191–215 (1996). https://doi.org/10.1007/BF02208489. Accessed February 10, 2018.
[21] Russ, A., Peters, S. J., Krasny, M. E., Stedman, R. C.: Development of ecological place meaning in New York City. J. Environ. Educ. 46, 73–93 (2015). https://doi.org/10.1080/00958964.2014.999743. Accessed February 10, 2018.
[22] Schiesari, L., Waichman, A., Brock, T., Adams, C., Grillitsch, B.: Pesticide use and biodiversity conservation in the Amazonian agricultural frontier. Philos. Trans. R. Soc. Lond. B, Biol. Sci. 368, 1619 (2013). https://doi.org/10.1098/rstb.2012.0378. Accessed February 10, 2018.
[23] Siqueira, M.: Jardins do Cerrado: potencial paisagístico da Savana brasileira. Rev. CAU/UCB 4, 33–47 (2016). https://portalrevistas.ucb.br/index.php/CAU/article/view/7065/4384. Accessed February 10, 2018.
[24] Standish, R. J., Hobbs, R. J., Miller, J. R.: Improving city life: options for ecological restoration in urban landscapes and how these might influence interactions between people and nature. Landsc. Ecol. 28, 1213–1221 (2013). https://doi.org/10.1007/s10980-012-9752-1. Accessed February 10, 2018.
[25] Tidball, K. G., Krasny, M. E.: Urban Environmental Education From a Social-Ecological Perspective: Conceptual Framework for Civic Ecology Education. CATE 3, 11 (2010). http://digitalcommons.lmu.edu/cate/vol3/iss1/11/. Accessed February 10, 2018.
[26] Tratalos, J., Fuller, R. A., Warren, P. H., Davies, R. G., Richard, G., Gastonet, K. J.: Urban form, biodiversity potential and ecosystem services. Landsc. Urban Plan. 83, 308–317 (2007). https://doi.org/10.1016/j.landurbplan.2007.05.003. Accessed February 10, 2018.
[27] Wisnik, G.: Espaço Público: Para ser 'público' de fato, é preciso conflito. Cidades em disputa. J. Nexo. 10, 2018 (2018). https://www.nexojornal.com.br/especial/2018/01/26/Cidades-em-disputa?utm_source=socialbttns. Accessed February 10, 2018.

Part III: **Renewable energy**

Antonio Pralon F. Leite and Douglas B. Riffel

The challenges of the new energy revolution

Abstract: The new energy revolution is about a change from a resource based, centralized energy source to a technology based, decentralized topology. There is strong evidence that the energy sources that have driven economic development since the beginning of the Industrial Revolution will be not available for future generations. The humankind is tracing a clear path from a high- to low-carbon energy source. The aim of this paper is evaluate the perspectives of this new energetic era and the relations of our demands. It is clear that is possible to build a more suitable society, but it will be strongly dependent how we will working together as a global economy, more integrated with a clear climate focused policy. Energy security, reduction of greenhouse gas emissions, sustainability and social inclusion are the main factors that should guide a change in the route of the world energy profile. Once this set of essential elements has been prioritized, it is possible to minimize climate imbalances on the planet and to improve the conservation of natural resources, a wise use of land for the production of bioenergy (without compromising food security), electricity generation based on renewable sources, the adoption of energy efficient strategies and a new conception of the city. However, none of this effort will be enough without the reduction of our hungry energy demand.

Keywords: New energy revolution, renewable energy, energy demand

1 Introduction

There is strong evidence that the energy sources that have driven the economic development since the beginning of the Industrial Revolution will be not available for future generations. The replacement of firewood by fossil fuels characterized a brief period in the history of humankind, guaranteeing the bases of industrialization and the development of cities. At that time, this resource was abundant and with a relatively low price. Nowadays, the combination of limited oil reserves, increasing energy consumption and climate change have made the fossil fuel market uncertain for the next decades. By 2050, the world's population is projected to increase by 50 %, with more than two-thirds of the population living in emerging countries – mainly in China, India and Southeast Asia - whose energy consumption per capita will have almost tripled [1]. Meanwhile, to guarantee their energy security, the industrialized countries are using coal – a fuel that emits large amounts of greenhouse gases – a mineral resource with

Antonio Pralon F. Leite, Universidade Federal de Sergipe, São Cristóvão, Brazil
Douglas B. Riffel, Universidade Federal de Sergipe, São Cristóvão, Brazil, e-mail: dougbr@academico.ufs.br

https://doi.org/10.1515/9783110584455-009

reserves for two more centuries. On the other hand, ever since the first decade of the twentieth century, there has been scientific evidence for excess greenhouse gases in the atmosphere, which has been related to the climate. These gases are released during the combustion of fossil fuels and can affect life forms and increase the impact of natural disasters as well as their frequency, threatening the survival of human civilization. These catastrophic scenarios indicate the urgency of changing current patterns of energy production and consumption, to ensure the sustainability of the environment, ensuring a stable and long lasting supply of energy. This means replacing fossil fuels by renewable energy and adopting strategies for the efficient generation and use of this energy. In other words, it is necessary to learn from nature a way to make use of natural resources wisely. A transformation of urban centers into "light" spaces, where the energy demand is greatly reduced, is compulsory.

The new energy revolution is about a change from a resource based, centralized energy source to a technology based, decentralized topology. It is possible to have an idea of that in terms of jobs. Worldwide, a small number of enterprises employ thousands of people in the coal industry. When something affects this industry, politicians and the media promptly become alarmed. Of course, this is a lot of jobs. On the other hand, in the last few years in Germany alone, hundreds of thousands of small enterprises in the photovoltaic (PV) industry have closed or at least reduced their staff, yet the effect on the media is not the same, despite the fact that in terms of the quality and the quantity of jobs, it is almost the same. The main reason is that the PV industry is decentralized [2].

2 Energy in the twenty-first century

According to the International Energy Agency (IEA), over the last 30 years, the world's energy consumption increased by almost 75 %. Currently, about 80 % of the world's energy supply comes from fossil fuels and the remainder comes from renewable sources and nuclear fuel. The production of both oil and gas will continue to rise, up to 2040 in the case of global annual oil production and around 2060 for natural gas [3]. Measuring nonrenewable energy reserves (fossil and nuclear fuels) is not an easy task: besides the difficulties and costs in computing the reserves of mineral deposits, their exploitation depends directly on the state of the global economy. The most optimistic estimates do not put this past the end of this century. However, as a former Saudi minister said, "just as the Stone Age did not end because of a shortage of stone, the Oil Age will end well before the oil reserves are exhausted." In fact, the first energy revolution was from wood to coal, the second, from coal to hydrocarbon sources and now, the third is the transition to the new energy sources, which is a clear path from high- to low-carbon sources of energy.

Furthermore, it is necessary to develop renewable energy technologies to replace the mass of technological equipment based on fossil fuels. For instance, in most industries sectors, there has been observed a gradual replacement of coal by natural gas. The 2014 report by the Intergovernmental Panel on Climate Change (IPCC) proposes the need to reduce greenhouse gas emissions by 50 %–80 % by 2050 to avoid environmental catastrophe [4]. To meet the energy demand by 2030, which is estimated to increase by 45 % over that of 2008, oil will account for 30 %, coal for 29 % and natural gas for 22 %, maintaining the total share of these sources since the 1970s at around 80 % [5]. Doubling the share of renewables and assuming that there is a reduction in the demand for electricity by 2030, fossil and nuclear fuel will still make up 64 % [6]. As the cheapest fossil fuel, coal will continue to play an important role in the world energy mix, constituting 25 % of the total energy supply in 2050. A cleaner utilization of coal is an ongoing and important trend. At present, coal-fired power generation with large capacity and high parameters, power generation with large-scale circulating fluidized beds, and integrated gasification combined-cycle power generation can increase the thermal efficiency of coal generators to be around 50 % [3]. According to the IPCC, global energy consumption generates an annual emission of about 8 billion tons of greenhouse gases, mainly CO2. The amount of carbon dioxide absorbed from the air by plants through photosynthesis is 60 billion tons per year. As CO2 emissions have increased in the last 150 years the projection is that the concentration of CO2 in the atmosphere will reach levels that can unbalance the climate [7]. The scenarios outlined by the IPCC suggest that is drastically essential to intensify the use of renewable sources and improving efficiency in the generation and final use of energy. This change in the world's energy base has become imperative and constitutes a major challenge in ensuring reduce of greenhouse gas and the conservation of the natural resources.

The "new energy revolution" may come earlier than expected. Indeed, a new route to a renewable energy future – independence from fossil fuels – was already phased in 2012, from which date the capacity of renewable power installations has exceeded that of nonrenewables by a rising margin [3]. In 2016, renewables reached 62 % of the new power capacity [6]. This marks the beginning of the renewable era, or the great change. It will only be a matter of time before the current global energy scenario based on fossil fuels will be over. The sun, the wind, the heat of the earth, waterfalls and plants will make up the sources of renewable energy, defined by its inexhaustible character. They are clean sources, or at least produce only a minimal amount of pollutants, unlike fossil (petroleum, coal and natural gas) or nuclear (uranium) fuels. Thus, these renewable resources give rise to solar, wind, geothermal, hydropower and bioenergy energies. The renewable sources just mentioned are those with the highest level of technological development and are already responsible for a significant amount of electricity generation worldwide. But there is a still promising source of renewable energy: the energy of the seas and oceans, with the potential to supply much of the world's electricity demand, is coming in the near future. Depending on its characteristics, a given source of renewable energy can be used to generate heat (either directly, e. g., solar thermal or via combustion, e. g., bioenergy) or to generate electricity directly (e. g., hydropower, wind, solar photovoltaic, geothermal and oceanic/tidal).

The global capacity for clean energy generation is currently 2,017 GW, with the following distribution [8]: hydropower (1,096 GW), wind (487 GW), solar PV (303 GW), biomass and thermal rejects (112 GW), geothermal (13.5 GW), CSP (4.8 GW) and tidal (0.5 GW). Several factors determine whether renewable energy source is used, such as local availability, access to the grid, raw material and equipment costs, labor qualification and regulations.

The World Energy Council projected some implications for the energy sector [9]:

1. The world's primary energy demand growth will slow and per capita energy demand will peak before 2030 (reaching 1.9 ton of oil equivalent) due to unprecedented efficiencies created by new technologies and more stringent energy policies. Energy intensity will decline three times faster dependent to the scenario. Substantial efficiencies will be gained through the deployment of solar and wind electricity generation capacity. Conversion rates for these renewable energy sources are much higher than those for fossil fuel plants, meaning less energy will be needed from the primary source.
2. Demand for electricity will double to 2060. Meeting this demand with cleaner energy sources will require substantial infrastructure investments and systems integration to deliver benefits to all consumers. New cleaner generation is needed to meet climate targets and utility business models are pushed to the limits by stringent policies and shifting consumer demands. The industry must find a way to navigate shifting dynamics. More stringent regulatory requirements for a low-carbon future will force companies everywhere to make significant changes in their business models or face collapse. This change is particularly pronounced for utilities who must respond quickly to changing demand patterns.
3. The phenomenal rise of solar and wind energy will continue at an unprecedented rate and create both new opportunities and challenges for energy systems. Solar and wind energy account for only 4 % of power generation in 2014, but by 2060 it will account for 20 % to 39 % of power generation. Large-scale pumped hydro and compressed air storage, battery innovation and grid integration provide dependable capacity to balance intermittency.
4. Demand peaks for coal and oil have the potential to take the world from "stranded assets" to "stranded resources." Fossil fuel share of primary energy has shifted just 5 % in the last 45 years from 86 % in 1970 to 81 % in 2014. To 2060, the momentum of new technologies and renewable energy generation results in the diversification of primary energy. Again, the more optimistic scenario fossil fuel share of primary energy will fall to 50 %.
5. Transitioning global transport forms one of the hardest obstacles to overcome in an effort to decarbonise future energy systems. Oil share of transport falls from 92% in 2014 to 60 % in the more optimistic scenario. Advances in second and later third generation biofuels make substantial headway in all scenarios.
6. Limiting global warming to no more than a 2 °C increase will require an exceptional and enduring effort, far beyond already pledged commitments, and with

very high carbon prices. Depending on the scenario, this reduction can fall by 61 % from 2014 to even increase of 5 % in a more fragmented global economic and political system.

7. Global cooperation, sustainable economic growth, and technology innovation are needed to balance the energy trilemma: energy security, energy equity and environmental sustainability.

Some scientists have trace a roadmap for a rapid decarbonization, like Rockström et al. [10], Wiseman [11], Tollefson [12], Capros et al. [13] and Cash and McCormack [14]. In most cases, the main roadblocks include the lobby of the fossil fuel industry; the financial, governance and implementation constraints; the unsustainable endless rise of consumption and inequitable wealth distribution. It is clear that the challenge is build a worldwide goal to instigate technological and institutional breakthroughs to overcome the human impact in the environment. Tollefson, looking from inside the actual COVID-19 pandemic, says be optimistic with the possibilities of change in economy and in the direction to avoid a potential climate catastrophe. As studied by Griffiths [15], diplomacy will play an key role on that.

2.1 Energy demand

Haley et al. [16] remember that energy efficiency has "nonenergy benefits," such as energy poverty reductions and economic stimulus. The policymakers should maintain support for aggressive efficiency programs, learning from the traditional demand side management strategies and looking for another funding streams, as the greenhouse gas emission reduction policies. Strategies to reduce energy consumption include the use of technological innovations in the process of production and end-use of energy, as well as a change in people's individual behavior. One example of this is the Danish wind turbine manufacturer, Vestas, that uses big data to optimize the performance of their wind turbines worldwide [17].

It is observed, however, that unequal efforts are being made in the search for better efficiency in the use of energy resources, and this is reflected in the relation between the energy consumed and the gross domestic product (GDP) of each country, the so-called "energy intensity." Currently, the United States has an energy intensity 50 % higher than that of the European Union countries, while Japan is 20 % lower than the USA [1]. The GDP is an index that measures the efficiency of the production system and also the lifestyle. Manufacturing productivity in economically developing countries is generally inefficient while their living standard is lower and energy consumption is smaller. In other words, they have energy-intensive production systems and a nonenergy-intensive lifestyle (not the same thing as energy conservation) at the same time. An industrialized country has nonenergy-intensive production systems and an energy-intensive lifestyle based on large-scale energy consumption. As an aggregate index, the GDP can hide some internal inequality between industry and lifestyle energy efficiency.

Although the ratio between energy consumed and goods produced has declined in recent decades, this has not meant a reduction in primary energy consumption. This is because the GDPs of the developed countries grew rapidly in this period while gains in energy efficiency did not follow this growth. Four major sectors are responsible for primary energy consumption: electricity generation, industry, transportation and "other sectors," which include the residential, tertiary and commercial sectors, as well as agriculture. According to the IEA, 36 % of the world's primary energy consumption goes to electricity generation. Industry consumes 36 %, buildings 30 % and transport, 28 % of the global end-use energy consumption in 2016. Another point to be in account is the currency units that must be the same, and the evaluated levels of prices must be the same. Suehiro [18] faced this problem and present a GPD evaluated with market exchange rates and another evaluated with purchasing power parity. One second improvement was the calculation of two index, one for industrial sector and another for nonindustrial sector. The idea is to separate the energy efficiency in the industry from the lifestyle. The more developed countries show a low index, with the industrial energy efficiency lower than from the lifestyle. The opposite is observed in economically developing countries.

Since 2010, energy intensity has declined at an average rate of 2.1 % per year [19]. The energy saved is equivalent to adding another European Union to the global energy market. Although the ratio between energy consumed and goods produced has declined in recent decades, it has not meant a reduction in primary energy consumption. This is because developed countries' GDP grew rapidly in this period and gains in energy efficiency did not follow this growth. In all of the 34 countries that make up the Organization for Economic Co-operation and Development (OECD), this energy consumption increased by 58 % between 1971 and 2002 [20]. Four major sectors are responsible for primary energy consumption: electricity generation, industry, transportation and "other sectors," which include the residential, tertiary and commercial sectors and agriculture. The industry consumes 36 %, buildings 30 % and transport, 28 % of global final energy consumption in 2016 [21]. More than 90 % of transport are powered by fossil-fuels, which means that efficiency improvements can significantly reduce emissions of pollutants and GHGs.

2.2 Electricity

The use of electric power is fundamental to ensure economic development and provide comfort to people. Extremely versatile, this form of energy has sustained practically all the technological advance, which is why its consumption has grown globally, despite the efforts made by the most industrialized countries to reduce their energy intensities. Projections from the World Energy Council indicate that by 2060 global electricity production is expected to grow 100 % compared to 2014 [9]. Given this framework, how to improve efficiency in electricity generation? Overall, it is reasonable to attribute an average value of 1/3 to the ratio of electric power produced to primary energy consumption. This fraction means that for every 3 units of energy supplied by the fuel, only 1 unit turns into electricity; the rest is released into the atmosphere in the form of heat. This inefficient relationship between final and primary energy can be significantly improved by two technological approaches: the combination of thermodynamic cycles to increase the heat/electricity conversion efficiency, and the large-scale use of cogeneration systems. The use on a global scale of high-efficiency power plants and cogeneration units would represent a saving of 15 % in total primary energy consumption by 2030, according to IEA data. This economy obviously translates into the lower consumption of CO2-emitting fuels, mainly oil and natural gas. Another alternative to increasing the global supply of electricity is small power plants, which can use different sources of renewable energy, such as hydropower, wind, solar thermal and photovoltaic.

2.2.1 Combined cycle

In the case of thermoelectric plants, which are normally installed far from the sites of consumption, the best way to increase the conversion efficiency is to use "combined cycles," the principle of which is to associate a gas turbine with a steam turbine, taking advantage of the heat rejected in one to power the other. Result: you can get up to 70 % higher yields. According to the IEA, if all new plants were built with the combined cycle technology, it would be possible to produce the same amount of electricity generated today, with a reduction of 36 % in primary energy consumption. This refers mainly to emerging countries, such as Brazil, China and India, with projections of strong growth in the coming decades, where most of the new thermoelectric power plants will be built. In almost all of these countries (except for Brazil), electric power plants are fueled by fossil fuels; by 2020 global consumption of fossil oil, natural gas and coal in electricity generation will have increased by 80 %, while the use of renewable energy will account for 20 % of global production [5]. With the start-up of high-yield plants, world consumption of primary energy would be reduced by 6 %, which is significant. The challenge is how to finance technology transfer to these countries – in accelerated processes of industrialization and economic growth – to allow the introduction of combined cycles in their electric generation matrix? One possibility could be the "Clean Development Mechanism" (CDM) provided for by the Kyoto Protocol. By the CDM, a country that reduces its GHG emissions can get carbon credits equivalent to the emissions avoided and use those resources to finance the technology transfer needed to meet its own targets or to help other countries achieve their reduction targets. Latin America and Asia share 90 % of the CDM projects, which began to be certified in 2005; Brazil leads with more than 15 % of them, followed by Mexico. By this mechanism, developing countries may have access to new technologies to make their energy matrixes cleaner.

2.2.2 Cogeneration

Cogeneration – another technological alternative to reduce waste in electricity generation – is also based on the use of rejected heat for the environment in thermodynamic energy conversion cycles. In this case, however, it is necessary that the production unit should be closed to the consumption to avoid losses in transportation. Ideally, it is possible to achieve a recovery ratio close to unity, with the useful energy now representing simultaneously heat, electricity and mechanical energy. In this way, thermal energy waste (as in thermal power plants) is avoided, as this energy can be used for various purposes, but especially in industrial processes such as drying, heating and distillation, among others. Currently, the efficiency of cogeneration systems is between 0.8 and 0.9, which means a considerable saving of primary energy used to generate electricity. The most commonly used techniques in cogeneration are diesel engines, steam and gas turbines. The sources of energy are fossil fuels and bioenergy, such as biomass of reforestation wood, biogas or vegetable waste. Bagasse and sugarcane straw are already widely used in the Brazilian sugar and ethanol industry, in which demand for thermal, electrical and mechanical energy is high. Another plant residue used intensively in cogeneration systems in Brazil is the bark of eucalyptus, in the pulp and paper industry. A decentralized conception of cogeneration would limit its application to plants with installed capacity of up to 1 GW, since units with higher powers would be located away from consumption centres, just where thermal energy is required. Thus, with small generating units, it would be possible to meet the needs of heat and electricity directly in the places of consumption, such as in factories, in large residential condominiums, in commerce and services sector. In Brazil, where the electric demand to provide air conditioning is high, cogeneration units could supply a good part of this consumption by the use of thermal cooling cycles – in which the main energy input is heat – in large spaces such as a commercial mall, hypermarkets, hospitals, airports and industries. In addition, since decentralized generation does not require the transport of electricity over long distances, the energy losses in the transmission are avoided, which are not negligible. In Brazil, these losses are estimated between 15 and 17 %.

2.2.3 Use of methane

The gas released in the decomposition of organic matter from landfills contains high concentrations of methane (CH4), which can be used to generate electricity. CH4 is highly deleterious to the atmosphere and its burning can attenuate the greenhouse gas effect of anthropogenic origin and at the same time generate energy. The energy potential of methane released in Brazilian cities is estimated at up to 3,600 GWh per year, or electricity to supply up to 18.3 million houses with average monthly consumption of 200 kWh [22]. In some European countries, biogas from urban waste drives powered plants and is also used as a source of thermal energy (heat) for space heating. In Copenhagen, 97 % of ambient heating comes from recovery heat of waste incineration plants and cogeneration, saving energy and reducing CO2 emissions and polluting gases. Sweden has a solid waste recycling plant – in Boras (350 km southwest of Stockholm) – with 100 % recovery and reintroduction, which generates 10 MW of electrical power, sufficient to feed 100 thousand inhabitants of the city.

2.3 Industry

The secondary sector has a major share in the consumption of primary energy and electricity. In this regard, the establishment of energy efficiency criteria in the industry is an important strategy in combating energy waste. Companies promoting energy sustainability and energy efficiency policies will receive international certification, based on ISO 50.001, a joint initiative of Brazil and the United States. The new ISO (International Organization for Standardization) was made possible by a partnership between the Energy Management Commission of ABNT (Brazilian Association of Technical Standards) and Procobre Brasil (entity linked to copper producers), with the objective of prioritizing energy consumption, leading to three aspects: qualitative (type of use), quantitative (rational use) and technological (efficiency). The copper industry can serve as an example as it participates in an international program that promotes actions to optimize the use of copper in the electrical sector, to boost energy efficiency, protection of the environment, safety and reliability in the generation, transmission and distribution of electricity. The new standard should contribute to PNE 2030, which considers energy efficiency to be critical to securing energy supply to the secondary sector in the coming years. It is estimated that, with the adoption of energy efficiency strategies, the Brazilian industry can save 440 thousand barrels of oil by 2020 [23]. ISO 50.001 should become an important tool for the dissemination of the concepts of energy efficiency in the country, as well as in the formulation of policies for the industrial sector, such as tax incentives for modernization and energy efficiency, energy performance criteria for obtaining financing, consumption reduction certificates and efficiency targets – based on benchmark indices – across a variety of industry sectors. In addition, it should foster the use of efficient products and equipment, such as electric motors certified with the Procel (National Energy Conservation Program) seal, together with the productive sector and consumers. The new certification will also contribute to expanding the market for technologies that generate or save energy with low environmental impact, so-called "clean technologies." Most companies are expected to adopt the new ISO as a way of demonstrating to the market their commitment to sustainable development, enabling society to more clearly identify the differences between responsible and nonresponsible companies.

2.4 Bioenergy

Bioenergy is defined as any energy produced from organic matter from animal or vegetable origin. Thus, it includes biogas, one of the products of the anaerobic decomposition of organic matter, which can be an important source of energy both in the countryside and in cities. The Earth's biomass potential is estimated to be 2 trillion tons, which in energy terms corresponds to eight times the world's primary energy consumption in 1995 [24]. The direct combustion of biomass (in furnaces, boilers, etc.) could also be a source of renewable energy when not derived from an extractive activity. The traditional and modern uses of bioenergy provide the largest amount of renewable energy. The total primary energy supplied in 2016 from biomass was approximately 62.5 Exajoules (EJ). The supply of biomass for energy has been growing at around 2.5 % per year since 2010. This is 10.5 % of all world primary energy consumption. In developing countries, this share reaches 34 %. In Brazil, where biomass makes up 26 % of the total, it is already widely used in several industries, such as the sugar, alcohol, paper and pulp industries, which generate a large amount of waste used to generate electricity, especially in cogeneration systems [25]. The production of firewood, coal and logs also generates large volumes of waste, which can be used to generate electricity. Road transport accounts for the largest share of energy consumption in the world to move people and products: about 80 % of the world's population uses roads, and petroleum products (diesel and gasoline) account for more than 90 % of total fuels. Green fuels (bioethanol and biodiesel) may play a significant role in the transportation sector on a global scale, as is the case in Brazil, where sugarcane ethanol is mixed with gasoline, making up a significant share (25 %) of the fuel of the national automotive fleet. On the other hand, a "large and long transition" points to an energy and technological race to supply the automobile of the future [26]. In this race, Brazil came out ahead with the flex-fuel engine, but electric cars and plug-in hybrids will enter this competition. The combination of green fuel with electric vehicles is expected to contribute heavily to the planet's new energy route, toward a greener world with lighter cities.

2.4.1 Bioethanol

Bioenergy is now among the renewable energies, and is the one with the greatest prospects of reaching a large scale of production. In 2015/16, Brazil produced 666.8 million tons of sugarcane, which yielded 33.8 million tons of sugar and 30.2 billion liters (8 billion gallons) of ethanol. That makes Brazil the world's largest sugar producer and second largest ethanol producer, behind the United States. Most of this production is absorbed by the domestic market, where it is sold as either pure ethanol fuel or blended with gasoline. All gasoline sold in Brazil includes 18 % to 27.5 % ethanol. The bioethanol produced in Brazil is largely advantageous in terms of its ability to reduce emissions of pollutants, compared to other countries. While the ratio between renewable energy and fossil energy used in the production of sugarcane ethanol in Brazil is 8.9, that of corn ethanol production in the United States is 1.3 and of wheat ethanol (Europe) is 2.0. Although Brazilian ethanol is highly environmentally beneficial, it is possible to improve the productivity in planting and harvesting cane with the introduction of innovative techniques, such as one being tested in São Paulo (within the framework of the FAPESP Bioenergia – BIOEN), which consists of the use of a versatile machine, which produces less soil compaction (reducing the loss of seedlings) and provides greater access to steep terrain while maintaining the stability of the equipment. With the total and efficient mechanization of the harvest, besides the absence of the release of greenhouse gases, the part of the straw that is deposited in the soil acts as a carbon sink; the average annual rate of accumulation of this residue is equivalent to 1.5 tons of carbon per hectare [27]. Brazilian bioethanol can also play a leading role in the air transport sector, as the Association of International Air Transport established that airlines should, by 2050, reduce by 50 % their CO2 emissions from 2005 levels. For this, biofuels need to meet the technical specifications required to replace aviation kerosene, which will involve research into new raw materials and further technological development of green fuels. Another plant input that can be used to produce ethanol efficiently is eucalyptus bark, a biomass discarded in large quantities in the manufacture of paper and cellulose. In Brazil, the area planted with eucalyptus (about 4.5 million hectares) represents an annual production potential of almost 12 billion liters of ethanol. The relationship between the volume of ethanol produced and the amount of wood is similar to that obtained with sugarcane. According to a survey by USP Luiz Queiroz School of Agriculture, with the 5 million tons of bark discarded annually by the Brazilian paper and pulp industry, it would be possible to produce 1 billion liters of ethanol. However, if this alternative input proves to be technically feasible, a large scale productive model based on extensive monoculture would be highly questionable.

2.4.2 Extensive monoculture for energy production: the two sides of the coin

Over the last 30 years, the increase in the efficiency of the Brazilian sugarcane industry has been very significant, thanks to an extensive technological development – including the generation, import, adaptation and transfer of technologies in production (agricultural and industrial) – in logistics and end uses. Ethanol production in Brazil has grown vigorously in recent decades, with the lowest costs in the world. Some of the main technological advances in the sugar and alcohol industry can serve as examples to boost the development of other biofuels, although there is room for further improvement in ethanol production and use in the coming years, as [28] states: "there is still exploring areas with great potential in competitiveness." On the other hand, bioethanol has always generated controversy among specialists and energy policy-makers, because its production usually takes a large and continuous field of land, whatever the vegetable cultivated. Possible damage to the environment, arising from the centralization of the production processes, and subsidy policies are also issues raised by specialists. About 35 % of all greenhouse gases emitted by agrochemical activities are due to farming, burning and excessive use of fertilizers [29]. Researchers at UNICAMP (University of Campinas) warn that the Brazilian option for large-scale ethanol production means reproducing a model of "harmful monoculture with little possibility of interaction with livestock, which can lead to the destruction of ecological diversity and small economies where large plants are installed" [30]. For these specialists, the economy of scale of extensive monoculture is only possible at the expense of huge subsidies to big producers; it disappears if the losses of environmental services and the "negative externalities" (costs of the environmental impact of agrochemicals applied to the crop) resulting from the alcohol industry are taken into account. They argue that agro-ecological systems integrated into microdistilleries can be economically viable in small-scale facilities, so-called microplants and mini-plants. Another pertinent question concerns the use of the land, the importance of good conservation practices in energy crops, to create favorable conditions for sowing and the development of cultivated plants. The ideal would be to make possible the coexistence of plantations devoted to the production of energy and food, through their appropriate management, allowing the plants and animals to coexist, the conservation of biodiversity and the presence of the countryman on the site, avoiding its massive migration to the urban centers.

2.5 Photovoltaic and wind power

In 2007, the most optimistic forecasts predicted 30 GWp of installed photovoltaic power in 2020 [31]. Only in 2016, 75 GWp were added, making a total of 303 GWp already installed [8]. Almost the same thing happened in the wind power sector, which added 55 GW during 2016, increasing the global installed power capacity by about 12 %, to nearly 487 GW. By the end of 2016, over 90 countries had seen commercial wind power activity, and 29 countries had more than 1 GW in operation. Of the total of 487 GW, almost 14.4 GW are installed offshore [8]. They are now the first-choice option for expanding, upgrading and modernizing power systems around the world. Wind and solar power, which in 2015 constituted about 90 % of the investment in renewable power, are now competitive with conventional sources of electricity, as their costs have plunged in recent years. The cost of wind turbines has fallen by nearly a third since 2009 and that of solar photovoltaic (PV) modules by 80 %. These developments are reflected in the equalization of the cost of electricity with some renewable technologies having reached grid parity. Currently, onshore wind, hydropower, geothermal and biomass are all more competitive than coal, gas and oil-fired power stations, even without subsidies and notwithstanding relatively low barrel prices. In regard to photovoltaic generation, for the fourth consecutive year, the Asian countries eclipsed all other markets, contributing about two-thirds of the world's total PV energy growth. There are some markets where solar PV now is considered a cost-competitive source for increasing electricity production and for providing energy access. Nevertheless, most markets are driven largely by government incentives or regulations.

2.6 Intermittence of power supply

Most renewable energy cannot be stored, and should be converted as it is available. This brings about two problems to solve: long periods with small renewable production (lulls) and short-term changes in either supply or demand (slews). There are two solutions: acting on the supply or on the demand. A storage system can also be used, such as: pumped storage, batteries or electric vehicles. It is possible to use cheap storage in other sectors (heating, gas and mobility) to solve the storage tasks in the electricity sector [2]. Systems integration technologies, such as energy storage, are being driven by decreasing costs, increasingly favorable regulatory treatment, and an improved understanding of their value [32].

Storage balances the timing between the supply and the demand. However, it is not profitable to store too much energy. The solution passes through a mix of different resources spread through the region, which helps to reduce but not eliminate the problem. Another option is the use of hybrid systems. There are already gas turbine power stations capable of fast loading for peak shaving installations, producing power already within 30 seconds and giving full power in less than 2 minutes.

One interesting approach is the concept of the swarm electrification, proposed as a solution for solar home systems [33]. A. This employs a smart low-voltage DC grid where the connected parties can act as producers and consumers at the same time (prosumers). It is a user-centric system, which is crucial to engage communities and create ownership that is reflected in both the technical care users take of the system as well as the experience of an economic benefit. Any person connected to the grid can consume, produce and store electricity depending on their own technical capacities installed and their own preferences. The community operates or owns the power grid and trades electricity with itself. The users make their economic decisions regarding energy use and grid structure. The grid is quite dynamic and can grow organically when more people connect. One example of this system is in Bangladesh, another in Kenya, where solar mini-grids are changing the lives of people in remote rural areas [34]. The same happening in Kenya [17]. In this kind of system, the users have to think about the production and demand together. As a result, people have the opportunity to see the impact of their demand profile on the system, choosing how and when such loads should be used. It is clear that an intercontinental electrical system cannot operate on such an intermittent grid, but with the use of smart grids and smart homes, consumers will be empowered to control their demand profiles and use energy more wisely.

> *Smart energy systems can enable demand-response measures. Technologies such as advanced metering infrastructure, smart appliances or bidirectional smart meters allow demand management and provide incentives for consumers to play an active role in energy systems. These approaches can stimulate more efficient energy use and contribute to load management and system flexibility* [19].

3 Sustainable cities, "white" cities

Based on conceptions established by several authors [35–37], a sustainable city can be defined as one that follows a developmental path without compromising the ability of future generations to supply their own needs, providing a suitable environment for the realization of desirable social aspirations, based on a rational management of biotic, abiotic (soil, water, wind and sun) and man-made inputs. However, it would be more appropriate to use the plural term "sustainable cities," since different stages of urbanization, geographical location, socioeconomic context and cultural aspects, among others, need to be taken into account. Whatever the case, the search for urban sustainability involves considering the interrelationship between several factors involved, although our approach focuses on the energy issue. Sustainable urban development depends on the buildings – the use of recycled, biodegradable and low environmental impact materials – the use of thermally appropriate materials in buildings and public roads, the use of efficient equipment and electrical devices, the implementation of low-energy buildings, the large-scale use of electric or biofuel-based vehicles and the use of public transportation.

> *Public transport (especially electric trains, trams and buses) seems a promising way to deliver passenger transportation – better in terms of energy per passenger-km, perhaps five or ten times better than cars. However, if people demand the flexibility of a private vehicle, what are our other options? My favorite suggestion is the provision of excellent cycle facilities, along with appropriate legislation (lower speed-limits, and collision regulations that favor cyclists, for example)* [38].

The use of solar energy in buildings to heat water (residential, hospital, industrial, hotels, etc.) is an alternative that has already been widely used in cities, with the potential to be applied in air conditioning systems. This strategic framework, once implemented, would complete the concept of "light cities." This term reinforces a relatively simple idea, the surface is brighter and more reflective of light, but one of great effectiveness, to reduce the amount of solar energy absorbed by the enclosures of the buildings and by the pavements of the public roads, which contributes significantly to the heating of the atmosphere of urban areas. The rising air temperature of cities is intensified by the use of air conditioners, which release even more heat into the urban space, creating a vicious circle that contributes to the formation of "heat islands." The Iranian-American expert, Dr. Hashem Akbari (professor at the Concordia University), who has been researching solutions to mitigate heat islands for more than 20 years, led a survey in Sacramento, a city of 500,000 inhabitants, whose results showed that buildings with lighter facades consume on average 40 % less energy in their summer air conditioning, in relation to those with darker facades [39].

3.1 The "colors" of the new energy revolution

Renewable energy and energy efficiency are emerging as essential factors for making the world "green" and creating "light" cities, laying the foundations for sustainable development. It is salutary that the world should become greener because renewable energy does not affect the climate, with zero or little environmental impact, and thus provides an effective conservation of natural resources and biodiversity. Another relevant aspect of the new energy revolution concerns urban centers. Currently, more than half of the world's people lives in cities, and it is estimated that by 2040 this proportion will reach 70 %. In Brazil, 88 % of the population will be living in cities in 2030, according to the National Energy Plan [40]. Energy consumption in buildings plays an important role in the energy matrix of the most industrialized countries and, consequently, has a major impact in global terms. Residential and commercial buildings in the United States, the European Union, Japan, China, India and Brazil together account for 2/3 of the world's energy consumption [41]. Roofs and pavements, usually made of dark colors, are responsible for excessive absorption of solar radiation, which is an important factor in increasing the demand for energy due to air conditioning. The replacement of dark colors with lighter ones in 100 m^2 of a roof implies a reduction in the consumption of air conditioning equivalent to 10 tons of CO2 each year in the atmosphere. As they constitute over 60 % of urban surfaces, increasing the worldwide albedos of urban roofs and paved surfaces will induce a negative radiative forcing on the earth equivalent to offsetting about 44 billion tons of carbon dioxide [42]. Thus, as cities become lighter, they become colder, which implies reducing the amount of energy consumed to provide thermal comfort, giving urban citizens healthier air and spending less energy during the summer. But the term "light cities" is much broader. It is a set of strategies, including, but not limited to, the use of thermally-appropriate materials in buildings, efficient use of energy in transportation and buildings, and the use of biogas from landfills, as well as a change in people's behavior, which will lead to saving energy and water as a daily practice, recycling materials, and facilitating the selective collection of waste.

In recent years, there has been observed an improvement in the energy efficiency of buildings, but much more is possible. First, policies have focused on the envelope, rather than cooling and heating equipment. There is a huge potential to achieve further energy savings by strengthening or establishing standards. In most countries, improvements of efficiency of up to 20 % are possible by using appliances, equipment and lighting products that are already commercially available, such as compact fluorescent lamps and light-emitting diodes [19].

3.2 The oasis effect

The "oasis effect" is a concept introduce by the Frenchman, Dr. Francis Meunier, a physicist and member of the IPCC, to soften the thermal waves in urban areas that lead to the formation of heat islands. According to Professor Meunier, the main factors that cause heat islands, besides the excessive absorption of solar energy by roofs and pavements, are: thermal waste derived from the anthropogenic consumption of energy, including transport; the shrinking of natural areas of evaporation and vegetation, caused by the urbanization process and the increase of the greenhouse effect due to the emission of gases resulting from industrial activity and means of transport [43]. In order to produce an oasis effect, Meunier proposes two technological alternatives to conventional air conditioning (A/C), which is largely responsible for the intensification of heat islands [44]. The first is to transfer to the subsoil or natural watercourses (rivers, groundwater, etc.) the waste heat from A/C appliances, mainly from large facilities. The second alternative is to use solar-powered A/C systems (thermal or photovoltaic): the sun's rays, previously directed onto the roofs and facades, helping to heat the buildings, once captured in this equipment, would be used as a source of energy to cool the enclosures. If the above alternatives were implemented on a large scale, simultaneously with the use of thermally appropriate building materials, the implementation of low-energy buildings and the massive use of electric vehicles, it would be possible to create an oasis effect: urban areas colder than the rural outlying areas. Today, there are 2 million electric cars around the world. By 2030, this is projected to increase by a factor of 28 [32].

3.3 Solar energy for air conditioning

Solar irradiance has the same temporal pattern of variation as two major contributors to peaks in the demand for electricity, namely refrigeration and air-conditioning: sunlight is itself directly related to this demand [45]. Air conditioning and refrigeration are important and increasing components of the peak electricity demand in buildings. Solar energy can be used in air conditioning equipment as an alternative, which will mitigate the heat islands in large cities. There are systems that use the photovoltaic effect of solar radiation (direct conversion of light into electricity) to drive conventional appliances (vapor compression) or make use of low-temperature heat rejection or solar energy to drive thermally driven chillers, in which the energy consumption of mechanical and electricity is minimal. In addition to these technological possibilities, the solar energy accumulated in the soil can be used as a cold or hot reservoir to cool ambient air in the summer and warm it in the winter by means of a device known as a geothermal heat pump that takes advantage of the temperature difference between the soil and the ambient air. Heat pumps consume electrical energy and, depending on their mode of operation, can serve either to heat and to cool. The lower the temperature difference between the ambient air and the room air to be conditioned, the better the performance of the equipment and, consequently, the lower the power consumption. This has suggested the idea of using a natural reservoir, the soil at a depth of a few meters deep, which is naturally maintained at a constant temperature, sometimes smaller (in the summer) or bigger (in winter) than the ambient air temperature. The use of this type of equipment has increased significantly in Europe, due to its low electrical consumption (in relation to transferred energy) and its "reversible" operation. In Germany, the current rate of new installations is 66 thousand units per year. In both cases, the use of the ground provides smaller temperature differences than those existing when "pumping heat" from (or to) the space to be air-conditioned with a common heat pump. From the energy point of view, this is the main advantage of using geothermal energy for cooling and heating.

3.4 Smart heating

One way is the direct use of solar thermal technology, which is widely used in many regions of the world to heat and cool space, to provide hot water, to dry products and to produce process heat, steam or cooling for industrial processes or for cooking. In 2016, solar heating and cooling technologies had been sold in more than 127 countries. Approximately 375 TWh (1,350 PJ) of heat is provided annually, equivalent to 221 million barrels of oil.

Another way is provided by [38]:

1. Reduce the average temperature difference. This can be achieved by turning thermostats down.
2. Reduce the leakiness of the building. This can be done by improving the building's insulation – think triple glazing, draught-proofing and fluffy blankets in the loft – or, more radically, by demolishing the building and replacing it with a better insulated building; or perhaps by living in a building of smaller size per person. (Leakiness tends to be bigger, the larger a building's floor area, because the areas of external wall, window and roof tend to be bigger too.)

3. Increase the efficiency of the heating system. You might think that 90 % sounds hard to beat, but actually we can do much better with a ground-source heat pump, that can reach 400 %.

An energy efficient house can reduce from the European average of 33 W/m^2 to lower than 7 W/m^2. The German Passivhaus standard aims for power consumption for heating and cooling of 15 kWh/m^2/y, which is 1.7 W/m^2; and total power consumption of 120 kWh/m^2/y, which is 13.7 W/m^2 [38]. As another example:

> *The National Energy Foundation built themselves a low-cost low-energy building. It has solar panels for hot water, solar photovoltaic (PV) panels generating up to 6.5 kW of electricity, and is heated by a 14 kW ground-source heat pump and occasionally by a wood stove. The floor area is 400 m^2 and the number of occupants is about 30. It is a single-story building. The walls contain 300 mm of rockwool insulation. The heat pump's coefficient of performance in winter was 250 %. The energy used is 65 kWh per year per square meter of floor area (7.4 W/m2). The PV system delivers almost 20 % of this energy.*

4 Conclusion

The stability of the energy supply, the reduction of greenhouse gas emissions, sustainability and social inclusion are the main desiderata that should guide a change in the world's energy profile. These necessitate the adoption of energy efficient strategies and a new conception of the city, the "light" city. Fulfilling these desiderata will make it possible to minimize disturbances to the planet's climate, improve the conservation of natural resources, use land more wisely for the production of bioenergy (without compromising the food supply) and base the generation of our electricity on renewable sources. To guarantee its sustainability, the ideal city should include, among other things: public roads and pavements made with thermally appropriate materials of low environmental impact, solar energy systems for water heating and energy-efficient equipment. However, none of this will be enough without reducing our hunger for energy.

It is necessary to overcome the technological barriers that still prevent the more effective use of some renewable energy whose supply depends on natural events, while at the same time making them more competitive in the energy market. Solar irradiance, winds, tides and rainfall are natural resources that can be significantly affected by global climate change, making these energy sources even more unpredictable than they are already. The new energy revolution must be guided looking for reducing the use of fossil energy resources and their environmental impact, and at the same time it must meet the demands of society.

The main externalities resulting from energy production should be taken into account in their total cost, to stimulate the efficient use of resources and avoid possible environmental impacts that may affect the quality of life in cities as well as in the countryside and destabilize the global climate. It is indisputable that bioregions such as the Amazon rainforest exert a strong influence on hydrological and fluviometric regimes, and are responsible for the climatic, fluvial stability and rainfall regime of entire continents. Hydroelectric potential should be used to increase the supply of clean energy, even in places like the Amazon in Brazil, although its exploitation must include effective measures to minimize socioenvironmental impacts.

Lastly, although the issues raised here as challenges of the new energy revolution argue in favor of a gradual and consistent reduction of anthropogenic CO2 emissions, there is no way to dissociate environmental problems from the predominance of an economic and productive system that has ignored its insertion into a planet, where natural resources are limited. To conclude, there is nothing better than the quotation made more than 30 years ago but remains actual [46]:

> *The energy transition cannot be reduced to simple technical improvements or the development of new energy supplies: it necessarily implies a complete change of societies on a world scale. Whatever its duration and rhythm, this mutation will be global. No revolution, until today, has really or durably questioned the material bases of social organization; however, no social alternative will be conceivable, from now on, unless it involves the establishment of a new energy system.*

Bibliography

[1] Meunier, F.: Les énergies renouvelables. Le Cavalier Bleu, Paris (2008).

[2] Stadler, I.: Fear of Change. Presentation given at The 1st Latin American Conference on Sustainable Development of Energy, Water and Environment Systems, 1st LA SDEWES, Rio de Janeiro (2018).

[3] Zou, C., Zhao, Q., Zhang, G., Xiong, B.: Energy revolution: From a fossil energy era to a new energy era. Natural Gas Ind. B 3, 1–11 (2016).

[4] UN: Relatório Global Cities and Climate Change: Global Report on Human Settlements (2011).

[5] IEA: Solar Heating & Cooling Programme. Solar Heating Worldwide: Markets and Contribution to the Energy Supply (2011).

[6] IRENA: REthinking energy report 2017: Accelerating the global energy transformation, International Renewable Energy Agency (2017).

[7] IPCC: Contribution of Working Group III to the Fourth Assessment Report (2007).

[8] REN21: Renewables 2017 Global Status Report. Renewable Energy Policy Network for the 21st Century (2017).

[9] WEC: World Energy Scenarios | 2016 — The Grand Transition. World Energy Council in collaboration with Accenture Strategy and Paul Scherrer Institute (2016).

[10] Rockstroem, J., Gaffney, O., Rogelj, J., Meinshausen, M., Nakicenovic, N., Schellnhuber, H. J.: A roadmap for rapid decarbonization. Science 355, 1269–1271 (2017).

[11] Wiseman, J.: The great energy transition of the 21st century: The 2050 Zero-Carbon World Oration. Energy Res. Soc. Sci. 35, 227–232 (2018).

[12] Tollefson, J.: How hot will Earth get by 2100? Nature 580(7804), 443–445 (2020).

[13] Capros, P., Kannavou, M., Evangelopoulou, S., Petropoulos, A., Siskos, P., Tasios, N., Zazias, G., DeVita, A.: Outlook of the EU energy system up to 2050: The case of scenarios prepared for European Commission's "clean energy for all Europeans" package using the PRIMES model. Energy Strat. Rev. 22, 255–263 (2018).
[14] Cash, D. W.: Choices on the road to the clean energy future. Energy Res. Soc. Sci. 35, 224–226 (2018).
[15] Griffiths, S.: Energy diplomacy in a time of energy transition. Energy Strat. Rev. 26 (2019).
[16] Haley, B., Gaede, J., Winfield, M., Love, P.: From utility demand side management to low-carbon transitions: Opportunities and challenges for energy efficiency governance in a new era. Energy Res. Soc. Sci. 59 (2020).
[17] Earley, K.: Why renewables are winning the 'carbon war'. Renew. Energy Focus 19, 117–120 (2017).
[18] Suehiro, S.: Energy intensity of GDP as an index of energy conservation. Institute of Energy Economics Japan Report (2007).
[19] IEA: Energy Technology Perspectives. International Energy Agency (2017).
[20] Meunier, F., Meunier-Castelain, C.: Adieu pétrole... Vive les énergies renouvelables!. Dunod, Paris (2006).
[21] IEA: Energy Efficiency. International Energy Agency (2017).
[22] Andrade & Canellas Energia S.A.: Lixo descartado em 2010 poderia ter gerado energia para atender 18,3 milhões de moradias (2011).
[23] bsnmEPE/MME: PNE 2030 Eficiência Energética, Brazil (2008).
[24] Ramage, J., Scurlock, J.: Biomass. In: Renewable energy: power for a sustainable future. Oxford University Press, New York (1996).
[25] MME: Balanço Energético Nacional. Brazil (2017).
[26] Sohn, I.: In: Zoom: The Global Race to Fuel the Car of the Future – By Vijay Vaitheeswaran and Iain Carson. Natural Resources Forum, vol. 32, pp. 348–349 (2008).
[27] Ereno, D.: In: Máquina Versátil, PESQUISA FAPESP – Tecnologia/Biocombustíveis, pp. 68–71 (2011).
[28] Macedo, I. C.: Situação atual e perspectivas do etanol. Estud. Av. 21, 157–165 (2007).
[29] Foley, J. A., Ramankutty, N., Brauman, K. A., Cassidy, E. S., Gerber, J. S., Johnston, M., Mueller, N. D., O'Connell, C., Ray, D. K., West, P. C., et al.: Solutions for a cultivated planet. Nature 478, 337–342 (2011).
[30] Ortega, E., Watanabe, M., Cavalett, O.: Production of ethanol in micro and mini-distilleries. Unicamp, Campinas (2007).
[31] Marigo, N.: The Chinese silicon photovoltaic industry and market: a critical review of trends and outlook. Prog. Photovolt. 15, 143–162 (2007).
[32] IEA: World Energy Outlook. International Energy Agency (2017).
[33] Koepke, M., Groh, S.: Against the odds: The potential of swarm electrification for small island development states. Energy Proc. 103, 363–368 (2016).
[34] Laursen, L.: Grids of all sizes, Nature Publishing Group Macmillan Building, 4 Crinan St, London N1 9XW, England, (2017).
[35] Bremer, U. F.: Por nossas cidades sustentáveis, em V CNP. 61a SOEAA, São Luiz (MA) (2004).
[36] Dias, L. E.: Conceitos e termos relativos a estudos de recuperação ambiental. In: SOL646 (2006).
[37] Pearce, D. W., Barbier, E., Markandya, A.: Sustainable development: economics and environment in the Third World. Edward Elgar Publishing, Aldershot (1990).
[38] MacKay, D.: Sustainable Energy-without the hot air. UIT, Cambridge (2008).
[39] Levinson, R., Akbari, H.: Potential benefits of cool roofs on commercial buildings: conserving energy, saving money, and reducing emission of greenhouse gases and air pollutants. Energy Effic. 3, 53–109 (2009).

[40] EPE/MME: PNE 2030 Seminários Temáticos – Projeções do Consumo Final de Energia, Brazil (2006).
[41] Stigson, B.: What is the real potential for energy efficiency? Energy Efficiency Global, Paris (2009).
[42] Akbari, H., Menon, S., Rosenfeld, A.: Global cooling: increasing world-wide urban albedos to offset CO2. Clim. Change 94, 275–286 (2008).
[43] Meunier, F.: Effect oasis to mitigate heat island. In: Proc. of 22nd IIR International Congress of Refrigeration, Beijing (2007).
[44] Tremeac, B., Bousquet, P., De Munck, C., Pigeon, G., Masson, V., Marchadier, C., Merchat, M., Poeuf, P., Meunier, F.: Influence of air conditioning management on heat island in Paris air street temperatures. Appl. Energy 95, 102–110 (2012).
[45] Barnham, K. W. J., Mazzer, M., Clive, B.: Resolving the energy crisis: nuclear or photovoltaics?. Nat. Mater. 5, 161 (2006).
[46] Pestre, D., Debeir, J.-C., Deleage, J.-P., Hemery, D.: Les servitudes de la puissance. Une histoire de l'energie. Vingtième Siècle. Revue d'histoire, p. 137 (1987).

Christina S. Birkel

Synthesis of inorganic energy materials

Abstract: The development of new – cheaper, more efficient, more sustainable, more reliable – functional materials with useful properties calls for ever-improving, smart and innovative synthesis strategies. A multitude of inorganic compounds are already used as energy materials, i. e., electrodes, catalysts, permanent magnets and many more are considered highly promising for these and similar applications. For the optimization of the preparation of the respective materials, a broad knowledge of possible synthesis techniques is required and some of them are highlighted in this chapter. Furthermore, (micro)structure-determining processing techniques are addressed and discussed in the context of thermoelectric materials.

Keywords: Inorganic compounds, energy-related materials, traditional synthesis, nonconventional field-assisted synthesis, wet chemical synthesis, thermoelectrics

1 Introduction

A plethora of inorganic – nonmolecular – compounds are associated with particular native properties and hold functions that can potentially be used in a wide variety of energy-related applications. Prominent materials that are already in use and commercially available include, for example, $Nd{:}Y_3Al_5O_{12}$ (solid-state lasers), $Nd_2Fe_{14}B$ (permanent magnets) and Bi_2Te_3 alloys (thermoelectrics). The three compounds mentioned here not only differ in chemical composition but also in their electronic and crystallographic structure. Intrinsically, these three factors determine the chemical, physical, mechanical behavior – and ultimately the properties – of the respective inorganic material. Furthermore, extrinsic aspects also play a key role for the final materials' properties. We have to ask ourselves in which form we are using the functional material – is it a single crystal, a loose (nano) powder, a thin/thick film or even a dense pellet? For the latter, the microstructure – morphology, density/porosity, elemental distribution – becomes a major concern that needs to be addressed when discussing the properties. For thin/thick films as well as nanoparticles, the surface chemistry (oxide layer, functional groups, etc.) are important aspects in understanding and eventually controlling the compound's behavior. The presence and influence of possible side phases also need to be considered. Therefore, the overall material's properties are always the combination of the intrinsic and the extrinsic properties of the respective inorganic compound. A large number of inorganic compounds are obtained as the thermodynamically stable product at elevated temperatures based on a given stoichiom-

Christina S. Birkel, Arizona State University, Tempe, USA, e-mail: christina.birkel@asu.edu

https://doi.org/10.1515/9783110584455-010

etry. However, kinetically controlled products, metastable phases, are also accessible using suitable synthetic techniques. In this chapter, a variety of synthesis methods will first be described including the distinction between traditional solid-state techniques, nonconventional field-assisted solid-state and softer wet chemical methods. Besides, processing procedures, e. g., for the densification of loose powders, will be addressed with a focus on how they influence the extrinsic materials' properties. Thermoelectric materials will then be discussed in greater detail providing a platform to point out the strengths and weaknesses/pitfalls of the different syntheses.

2 General synthesis and processing techniques

The field of synthetic chemistry is indeed quite extensive and substantial for the development of a large number of industrial processes and functional (energy) materials. It comprises different areas, such as inorganic, organic and biochemistry with different subareas like solid-state, bioinorganic, (macro)molecular and polymer chemistry. Here, the focus will exclusively be the field of synthetic inorganic, mostly solid-state, chemistry that allows the preparation of a variety of inorganic compounds that are of interest for energy-relevant applications. Of course, there exist a number of books that deal with this exact topic and the reader is referred to these sources for further reading [1–4]. One book in particular covers most techniques in synthetic inorganic chemistry in great detail while also providing information about different inorganic materials [5]. It is not the intention of this present chapter to fully cover this wide topic, but rather to highlight selected synthetic methods. Besides, thermoelectric materials are chosen as an example of energy materials to present different synthesis methods and discuss their advantages and disadvantages. Of course, these methods can also be used to obtain further energy materials, such as photocatalytic metal oxides, battery materials, catalysts, catalytic-support materials and metal organic framework structures for gas storage.

Chemical solid-state reactions usually require energy converting precursors A and B into an inorganic compound AB, for example. This energy can be delivered by applying heat, therefore many classical solid-state reactions are performed at elevated – and oftentimes very high – temperatures. Figure 1 depicts different temperature regions and corresponding synthesis techniques. They span the large temperature regime from below 0 °C or around room temperature to temperatures exceeding 3000 °C. Please note that the actual temperatures during ball milling are highly dependent on the specific process parameters, such as precursors, milling container/balls, type of ball mill, milling speed and time. This mechanochemical method can be very beneficial for the synthesis of inorganic (and also organic) compounds and is therefore the subject of many ongoing research efforts. Temperature measurements during the synthesis are very challenging. While many reactions are believed to proceed at

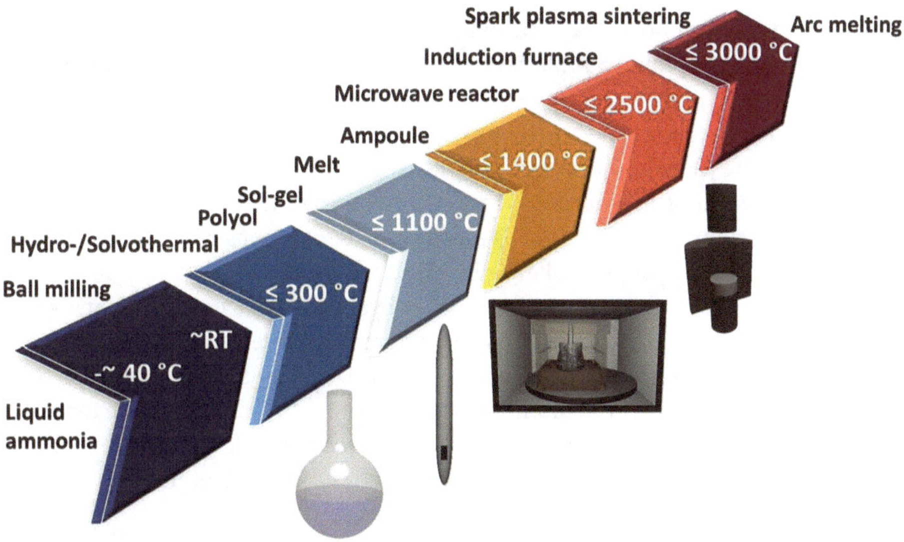

Figure 1: Different temperature regions and corresponding synthesis techniques to be used for the preparation of inorganic compounds.

around room temperature, many arguments exist that the real local temperatures are much higher. Due to the complexity and volume of this topic, the reader is referred to some excellent (review) articles that cover the mechanochemical synthesis/mechanical alloying including many inorganic compounds as examples [6–8].

In the following, the subcategories of (i) wet chemical methods, (ii) classical/conventional solid-state and (iii) nonconventional/field-assisted solid-state techniques will be discussed. The different techniques will be addressed in the order of increasing reaction temperatures starting with reactions in water and other solvents (hydrothermal/solvothermal).

2.1 Wet chemical synthesis methods

Wet chemical techniques – also referred to as low temperature or *chimie douce* methods – offer the huge advantage of intimately mixed precursors on the atomic scale. Hence, diffusion does not become the rate-limiting factor that often leads to kinetically controlled reactions rather than those thermodynamically controlled. As a consequence, metastable phases may be accessible that would otherwise not crystallize under high temperature conditions where the most thermodynamically stable product forms. On the other hand, researchers also need to deal with certain drawbacks associated with wet chemical techniques: The suitable precursor might be costly, toxic or not even available at all; solvents are needed that can have a negative environmental impact; optimization of the synthesis parameters can be time, energy and money consuming and might not be applicable to the upscaling of the process or to the synthesis of similar compounds. A plethora of wet chemical methods, i. e., syntheses in the presence of a liquid medium, exist and it is not the claim of this section to cover all of these. Instead, a few selected methods will be presented that are used to synthesize energy materials: (i) hydrothermal/solvothermal syntheses, (ii) polyol synthesis, (iii) wet chemical synthesis in a microwave reactor, (iv) sol-gel methods.

Solvothermal processes are homogeneous or heterogeneous reactions in a liquid medium, whereas the liquid can be water, glycols, alcohols or other solvents. In the case of water, the reaction is also referred to as hydrothermal synthesis. Usually, hydrothermal and solvothermal reactions are performed in a sealed container or high-pressure autoclave under subcritical or supercritical conditions of the solvent. In both cases, the solvent plays an important role by possibly acting as part of the reactant, changing the chemical and physical properties of reactants and products and accelerating reactions [2]. The most important applications of the solvothermal syntheses are

(i) the preparation of bulk single crystals;
(ii) zeolites and other open framework systems as well as (ii) inorganic-organic hybrid materials.

Another highly versatile wet chemical technique is the polyol synthesis that takes place in high-boiling, multivalent alcohols, such as ethylene glycol (EG), diethylene/triethylen/ tetraethylen glycol (DEG, TrEG, TEG) and polyethylene glycol (PEG). This approach is mostly directed at the synthesis of nanoparticles and was first described by Fievet, Lagier and Figlarz in 1989 [9, 10]. Initially, the method was used to synthesize metal particles, such as Co, Ni, Cu and Pt, but was quickly extended to further metals as well as intermetallic compounds. The general technique is highly flexible given by the vast number of possible polyols with different viscosities, polarities and boiling points. The latter enable processes at different temperatures of up to 320 °C. Besides, their chelating effect leads to suppressed grain growth and, therefore, formation of nanoscale compounds. Oftentimes, the shape of the resulting particles can be controlled by using further surfactants, resulting in sphere-, cube-, wire-, flower-, star-like morphologies [11]. Quite similarly, these reactions can also take place in other high-boiling solvents, e. g., dioctyl, phenyl or benzyl ether that offers an additional degree of freedom in the choice of reaction parameters. Due to the huge number of publications in the field, the reader is referred to these review articles/books for further information [12–15].

Wet chemical syntheses inside a microwave reactor have become standard processes in inorganic and organic laboratories, with impacts in various fields, such a polymer chemistry, medicinal chemistry, drug discovery, nanoparticle synthesis, green and sustainable chemistry, flow chemistry, just to name a few [16]. Microwave radiation falls between radio and infrared frequencies and most commercially available instruments operate at 1.45 GHz. The corresponding wavelengths lie between 1 m and 1 cm showing that microwaves do not interact at the atomic or molecular level. They do interact, however, with different substances via different mechanisms, oftentimes leading to rapid heating effects. Water is, for example, a classic substance that couples strongly to microwave radiation. Here, dielectric heating is achieved by the reorientation of polarized molecules in a rapidly alternating electric field. Historically, this phenomenon was used for food preparation and heating and it took surprisingly long for it to be transferred to chemical reactions. One reason for this is that dielectric heating is a very complex issue and highly dependent on the respective physical properties of the reaction medium/reactant, etc. in contrast to conventional conductive heating. Clarifying the processes that happen during microwave syntheses is still a major research effort, yet the large gain brought by the use of microwave reactors has widely been accepted in the scientific community.

The last type of wet chemical technique that is included here is the sol-gel method in all its known varieties. Generally, during this process a *sol* is formed first from liquid precursors that is then transferred into a network structure, called a *gel* [17]. Broadly defined, a sol is a colloidal suspension with molecules or polymolecular particles being dispersed in a liquid medium. A gel is considered a nonfluid 3D network that extends through a fluid phase, whereas different types of gels exist depending on the type of gel-building substance and bonding [18]. A very common small molecule used for sol-gel syntheses is citric acid that is typically mixed with aqueous metal salts and consecutively heated slowly to form a viscous gel. A subsequent heating step – in air if oxide formation is desired – is applied to form the final inorganic solid. Due to the intimate mixing of the precursors on an atomic scale, lower temperatures can usually be applied in contrast to classical high temperature solid-state methods. Sol-gel chemistry is also highly flexible. For example, structure-directing agents can be added to obtain ordered porous structures or nanostructures and the process is applicable to the syntheses of a variety of systems, such as metals, oxides, chalcogenides, carbides and sulfides.

2.2 Conventional solid-state synthesis methods

Classical "dry" solid-solid reactions are typically slow and require high temperatures due to the necessary mass transport within the rather inhomogeneous reactant mixture. Ball-milling/high energy mixing of the reactants can be beneficial but might also introduce impurities due to traces of oxygen or wear of the container and ball material. A sort of intermediate process between wet chemical techniques at moderate temperatures and classical solid-solid processes at high temperatures are reactions in the melt. As in all kinds of solvents, diffusion occurs much faster in liquid melts. Different types of these flux syntheses utilize liquid metals [19] or metal salts [20] as the solvent and can be used for single crystal growth as well as preparation of polycrystalline powder materials. In these cases, the flux does not take part in the reaction and, therefore, must not react with the precursors and must be readily removable after the reaction.

Although many inconveniences of conventional solid-state synthesis techniques have already been mentioned, they are still the oldest, simplest and most widely used method to make inorganic solids. Since the synthesis protocol usually only involves mixing of the reactants and heating them for an extended amount of time, they are also referred to as *shake 'n bake* or *beat 'n heat* methods. In reality, additional aspects need to be considered for successful synthesis of high-quality materials, i. e., without side phases or unreacted precursors. In order to maximize the contact area between the precursors, dense pellets – obtained by cold pressing of precursor powders – rather than loose powders are used. The reaction container needs to be chosen carefully in order to avoid reaction between the container material and the reactants. The atmosphere needs to be controlled when inert conditions are, for example, necessary during the synthesis of intermetallic compounds.

Furthermore, the source of heat can vary. Furnaces with different temperature regions that can be programmed to meet the researcher's needs, are readily available and most common in inorganic solid-state laboratories. Oxide synthesis can be performed in air and take place in alumina crucibles that are usually inert to the reaction mixtures and high-temperature stable. For intermetallics, chalcogenides, carbides, etc., inert conditions are provided by using sealed quartz ampoules as the container where smaller tantalum crucibles can be embedded if side reactions between the quartz walls and the precursors could occur. Furthermore, reactions can be run under certain gaseous atmospheres, such as reducing H_2/N_2 mixtures or nitrogen gas for the formation of nitrides. Beside conventional furnaces, microwave ovens can be used for high temperature solid-state reactions (see Section 2.3). For even higher temperatures, an electrical furnace where the heat is applied by induction heating (induction furnace) or a movable electric arc inside an arc melter can be employed. Both devices allow for melting and alloying metals and refractories. Spark plasma sintering also relies on an electrical current running through a graphite die and the sample if it is conductive thereby heating it up very rapidly as a result of Joule heating effects. The simultaneous application of pressure leads to the densification of powder samples (see Section 2.3).

2.3 Nonconventional solid-state synthesis methods

In this section, two nonconventional synthesis/processing methods will be addressed: Microwave heating and spark plasma sintering. Both can also be considered field-assisted methods since microwave radiation and electric currents are involved, respectively.

Reactions inside a microwave oven are well established for solution-based syntheses and have been successfully used to access a variety of different organic as well as inorganic compounds [16]. Researchers benefit from reduced equipment costs, rapid processing, increased energy efficiency and high-yield, high-purity products [21]. These attractive prospects have also triggered research efforts in the field of solid-state chemistry. The corresponding studies allow the conclusion that microwave reactions are indeed highly promising for reactions in the solid-state and can be used to prepare a vast number of inorganic compounds. These solids include (but are not limited to) intermetallic compounds [22], carbides, e. g., SiC [23], (complex) oxides, such as $BaTiO_3$ [24] and different bronzes [25], chalcogenides [26–28], silicides [29], phosphates [30] and nitrides, such as AlN [31] and GaN, TiN and VN [32]. They are either obtained through direct reaction of the elements or with the help of an additional microwave susceptor, carbothermal reduction of the oxides or under a nitrogen/ammonia gas flow.

Microwaves are high-frequency oscillating electric and magnetic fields. The microwave photon corresponding to 2.45 GHz possesses energies close to 0.0016 eV that is several magnitudes lower than those of X-rays, for example, (~ 10^5 eV). Microwave heating in solution relies on the dielectric dipoles trying to align themselves with the oscillating electric field of the microwaves and as a result energy is lost in the form of heat through molecular friction and dielectric loss. Thus, the dielectric loss tangent determines the ability of a material to convert electromagnetic energy into heat at a given frequency and temperature. Although the interaction between microwave radiation and solids also relies on the dielectric properties, the mechanisms are more complex and not as well understood. Solids that are subject to microwave radiation can either act as reflectors (e. g., bulk metals and alloys), transmitters (e. g., fused quartz, zircon, some glasses and ceramics) or absorbers. It is important to note that the dielectric properties of solids are highly dependent on their chemical composition and on their physical state. One example is the different heating behaviour in response to microwave radiation of different forms of carbon where amorphous carbon heats up more rapidly than graphite. Another example are metals: While solid pieces of metals lead to large electric field gradients within a microwave cavity causing electric discharges, metal powders couple effectively with microwaves. The resulting rapid heating of the metal powders is primarily due to conduction mechanisms and possibly localized plasma effects [33, 34].

Additionally, impurities can give rise to high dielectric constants and losses resulting in strong coupling [34]. Aside from these factors, the dielectric properties themselves are temperature-dependent. While SiO_2 possesses a low dielectric loss and, therefore, does not heat up significantly under microwave radiation, NiO and Cr_2O_3 heat up very rapidly. The latter even exhibits thermal runaway due to increased dielectric loss at higher temperatures [35]. Many inorganic materials couple strongly to the microwave radiation and, as a result, heat up very rapidly. Consequences of this mechanism are direct, volumetric, instantaneous and selective heating. All these factors lead to highly energy efficient solid-state reactions. Often materials with improved properties are obtained as a result of their particular microstructure [36] as well as sample morphology and crystallinity [37].

Spark plasma sintering (SPS) describes a densification technique where a direct current is passed through the die – and the material if conducting – while a simultaneous pressure is applied [38–40]. The term seems misleading because neither spark nor plasma has been observed so far [41, 42]. However, "spark plasma sintering" will be used here since this term is currently most widely used by researchers working in that field [43]. Similar to microwave heating, it is a rapid technique that has been used successfully for the densification of a variety of materials with improved materials properties, such as carbides, nitrides, oxides, borides, metals and intermetallics [44]. Especially the field of thermoelectric research has benefited from this fast method since grain growth can be bypassed and pellets with low thermal conductivity are obtained [45–47]. Beside the consolidation, SPS has also been utilized to synthesize known [48, 49] and new [50, 51] materials.

Usually work in this area focuses on the thermal effect of the current that passes through the process setup. Due to well understood Joule heating effects of conducting materials, the die and/or the powder to be compacted are subject to very high heating rates of up to 1000 °C/min. These can be introduced by low voltages below 10 V that produce high currents between 1 kA and 10 kA. Additionally, pulse and pause durations of the applied pulsed current can be defined whereas typical pulse duration is in the order of a few milliseconds. The current effect depends on the current density, current direction, current pattern (continuous/pulsed), temperature and materials properties. It can have pronounced effects on the synthesis and microstructural development during SPS processing. For example, Friedman et al. have observed strong indications that high-density currents promote nucleation of powder phases, greatly increase growth kinetics and the reactivity in solid-solid interfacial reactions [52].

It is highly desirable to look into possible interactions between the electric current and the microstructure evolution during spark plasma sintering. Guillon et al. have summarized the occurring effects: (i) Percolation effects as a result of inhomogeneous current flow through the sample that is followed by Joule heating and phase and microstructure formation along those percolation paths (e. g., demonstrated in nanostructured silicon [53]). (ii) Peltier effect that is mostly relevant in semiconducting samples and leads to heating and cooling at the electrode-(die)/sample interface. (iii) Electrochemical reactions at the electrodes that can allow the synthesis of hardly accessible phases [54] and electromigration that can induce the formation of intermetallic phases [52].

All of the above mentioned phenomena enhance mass transport in current-activated processes. Electromigration effects might be negligible during spark plasma sintering due to low current densities and short reaction times. However, a number of additional fundamental current effects increasing solid-state reactions and sintering processes have been observed. In a recent review on the Pulsed Electric Current Sintering Process (PECS) [55], Munir et al. report on the reaction between Mo foils and Si wafers with and without passing a current through the multilayer ensemble. They concluded that (i) the reaction between Mo and Si was not affected by the pulse pattern [56], (ii) there is no effect of the current direction on the growth of the product layer, however, (iii) the reaction rate to produce the interface layer was noticeably influenced by the presence of the current.

3 Thermoelectric materials

Thermoelectric materials can convert between thermal and electrical energy and vice versa [57]. The underlying physical effect describes the occurrence of a voltage when a temperature gradient is applied to a material. Very simply, this can be understood as charge carriers behaving like gas molecules and moving faster on the hot side than on the cold side thereby creating a charge gradient. The Seebeck coefficient is a measure of the magnitude of an induced voltage that stems from a certain temperature difference across the material, and is an important part of the thermoelectric figure of merit that is directly related to the conversion efficiency. For the figure of merit zT to be as high as possible, the electronic conductivity σ and Seebeck coefficient α need to be large while the thermal conductivity κ needs to be minimized ($zT = \alpha^2\sigma/\kappa$). Since these physical properties are interdependent, there is a compromise around an ideal carrier concentration that can be found in semiconductors. State-of-the-art commercial materials are Bi_2Te_3 and doped versions of it that work in the intermediate temperature regime, while SiGe and $Yb_{14}MnSb_{11}$ are options for high-temperature applications [45].

In general, the performance of a thermoelectric device is dependent on many different factors (see Figure 2): (i) The target compound dictates its electronic and crystallographic structure as well as chemical composition (intrinsic properties). (ii) The processing conditions, e. g., during densification, influence the microstructure, i. e., the morphology, density and elemental distribution (extrinsic properties). (iii) Both result in a certain set of physical parameters that determine the thermoelectric properties. (iv) Before a thermoelectric material can fulfil its purpose as a thermoelectric device, contacts/wires as well as a cooling mechanism need to be added for the current flow and the necessary temperature gradient, respectively.

There are different approaches to maximize the thermoelectric figure of merit, such as band structure engineering to obtain large Seebeck coefficients and large electronic conductivities [58]. Another option is to target the thermal conductivity and to minimize it by introducing grain boundaries where heat-carrying phonons are scattered reducing the lattice thermal conductivity [59]. Although the electronic part of the thermal conductivity (directly connected to the electronic conductivity through the Wiedemann–Franz law) will not be affected, usually significant improvement in the figure of merit is achieved. Two model systems that consist of a large amount of grain

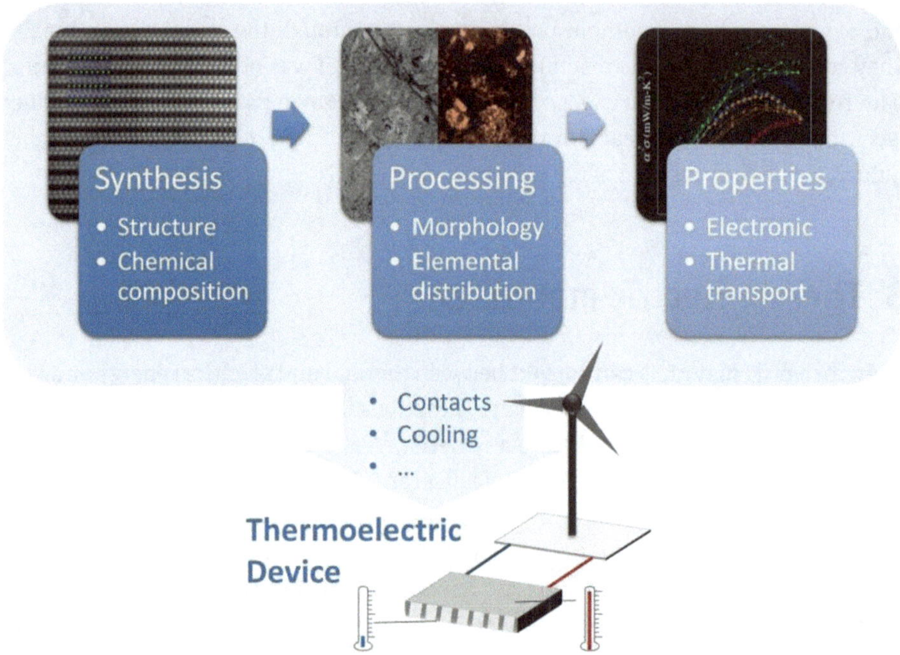

Figure 2: Factors that determine the performance of a thermoelectric device ranging from intrinsic to extrinsic materials properties. Also included are "engineering" considerations for the device fabrication, such as contacts and cooling to ensure the temperature gradient.

boundaries are (i) bulk nanostructures and (ii) heterostructures with micro-/nano-sized inclusions.

Synthetic methods to obtain nanoparticles are divided in "top-down" and "bottom-up" processes [60]. Ball milling is a "top-down" approach where bulk material is mechanically milled until grain sizes <100 nm are obtained. Many of the wet chemical techniques described in detail in Section 2.1 are considered "bottom-up" techniques or variation thereof. The following two are among the most important issues that have to be considered for the final thermoelectric performance: Organic components that are usually introduced by the capping agents necessary to limit grain growth or high-boiling solvents need to be avoided or removed since they limit the electronic transport through the material. On the other hand, unprotected surfaces are very reactive and prone to oxidation or decomposition that alters the overall chemical composition of the thermoelectric. Examples of nanostructured thermoelectric compounds synthesized by wet chemical methods include $FeSb_2$ [61], Zn-Sb [47, 62], Bi_2Te_3 [63] and PbTe-$PtTe_2$ [64].

The other crucial aspect is the need for dense/compact samples for thermoelectric characterizations since they cannot be performed on loose powders. Conventionally, powders are compacted by means of hot pressing; however, this leads to significant grain growth of the initially nanostructured compounds. Spark plasma sintering has been shown to be very valuable for the densification of nanoparticles due to rapid heating and cooling and overall fast densification processes. Still, detailed structural and microstructural investigations are necessary after the densification process. Particularly nanostructured compounds with high surface areas/energies can decompose or transform into different phases with different intrinsic physical properties. Hence, we cannot expect the dense material to be the same compound as initially synthesized. One lucid example is the densification of Zn-Sb nanoparticles that mainly decompose into a different Zn-Sb phase, ZnO and Sb even at rather low temperatures during the SPS process [47].

Possibly the more elegant option to enhance the thermoelectric figure of merit is the preparation of heterostructured materials, i. e., bulk structures that include nano- or micron-sized inclusions of a secondary phase. In the sense of "one stone – two birds," these configurations can exhibit reduced lattice thermal conductivity due to grain boundary scattering of the phonons, yet increased electronic conductivity due to additional charge carriers provided by metallic inclusions, for example. There are many examples of heterostructured thermoelectrics: Oxides [65], intermetallics [66, 67], chalcogenides [68, 69]. Their synthesis involves wet chemical as well as solid-state techniques. In the case of a half-Heusler intermetallic TiNiSn, microwave heating and subsequent spark plasma sintering was employed to prepare the thermoelectric material [70]. Moreover, significant enhancement of the figure of merit was achieved by the formation of respective heterostructures [71]. Spark plasma sintering can also be used to prepare thermoelectrics in a single-step process, as was shown for a WO_{3-x} Magnéli phase [72]. Again, an improved figure of merit was realized by the formation of Ta-rich heterostructures [73]. One of the highest *zT*s ever achieved (~2.2 @ 915 K) was reported for PbTe-SrTe heterostructures with all-scale hierarchical architectures on the atomic scale (alloy scattering), nanoscale and mesoscale (grain-boundary scattering) [74]. This material was synthesized at high temperatures, quenching to room temperature and subsequent densification by spark plasma sintering.

4 Conclusion and outlook

This chapter highlights a number of synthesis techniques that can be employed to prepare inorganic energy materials. In many cases, there exist multiple synthetic approaches for the preparation of one compound. Therefore, the information provided here can be used as a guideline to choose an appropriate synthesis method in regard to the target compound, temperature where it should form, desired morphology and potential application. Depending on these factors, the synthesis conditions need to be optimized based on the knowledge of how the synthesis parameters influence the structure, micro/surface structure and ultimately the properties of the material. For thermodynamically stable compounds, the conventional solid-state route is the most applicable since it is synthetically straightforward, does not involve any solvents and uses simple, often elemental, precursors. Flexibility is provided by applying varying heating rates, different reaction temperatures and the possibility to quench from high temperatures to room temperature. The latter mainly effects the microstructure that is a key factor in optimizing the thermoelectric properties, for example. In order to obtain different morphologies or nanostructures, mostly wet chemical techniques are used in a "bottom-up" approach. Here, one needs to consider the presence of solvents, organic molecules, possibly oxidized surfaces and methods to densify the obtained powders before the properties can be measured and assessed.

Ultimately, a broad knowledge of the synthesis techniques, together with their influence on the inorganic compound of choice, guarantees optimization of known processes and will lead to the development of future energy materials with improved properties.

Bibliography

[1] Schubert, U., Hüsing, N.: Synthesis of Inorganic Materials. Wiley VCH (2000).
[2] Xu, R., Xu, Y.: Modern Inorganic Synthetic Chemistry. Elsevier (2017).
[3] Hagenmuller, P.: Preparative Methods in Solid State Chemistry. Academic Press, New York (1972).
[4] West, A. R.: Synthesis, Processing and Fabrication Methods. In: Solid state chemistry and its applications. Wiley VCH (2014).
[5] Rao, C. N. R., Kanishka, B.: Essentials of Inorganic Materials Synthesis. Wiley VCH (2015).
[6] Suryanarayana, C.: Mechanical Alloying and Milling. Prog. Mater. Sci. 46, 1–184 (2001).
[7] McKissic, K. S., Caruso, J. T., Blair, R. G., Mack, J.: Comparison of Shaking versus Baking: Further Understanding the Energetics of a Mechanochemical Reaction. Green Chem. 16(3), 1628–1632 (2014).
[8] James, S. L., Adams, C. J., Bolm, C., Braga, D., Collier, P., Friščić, T., Grepioni, F., Harris, K. D., Hyett, G., Jones, W., et al.: Mechanochemistry: Opportunities for New and Cleaner Synthesis. Chem. Soc. Rev. 41(1), 413–447 (2012).
[9] Fievet, F., Lagier, J., Figlarz, M.: Preparing Monodisperse Metal Powders in Micrometer and Submicrometer Sizes by the Polyol Process. Mater. Res. Soc. Bull. 14(12), 29–34 (1989).
[10] Fievet, F., Lagier, J., Blin, B.: Homogeneous and Heterogeneous Nucleations in the Polyol Process for the Preparation of Micron and Submicron Sized Metal Particles. Solid State Ion. 32/33, 198–205 (1989).
[11] Sau, T., Rogach, A. (eds.): Complex-Shaped Metal Nanoparticles: Bottom-Up Syntheses and Applications. Wiley VCH (2012).
[12] Dong, H., Chen, Y.-C., Feldmann, C.: Polyol Synthesis of Nanoparticles: Status and Options Regarding Metals, Oxides, Chalcogenides, and Non-Metal Elements. Green Chem. 17(8), 4107–4132 (2015).
[13] Schmid, G. (ed.): Nanoparticles: From Theory to Application. Wiley VCH (2005).
[14] de Mello Donega, C. (ed.): Nanoparticles Workhorses of Nanoscience. Springer (2014).
[15] Altavilla, C., Ciliberto, E. (eds.): Inorganic Nanoparticles: Synthesis, Applications, and Perspectives. CRC Press (2010).

[16] Cravotto, G., Carnaroglio, D. (eds.): Microwave Chemistry. De Gruyter (2017).
[17] Brinker, C., Scherer, G.: Sol-Gel Science: The Physics and Chemistry of Sol-Gel Processing. Academic Press, New York (1990).
[18] Danks, A. E., Hall, S. R., Schnepp, Z.: The Evolution of "sol–gel" Chemistry as a Technique for Materials Synthesis. Mater. Horizons 3, 91–112 (2016).
[19] Kanatzidis, M. G., Pöttgen, R., Jeitschko, W.: The Metal Flux: A Preparative Tool for the Exploration of Intermetallic Compounds. Angew. Chem., Int. Ed. Engl. 44(43), 6996–7023 (2005).
[20] Liu, X., Fechler, N., Antonietti, M.: Salt Melt Synthesis of Ceramics, Semiconductors and Carbon Nanostructures. Chem. Soc. Rev. 42(21), 8237 (2013).
[21] Adam, D.: Out of the Kitchen. Nature 421, 571–572 (2003).
[22] Lekse, J. W., Stagger, T. J., Aitken, J. A.: Microwave Metallurgy: Synthesis of Intermetallic Compounds via Microwave Irradiation. Chem. Mater. 19(17), 3601–3603 (2007).
[23] Ramesh, P. D., Vaidhyanathan, B., Ganguli, M., Rao, K. J.: Synthesis of B-SiC Powder by Use of Microwave Radiation. J. Mater. Res. 9(12), 3025–3027 (1994).
[24] Zhang, H., Ouyang, S., Liu, H., Li, Y.: Kinetics and Mechanism of the Microwave Synthesis of Barium Titanate. In: MRS Proc., vol. 430, pp. 447–452 (1996).
[25] Rao, K. J., Ramakrishnan, P. A., Gadagkar, R.: Microwave Preparation of Oxide Bronzes. J. Solid State Chem. 148(1), 100–107 (1999).
[26] Whittaker, A. G., Mingos, D. M. P.: Microwave-Assisted Solid-State Reactions Involving Metal Powders. J. Chem. Soc., Dalton Trans. 18, 2751–2752 (1992).
[27] Whittaker, A. G., Mingos, D. M. P.: Microwave-Assisted Solid-State Reactions Involving Metal Powders. J. Chem. Soc., Dalton Trans. 12, 2073–2079 (1995).
[28] Vaidhyanathan, B., Ganguli, M., Rao, K. J.: Fast Solid State Synthesis of Metal Vanadates and Chalcogenides Using Microwave Irradiation. Mater. Res. Bull. 30(9), 1173–1177 (1995).
[29] Vaidhyanathan, B., Rao, K. J.: Microwave Assisted Synthesis of Technologically Important Transition Metal Silicides. J. Mater. Res. 12(12), 3225–3229 (1997).
[30] Vaidhyanathan, B., Rao, K. J.: High Microwave Susceptibility of NaH2PO4·2H2O :Rapid Synthesis of Crystalline and Glassy Phosphates with NASICON-Type Chemistry. J. Solid State Chem. 132, 349–354 (1997).
[31] Ramesh, P. D., Rao, K. J.: Microwave-Assisted Synthesis of Aluminum Nitride. Adv. Mater. 7(2), 177–179 (1995).
[32] Vaidhyanathan, B., Rao, K. J.: Synthesis of Ti, Ga, and V Nitrides: Microwave-Assisted Carbothermal Reduction and Nitridation. Chem. Mater. 9, 1196–1200 (1997).
[33] Whittaker, A. G., Mingos, D. M. P.: Microwave-Assisted Solid-State Reactions Involving Metal Powders and Gases. J. Chem. Soc., Dalton Trans. 3, 2541–2543 (1993).
[34] Rao, K. J., Vaidhyanathan, B., Ganguli, M., Ramakrishnan, P. A.: Synthesis of Inorganic Solids Using Microwaves. Chem. Mater. 11, 882–895 (1999).
[35] Mingos, D. M. P., Baghurst, D. R.: Br. Ceram. Trans. J. 91, 124–127 (1992).
[36] Binner, J., Vaidhyanathan, B., Paul, A., Annaporani, K., Raghupathy, B.: Compositional Effects in Nanostructured Yttria Partially Stabilized Zirconia. Int. J. Appl. Ceram. Technol. 8(4), 766–782 (2011).
[37] Yang, G., Wang, G., Hou, W. H.: Microwave Solid-State Synthesis of LiV3O8 as Cathode Material for Lithium Batteries. J. Phys. Chem. B 109(22), 11186–11196 (2005).
[38] Munir, Z. A., Anselmi-Tamburini, U., Ohyanagi, M.: The Effect of Electric Field and Pressure on the Synthesis and Consolidation of Materials: A Review of the Spark Plasma Sintering Method. J. Mater. Sci. 41, 763–777 (2006).
[39] Garay, J. E.: Current-Activated Pressure-Assisted Densification of Materials. Annu. Rev. Mater. Res. 40, 445–468 (2010).

[40] Guillon, O., Gonzalez-Julian, J., Dargatz, B., Kessel, T., Schierning, G., Räthel, J., Herrmann, M.: Field-Assisted Sintering Technology/Spark Plasma Sintering: Mechanisms, Materials, and Technology Developments. Adv. Eng. Mater. 16(7), 830–849 (2014).
[41] Hulbert, D. M., Anders, A., Dudina, D. V., Andersson, J., Jiang, D., Unuvar, C., Anselmi-Tamburini, U., Lavernia, E. J., Mukherjee, A. K.: The Absence of Plasma in "spark Plasma Sintering". J. Appl. Phys. 104(3), 33305 (2008).
[42] Hulbert, D. M., Anders, A., Andersson, J., Lavernia, E. J., Mukherjee, A. K.: A Discussion on the Absence of Plasma in Spark Plasma Sintering. Scr. Mater. 60, 835–838 (2009).
[43] Orrù, R., Licheri, R., Locci, A. M., Cincotti, A., Cao, G.: Consolidation/synthesis of Materials by Electric Current Activated/assisted Sintering. Mater. Sci. Eng., R Rep. 63(4–6), 127–287 (2009).
[44] Hungría, T., Galy, J., Castro, A.: Spark Plasma Sintering as a Useful Technique to the Nanostructuration of Piezo- Ferroelectric Materials. Adv. Eng. Mater. 11(8), 615–631 (2009).
[45] Snyder, G. J., Toberer, E. S.: Complex Thermoelectric Materials. Nat. Mater. 7, 105–114 (2008).
[46] Kieslich, G., Birkel, C. S., Veremchuk, I., Grin, Y., Tremel, W.: Thermoelectric Properties of Spark-Plasma Sintered Nanoparticular FeSb2 Prepared via a Solution Chemistry Approach. Dalton Trans. 43(2), 558–562 (2014).
[47] Birkel, C. S., Claudio, T., Panthöfer, M., Birkel, A., Koll, D., Kieslich, G., Schmidt, J., Hermann, R., Tremel, W.: Properties of Spark Plasma Sintered Nanostructured Zn1+xSb. Phys. Status Solidi A 208(8), 1913–1919 (2011).
[48] Kieslich, G., Birkel, C. S., Douglas, J. E., Gaultois, M., Veremchuk, I., Seshadri, R., Stucky, G. D., Grin, Y., Tremel, W.: SPS-Assisted Preparation of the Magnéli Phase WO2.90 for Thermoelectric Applications. J. Mater. Chem. A 1(42), 13050–13054 (2013).
[49] Kieslich, G., Veremchuk, I., Antonyshyn, I., Zeier, W. G., Birkel, C. S., Weldert, K., Heinrich, C. P., Visnow, E., Panthöfer, M., Burkhardt, U., et al.: Using Crystallographic Shear to Reduce Lattice Thermal Conductivity: High Temperature Thermoelectric Characterization of the Spark Plasma Sintered Magnéli Phases WO2.90 and WO2.722. Phys. Chem. Chem. Phys. 15(37), 15399–15403 (2013).
[50] Cabouro, G., Chevalier, S., Gaffet, E., Grin, Y., Bernard, F.: Reactive Sintering of Molybdenum Disilicide by Spark Plasma Sintering from Mechanically Activated Powder Mixtures: Processing Parameters and Properties. J. Alloys Compd. 465(1–2), 344–355 (2008).
[51] Zhang, F., Ahmed, F., Bednarcik, J., Burkel, E.: Diamond Synthesis through the Generation of Plasma during Spark Plasma Sintering. Phys. Status Solidi 209(11), 2241–2246 (2012).
[52] Friedman, J. R., Garay, J. E., Anselmi-Tamburini, U., Munir, Z. A.: Modified Interfacial Reactions in Ag–Zn Multilayers under the Influence of High DC Currents. Intermetallics 12(6), 589–597 (2004).
[53] Schwesig, D., Schierning, G., Theissmann, R., Stein, N., Petermann, N., Wiggers, H., Schmechel, R., Wolf, D. E.: From Nanoparticles to Nanocrystalline Bulk: Percolation Effects in Field Assisted Sintering of Silicon Nanoparticles. Nanotechnology 22(13), 135601 (2011).
[54] Beekman, M., Baitinger, M., Borrmann, H., Schnelle, W., Meier, K., Nolas, G. S., Grin, Y.: Preparation and Crystal Growth of Na24Si136. J. Am. Chem. Soc. 131(1), 9642–9643 (2009).
[55] Munir, Z. A., Quach, D. V., Ohyanagi, M.: Electric Current Activation of Sintering: A Review of the Pulsed Electric Current Sintering Process. J. Am. Ceram. Soc. 94(1), 1–19 (2011).
[56] Chen, W., Anselmi-Tamburini, U., Garay, J. E., Groza, J. R., Munir, Z. A.: Fundamental Investigations on the Spark Plasma Sintering/synthesis Process I. Effect of Dc Pulsing on Reactivity. Mater. Sci. Eng. A 407(1–2), 24–30 (2005).
[57] Rowe, D. M.: In: CRC Handbook of Thermoelectrics, Boca Raton (1995).
[58] Makongo, J. P. A., Misra, D. K., Zhou, X., Pant, A., Shabetai, M. R., Su, X., Uher, C., Stokes, K. L., Poudeu, P. F. P.: Simultaneous Large Enhancements in Thermopower and Electrical Conductivity of Bulk Nanostructured Half- Heusler Alloys. J. Am. Chem. Soc. 133(46), 18843–18852 (2011).
[59] Slack, G.: Solid State Physics, vol. 34. Academic Press, New York (1979).

[60] Kumar, N., Kumbhat, S.: Nanomaterials General Synthetic Approaches. In: Essentials in Nanoscience and Nanotechnology. Wiley VCH (2016).

[61] Kieslich, G., Birkel, C. S., Veremchuk, I., Grin, Y., Tremel, W.: Thermoelectric Properties of Spark-Plasma Sintered Nanoparticular FeSb2 Prepared via a Solution Chemistry Approach. Dalton Trans. 2014(43), 558–562 (2078).

[62] Birkel, C. S., Mugnaioli, E., Gorelik, T., Kolb, U., Panthöfer, M., Tremel, W.: Solution Synthesis of a New Thermoelectric Zn(1+x)Sb Nanophase and Its Structure Determination Using Automated Electron Diffraction Tomography. J. Am. Chem. Soc. 132(28), 9881–9889 (2010).

[63] Stavila, V., Robinson, D. B., Hekmaty, M. A., Nishimoto, R., Medlin, D. L., Zhu, S., Tritt, T. M., Sharma, P. A.: Wet-Chemical Synthesis and Consolidation of Stoichiometric Bismuth Telluride Nanoparticles for Improving the Thermoelectric Figure-of-Merit. ACS Appl. Mater. Interfaces 5(14), 6678–6686 (2013).

[64] Zhou, W., Zhu, J., Li, D., Hng, H. H., Boey, F. Y. C., Jan, M., Zhang, H., Yan, Q.: Binary-Phased Nanoparticles for Enhanced Thermoelectric Properties. Adv. Mater. 21(31), 3196–3200 (2009).

[65] Ou, C., Hou, J., Wei, T. R., Jiang, B., Jiao, S., Li, J. F., Zhu, H.: High Thermoelectric Performance of All-Oxide Heterostructures with Carrier Double-Barrier Filtering Effect. NPG Asia Mater. 7(5), 1–10 (2015).

[66] Maji, P., Makongo, J. P. A., Zhou, X., Chi, H., Uher, C., Poudeu, P. F. P.: Thermoelectric Performance of Nanostructured P-Type Zr0.5Hf0.5Co0.4Rh0.6Sb1-xSnx Half-Heusler Alloys. J. Solid State Chem. 202, 70–76 (2013).

[67] Douglas, J. E., Birkel, C. S., Miao, M., Torbet, C. J., Stucky, G. D., Pollock, T. M., Seshadri, R.: Enhanced Thermoelectric Properties of Bulk TiNiSn via Formation of a TiNi 2 Sn Second Phase. Appl. Phys. Lett. 101, 183902 (2012).

[68] Zhang, Y., Bahk, J.-H., Lee, J., Birkel, C. S., Snedaker, M. L., Liu, D., Zeng, H., Moskovits, M., Shakouri, A., Stucky, G. D.: Hot Carrier Filtering in Solution Processed Hetero-Structures: A Paradigm for Improving Thermoelectric Efficiency. Adv. Mater. 26, 2755–2761 (2014).

[69] Zhang, Y., Day, T., Snedaker, M. L., Wang, H., Krämer, S., Birkel, C. S., Ji, X., Liu, D., Snyder, G. J., Stucky, G. D.: A Mesoporous Anisotropic N-Type Bi 2 Te 3 Monolith with Low Thermal Conductivity as an Effi Cient Thermoelectric Material. Adv. Mater. 24, 5065–5070 (2012).

[70] Birkel, C. S., Zeier, W. G., Douglas, J. E., Lettiere, B. R., Mills, C. E., Seward, G., Birkel, A., Snedaker, M. L., Zhang, Y., Snyder, G. J., et al.: Rapid Microwave Preparation of Thermoelectric TiNiSn and TiCoSb Half-Heusler Compounds. Chem. Mater. 24(13), 2558–2565 (2012).

[71] Birkel, C. S., Douglas, J. E., Lettiere, B. R., Seward, G., Verma, N., Zhang, Y., Pollock, T. M., Seshadri, R., Stucky, G. D.: Improving the Thermoelectric Properties of Half-Heusler TiNiSn through Inclusion of a Second Full- Heusler Phase: Microwave Preparation and Spark Plasma Sintering of TiNi1+xSn. Phys. Chem. Chem. Phys. 15, 6990–6997 (2013).

[72] Kieslich, G., Birkel, C. S., Douglas, J. E., Gaultois, M., Veremchuk, I., Seshadri, R., Stucky, G. D., Grin, Y., Tremel, W.: SPS-Assisted Preparation of the Magnéli Phase WO2.90 for Thermoelectric Applications. J. Mater. Chem. A 1(42), 13050 (2013).

[73] Kieslich, G., Burkhardt, U., Birkel, C. S., Veremchuk, I., Douglas, J. E., Gaultois, M. W., Lieberwirth, I., Seshadri, R., Stucky, G. D., Grin, D. Y., et al.: Enhanced Thermoelectric Properties of the N-Type Magnéli Phase WO2.90: Reduced Thermal Conductivity through Microstructure Engineering. J. Mater. Chem. A 2, 13492–13497 (2014).

[74] Biswas, K., He, J., Blum, I. D., Wu, C. I., Hogan, T. P., Seidman, D. N., Dravid, V. P., Kanatzidis, M. G.: High Performance Bulk Thermoelectrics with All-Scale Hierarchical Architectures. Nature 489(7416), 414–418 (2012).

Part IV: **Sustainable smart materials**

Annelise Kopp Alves

Nature-inspired smart materials for multifunctional applications

Abstract: Smart materials are a group of materials that somehow respond to a certain stimulus. Nature is one of the best inspiring agents when one wants to develop smart materials. In the wild, life has to prosper under the Earth's harsh environment. In our daily lives, there are many examples of smart materials inspiring by nature: Velcro®, glass lenses that react to environmental light changes, materials that change their forms under load, heat or humidity for sensor and actuators applications. This field of research includes the development of new materials, the improvement of the characteristics of traditional materials, the fabrication of such materials and the integration of these materials in a useful component. It is a strong interdisciplinary field, involving materials scientists, designers, engineers, architects, medics and many others regarding to the component application and multifunctional integration.

Keywords: Biomimicry, smart materials, multifunctional materials

1 Introduction

Life on Earth first appears 4.28 billion years ago. Up to the present day, life has all this time to evolve and find solutions to prosper and grow, even in the harsher environments. Plants and animals have to find solutions to keep alive and to pass on their genes when facing the most different situations our ever-changing planet imposes upon them. Of course, some species have perished during this process, but the smarted and well adapted survive. From them, we can inspire ourselves to find more efficient, economic and sustainable solutions to our own daily struggle to survive.

Mother nature has few materials, time and space to develop fauna and flora. It also has to keep energy consumption to a minimal level to create sustainable beings. Their structures should also endure several challenges like drought, high humidity, heavy winds, salty waters, too much or too little sunlight, snow, etc. It was a development that leads to mistakes, and the price for those was extinction. But, the successful multifunctional structures developed by nature prosper and are with us today, as examples of what can be done under restricted conditions.

Traditionally, materials are selected to be used for their specific property, considering their function in the system. Some are rigid and then they are applied as structural material, some conduct electricity and then, used in electric devices, some have

Annelise Kopp Alves, Universidade Federal do Rio Grande do Sul, Porto Alegre, Brazil, e-mail: annelise.alves@ufrgs.br

https://doi.org/10.1515/9783110584455-011

specific optical characteristics and then, used to control light, and so on. But seldom the materials are combined by humans in a multifunctional way. It took the development of the human kind up to today for us to realize what nature already did from its beginning: multifunctional structures are a need to save resources (energy and raw materials) and to perform better on duty, combining more than one functionality in a single system using few materials. Biomimicry or biomimetics are the terms used in the field where solution inspiration come from nature.

Multifunctional materials can be of different kinds. One of the most interesting types is the stimulus-responsive materials (SRMs). They respond to a stimulus by changing their properties. A common SRM is the pieazoelectric materials, the ones that produced electricity under stress or vice versa. These types of materials are used in actuators, sensors and energy harvesting applications. Table 1 presents some of the most common SRMs applied in multifunctional devices and assemblies.

It is interesting to sort multifunctional materials according to the function they perform in the system they are imbedded in. Some examples of these classification are: autonomy (present self-healing/repairing properties, self-powered; self-diagnosing; self-assembling); structural; sound/vibration damping; shape-changing; isolation/conduction of electricity or heat; radiation protection; light emission; chemical functions (catalysts, selective permeation); electromagnetics, among others.

The biomimetic innovations are possible to be categorized based on the source of inspiration: how nature create things (materials), how the organisms respond to their environment (sensors), how they move in their environment (biomechanics and kinetics) and how they behave and function (processes).

Considering this context, this chapter propose to illustrate the state of the art of nature inspired multifunctional materials, structures and their integration into useful application. With examples inspired from nature, it will be exemplified how humans develop solutions to the most common or complex problems, and also indicates the perspectives of bioinspired applications.

2 Fauna inspired materials and applications

Earth's fauna is wonderful in its variety of beings with different forms, colors and sizes. From the tiniest animal, the Myxozoa species have no more than 20 µm when adult, up to the largest one, the blue whale that has approximately 30 m in length, all have to adapt to the environment to keep alive and pass on their genes. Coincidently, the smallest and the largest animal on Earth live in the water, where it is said life begins.

Humans use the seas and water bodies to navigate from one point to another, delivering goods and people around the globe. During the movement of vessels in the water, 40 up to 80 % of resistance to movement, depending on the vessel speed, is associated with friction and drag resistance [5]. Bio-fouling, the adhesion of a marine

Table 1: Common stimulus-responsive materials used in multifunctional assemblies.

SRM	Stimulus	Type	Answer
Thermo-responsive	Heat	Thermochromic	Color change
		Thermoelectric	Voltage change
		Shape memory alloy	Physical form
Mechano-responsive	Stress/pressure	Smart gel	Swell or deflate
		Piezoelectric	Voltage change
Electro-responsive	Electrical current/voltage	Electrochromic	Color change
		Dielectric elastomers	Large strain (up to 300 %)
		Electro-rheological fluids	Shape shift
		Electrostritctive	Produce striction
		Electro-active polymers	Shape shift
		Piezoelectric	Shape shift or produce stress
		Thermoelectric	Temperature change
Magneto-responsive	Magnetic field	Magnetorheological fluids	Shape shift
		Magnetostrictive	Produce striction
		Magnetocaloric	Temperature change
Chemo-responsive	pH	Smart gels	Imbedded into polymers to produce swelling or deflating
	Moisture		
	Solvent		
Photo-responsive	light	Photochromic	Color change
		Photomechanical	Shape shift
Molecular-responsive	Molecules	Gas sensing materials	Change color or voltage
pH-responsive	pH		Volume or color change

organism in the ships surfaces, is another important problem to solve in the marine industry. So, it is of great economical and system efficiency interest to prevent these issues.

One of the most traditional inspiration for biomimetic materials and applications in drag reduction and antifouling surfaces is the shark skin. This skin contains smooth scales made of dentin, resembling tiny teeth (Figure 1), the so-called dermal denticles. The scales can prevent the formation of vortices, reducing drag [5]. Shark skin also has the property of preventing barnacles and other organisms from glomming on due to its rough texture formed by the dermal denticles.

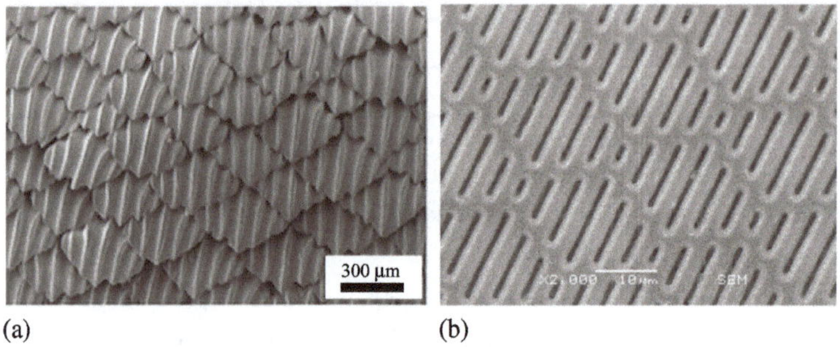

Figure 1: (a) SEM images of the shark-skin surface (adapted from [9]) and (b) SEM image of the developed biomimetic film based on shark-skin (adapted from [20]).

The group of researches of Fu et al. [5] studied the effect of the dimensions of riblets, mimicking the shark skin over a graphite plate (Figure 1b) and the influence of the water velocity, from 5 to 15 knots if considered a ship. They found out that the riblets work as an antifriction coating, since when the valleys are filled up with water, the flow is easier conducted over the surface. Choosing the ideal riblet configuration and dimensions, authors could measure a drag reduction up to 24 %.

The work of Liu et al. [9] developed a Lotus effect surface based on shark skin morphology replica and micro/nano-structures by thermal treating the surface of polydimethylsiloxane (PDMS). After heat treatment, the replica achieves a water contact angle of 160°, and increase of 40° compared to the nontreated replica.

Considering clean water sources, populated drought and desert regions suffer from a shortage that should be looked after. In the Namib desert, a region of the planet that receives up to 1.4 cm of rain a year, nature adapted a little desert beetle (Stenocara gracilipes) to live there. This beetle has the ability to collect water for its survival using the humidity of the atmosphere (dew, mist or ocean fog). It is possible because its body surface has special wetting properties, composed of hydrophobic and hydrophilic regions that form a droplet and absorb the water [13]. Through the grooves and smooth surfaces of the external upper wings, the water is collected and guided to the beetle mouth (see Figure 2).

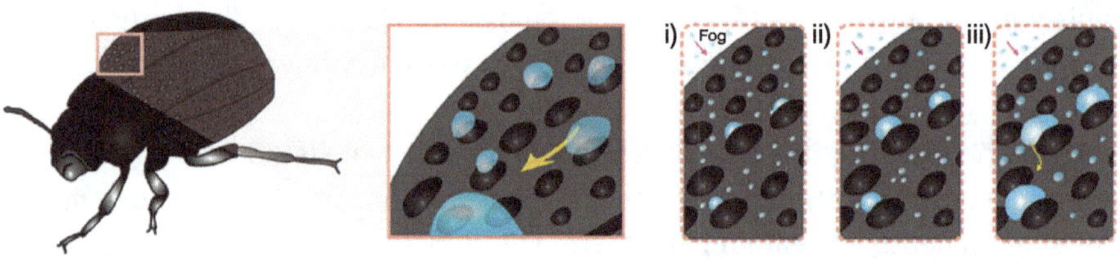

Figure 2: Schematic of the fog-harvesting process in a desert beetle. Beetle' back side surface consists of a microsized bumpy superhydrophilic structure that absorbs the fog and hardened forewings, which are superhydrophobic and helps collect and direct water droplets [13].

This insect has attracted interest of the researches that developed inspired water collected systems in a novel and functional way [6, 14, 17, 19]. Parker et al. [14] studied the mechanism that best simulates the water collection of the beetle by analyzing its wings surface characteristics. They have found that the wings are composed of regions that have a wax cover (hydrophobic region) and near-random array of bumps with 0.5 mm in diameter that have wax in their tips. The authors then developed a multifunctional system based on a hydrophobic flat membrane type PP (PPMM) with bumps made of a biopolymer of polydopamine (PDA) throughout several microfabrication processes. With this system, they collect water from a fog chamber at a rate up to 97 $mg\,cm^{-2}\,h^{-1}$ (3.28 L/m^2 day), using a plate of $2.5 \times 2.5\,cm^2$.

Sometimes, the animal kingdom made us curious to understand how animals sustain such harsh living conditions. For example, in the report of May et al. they observed that the drilling action of a woodpecker caused its head to move in straight trajectory, reaching maximum speeds of 5 to 7 $m \cdot s^{-1}$ and decelerations of 600 to 1500 g within 0.5 to 1.0 ms of impacting duration. In the same conditions, the human brain would be severely injured since its tolerance to deceleration is just 300 g. So, what is behind the antishock mechanism in the woodpecker's head?

The shape of the skull, the presence of a hyoid structure, the porous and dense periodical arrangement of bone structures, a strong beak and a nonuniform distribution of Young' modulus in the skull of the woodpecker (Figure 3a) hinders stress propagation [22]. Inspired by these studies, Yoon and Park [18], simplified the head structure as a mass-damper-spring model and made a bio-inspired shock-absorbing equipment consisting of close-packed microglass spheres (Figure 3b). The equipment can protect micromechanical and electronic devices from high-g damage. Also, based on these studies, mechanisms to protect instruments and the elements of inner structures, for example, light and high-performance helmets and spacecrafts were developed.

Construct stronger but lighter shelter systems is also an important necessity for our survival. Skulls have an extraordinary impact-resistant structure and are light at the same time as they protect the most important organ of an animal body (Figure 5a). Inspired by its performance and physical property, it is possible to applied these constructive characteristics in structure or architecture design. Considering song bird's skulls such as Carrion crows, their structure is based on elongated cells in a single or two-story air-package area, providing an acoustic function, distributing air at the inner layer but also determining a light and strong structure. Based on the bone structure of birds' skull, Harris A. have developed a biomimetic double-layer surface, that is strong and lightweight to be used as a roof for buildings (Figure 5b).

In the animal kingdom, the capacity to move from one place to another and to reach with limbs for food or support is an important tool in the daily survival fight. Think of the power of the elephant's trunk. It is flexible, it can bend in different directions, elongate and hold objects. A group of researchers from the German company Festo, developed the Bionic Handling Assistant inspired in the elephant's trunk particular characteristics (Figure 4) [4].

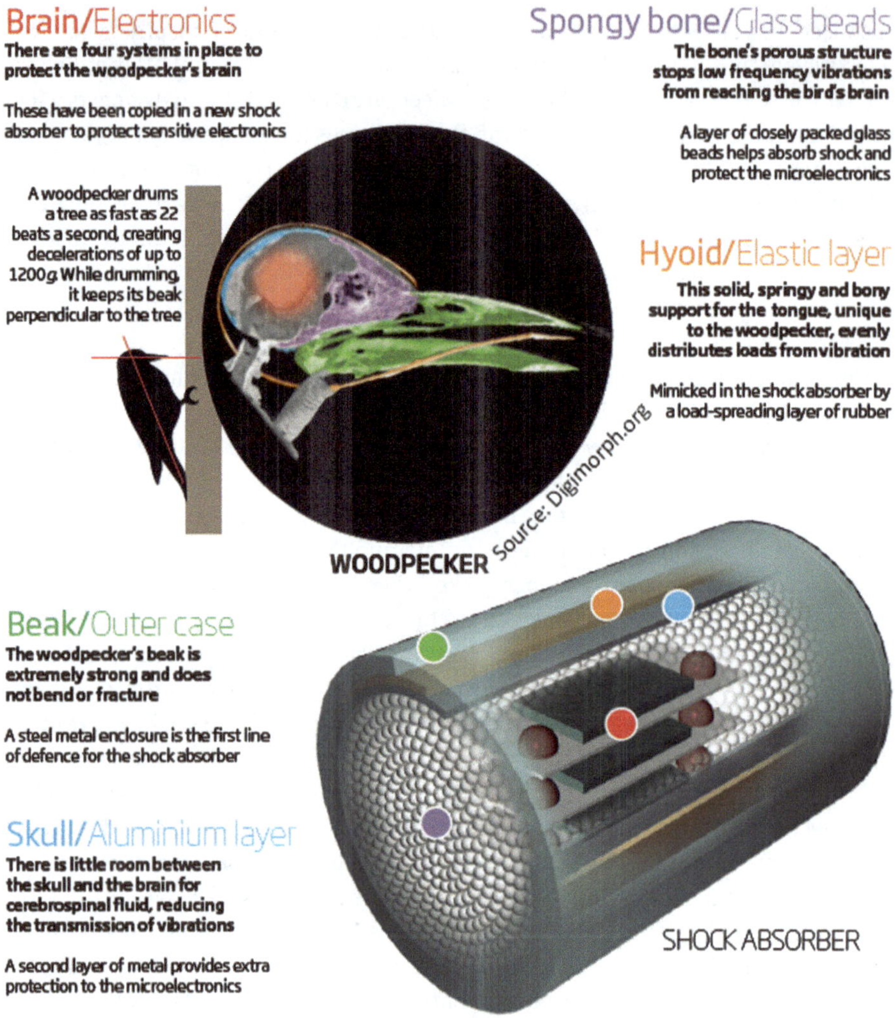

Figure 3: Bioinspired shock absorber based on the woodpecker (adapted from [11]).

The system has in its end a claw with sensors that can hold and transport even the more sensible objects with safety. The assistant arm is made of polyamide, which makes the structure flexible, but when needed, the pneumatic control unit can make it very rigid. The company used a 3D printing technology to make the soft and lightweight trunk parts that can move, expand and contract using pneumatically-powered artificial muscles.

Construct stronger but lighter shelter systems is also an important necessity for our survival. Skulls have an extraordinary impact-resistant structure and are light at the same time as they protect the most important organ of an animal body (Figure 5a).

(a)

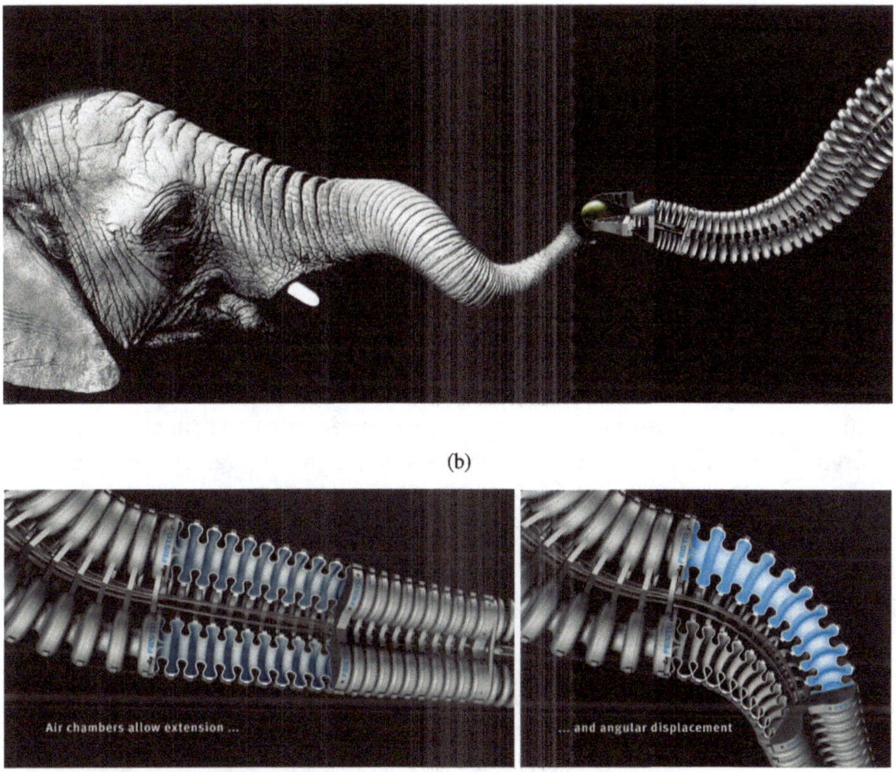

Figure 4: Bioinspired bionic handling assistant made of lightweight and flexible material. Adapted from Festo website (https://www.festo.com/group/de/cms/10241.htm), [4].

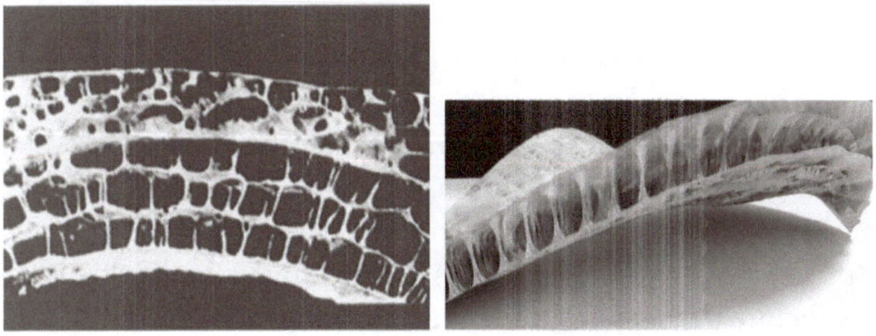

(a) (b)

Figure 5: (a) Bone structure of birds' skull and (b) Digital representation of a biomimetic double-layered surface (adapted from Harris [7], © Copyright Andres Harris).

Inspired by its performance and physical property, it is possible to applied these constructive characteristics in structure or architecture design. Considering song bird's skulls such as Carrion crows, their structure is based on elongated cells in a single or two-story air-package area, providing an acoustic function, distributing air at the inner layer but also determining a light and strong structure. Based on the bone structure of birds' skull, Harris A. have developed a biomimetic double-layer surface, that is strong and lightweight to be used as a roof for buildings (Figure 5b).

3 Flora inspired materials and applications

Plants are probably the most biomimicry inspiring species in terms of different applications (Figure 6). The large numbers of innovations inspired by flora can be attributed to diversity; there is approximately 300,000 plant species, and its survival abilities, since plants are unable to move they have to endure and adapt in situ in order to survive. Natural materials have numerous functionalities (e. g., stability, stiffness, toughness, self-healing, etc.), based on a limited number of components (i. e., cellulose, hemicellulose, lignin and pectin). These different properties can be attributed to the hierarchical organization of such components [12]. For example, Velcro® is the quintessence of application of biomimetics. It is found on everything from astronaut suits to the most mundane uses like shoes and bags. The Swiss engineer, George de Mestral, looked at plant burrs (Arctium lappa) (Figure 7a) that sticks on clothing or hair under a microscope and noticed they are made of tiny hooks that could catch on loops of the clothing, for example. He then developed a system (Figure 7b) based on these characteristics and called it Velcro®, from the French word "velour," velvet, and "crochet," hook [16].

As famous as Velcro® is in the biomimetic world, is the technology called the Lotus effect [2]. It is based on the surface behavior of the Lotus leaf (Nelumbo nucifera) that is water-repellant, superhy-drophobic and self-cleaning, due to the presence of hydrophobic coating and multiscale hierarchical roughness on its surface (Figure 8) [1].

Plants have also to protect its fruits to assure the continuation of the species. Like an overcaring mother that protects their children on their first skate ride with several layers of protection gear, the pomelo (Citrus maxima), shows a particular damping system for its fruits. They can weigh up to 6 kg and the higher branches of a pummelo tree reach 15 m high. They resist to a drop from this height without damage the pulp [3]. Analyzing their structure (Figure 9a, b, c, d), researchers have found that a complex hierarchical structure made of interconnected porous layers and fiber networks is able to dissipate more than 90 % of the impact energy [21]. Inspired by this, authors have developed crash absorbers based on metal foams [15] (Figure 9c). The work of [8], studied the properties of biomimetic surfaces structures based on the Calathea zebrina leaf surfaces to diminish the splash behavior of water droplets. The production of aerosol particles caused by among others, the splash phenomena of droplets, can be associated to virus and bacteria transportation and infection. Decreasing the

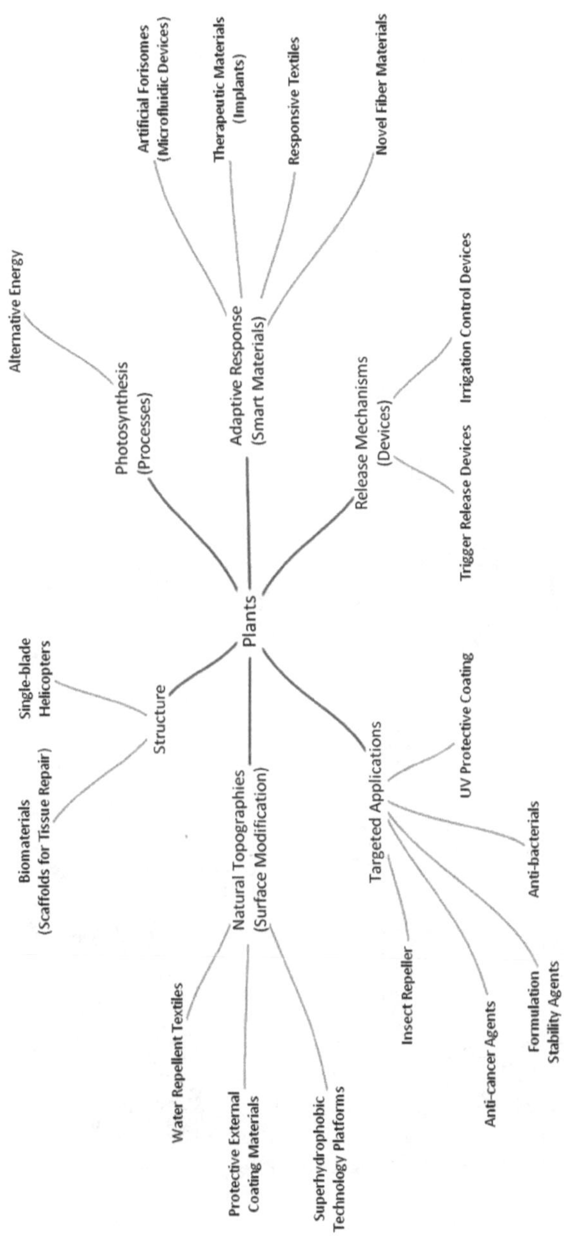

Figure 6: Variety of biomimicry applications inspired by plants [10].

amount of aerosol particles in the air has a significant impact in the human health. Also, on the economical point of view of agriculture, if a pesticide or other products spray is evenly distributed over the leaf, it can significantly reduce the amount of product used.

(a) (b)

Figure 7: Photography of the Arctium lappa plant and (b) SEM colorized image of Velcro®.

Figure 8: Superhydrophobic, self-cleaning surface of Lotus (Nelumbo nucifera). Lotus leaf and flower (a), Lotus leaf contaminated with clay (b) removal of the particles with water (c). SEM micrographs (d–f) of the lotus leaf surface in different magnifications: (d) papilla epidermis, (e) single cell papilla and (f) epicuticular nonacosan-10-ol tubules on the cell surface ([1], creativecommons.org).

The leaf of the C. zebrina has a hydrophilic behavior, and because of it, the tendency to form a homogeneous film rather than a splash when water droplets hit its surface. Analyzing more carefully the surface of the leaf, authors suggested a polymer surface

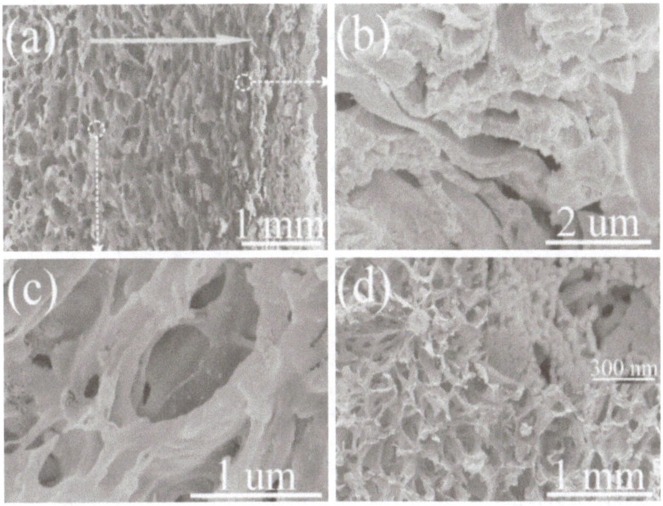

Figure 9: SEM images of pomelo peel: longitudinal section of the mesocarp, (b) magnification of the area within the white circle close to the exocarp in (a), (c) the magnification of the area within the white circle close to the endocarp in (a), (d) cross-section (adapted from [21]).

Figure 10: (a) Calathea zebrina leaves; (b) SEM micrograph of the leaf surface (tilt angle 45°); (c) SEM image of the replicated dark leaf areas and (d) bright leaf areas in side view (90° tilt angle) ([1] – creativecommons.org).

that mimic the topography of the leaf (Figure 10). They have found out that the conical structures of the C. zebrina leaf surface can make a polymer surface go from hydrophilic to superhy-drophilic and have a significant influence on its splashing behavior, by keeping drop impact locations drier due to capillary transport.

Considering all these examples, Nature is daily dealing with the harsh conditions of the Earth's surface, letting us know how to be creative, economical and sustainable by the strategies and solutions it founds to keep the many different organisms alive. Biomimicry is a fountain of inspiration for scientists, architects and engineers to create multifunctional systems. All it takes is a little patience and observation.

Bibliography

[1] Barthlott, W., Mail, M., Bhushan, B., Koch, K.: Plant Surfaces: Structures and Functions for Biomimetic Innovations. Nano-Micro Lett. 9, 23 (2017).
[2] Barthlott, W., Neinhuis, C.: Purity of the sacred lotus, or escape from contamination in biological surfaces. Planta 202, 1–8 (1997).
[3] Fischer, S. F., Thielen, M., Loprang, R. R., Seidel, R., Fleck, C., Speck, T., Bührig-Polaczek, A.: Pummelos as Concept Generators for Biomimetically Inspired Low Weight Structures with Excellent Damping Properties. Adv. Eng. Mater. 12, B658–B663 (2010).
[4] Festo: https://www.festo.com/group/de/cms/10156.htm. Accessed on March, 2018.
[5] Fu, Y. F., Yuan, C. Q., Bai, X. Q.: Marine drag reduction of shark skin inspired riblet surface. Biosurf. Biotribol. 3, 11–24 (2017).
[6] Guadarrama-Cetina, J., et al.: Dew condensation on desert beetle skin. Eur. Phys. J. E 37(11), 1–6 (2014).
[7] Harris, A.: http://www.andres.harris.cl/about/32-2. Accessed on March, 2018.
[8] Koch, K., Barthlott, W.: Superhydrophobic and Superhydrophilic Plant Surfaces: An Inspiration for Biomimetic Materials. Philos. Trans. Math. Phys. Eng. Sci. 367(1893, Biomimetics I: Functional Biosurfaces), 1487–1509 (2009).
[9] Liu, Y., Li, G.: A new method for producing "Lotus Effect" on a biomimetic shark skin. J. Colloid Interface Sci. 388, 235–242 (2012).
[10] Lurie-Luke, E.: Product and technology innovation: What can biomimicry inspire? Biotechnol. Adv. 32, 1494–1505 (2014).
[11] Marks, P.: Woodpecker's head inspires shock absorbers. New Sci. 21 (2011).
[12] Mazzolai, B., Beccai, L., Mattoli, V.: Plants as Model in Biomimetics and Biorobotics: New Perspectives. Front. Bioeng. Biotechnol. 2 (2014).
[13] Moazzam, P., Tavassoli, H., Razmjou, A., Warkiani, M. E., Asadnia, M.: Mist harvesting using bioinspired polydopamine coat-ing and microfabrication technology. Desalination 429, 111–118 (2018).
[14] Parker, A. R., Lawrence, C. R.: Water capture by a desert beetle. Nature 414, 33–34 (2001).
[15] Thielen, M., Schmitt, C. N. Z., Eckert, S., Speck, T., Seidel, R.: Structure–function relationship of the foam-like pomelo peel (Citrus maxima) – an inspiration for the development of biomimetic damping materials with high energy dissipation. Bioinspir. Biomim. 8, 2 (2013).
[16] Velcro, S. A.: Improvements in or Relating to a Method and a Device for Producing a Velvet Type Fabric Patent No. 721 338. Switzerland (1955).
[17] Wang, Y., et al.: Biomimetic water-collecting fabric with light-induced superhydrophilic bumps. ACS Appl. Mater. Interfaces 8(5), 2950–2960 (2016).
[18] Yoon, S-H., Park, S.: A mechanical analysis of woodpecker drumming and its application to shock-absorbing systems. Bioinspir. Biomim. 6, 1–12 (2011).
[19] Zhai, L., et al.: Patterned superhydrophobic surfaces: toward a synthetic mimic of the Namib Desert beetle. Nano Lett. 6(6), 1213–1217 (2006).

[20] Zhao, D., Tian, Q., Wang, M., Jin, Y.: Study on the Hydrophobic Property of Shark-Skin-Inspired Micro-Riblets. J. Bionics Eng. 11, 296–302 (2014).
[21] Zhao, S., Wang, L., Wang, Y., Li, X.: Hierarchically porous LaFeO3 perovskite prepared from the pomelo peel bio-template for catalytic oxidation of NO. J. Phys. Chem. Solids 116, 43–49 (2018).
[22] Zhu, Z., Wu, C., Zhang, W.: Frequency Analysis and Anti-Shock Mechanism of Woodpecker's Head Structure. J. Bionics Eng. 11, 282–287 (2014).

Michael Koerdt, Martina Hübner, Maryam Kahali Moghaddam, Walter Lang, and Axel Siegfried Herrmann

Smart fiber-reinforced polymer composites and their resource-efficient production by means of sensor integration

Abstract: The publication gives an overview about the latest investigations of smart fiber-reinforced polymer composites and their resource-efficient production by means of sensor integration at the Institute for Microsensors, -actuators and -systems (IMSAS) and the Faserinstitut Bremen e. V. The microsensors fabricated by the IMSAS were designed with the goal of avoiding any significant weakening of the composite material structure due to their embedment, allowing the sensors to remain in the composite structure throughout the product life cycle. The integrated sensors can provide useful information about the condition of the materials in real time during production and in service. Results for the verification of this approach are presented. The effects of the sensors on the mechanical properties of the composite are evaluated by mechanical tests. In addition, the composite samples with integrated sensors are examined by X-ray microcomputed tomography (X-μCT). The focuses of the studies are on the influence of the embedments on the structural integrity.

Keywords: Sensor integration, online process monitoring (OPM), structural health monitoring (SHM)

1 Introduction

Fiber-reinforced polymer composites are known as the materials of choice in many areas such as wind power and transportation where weight reductions are needed for lighter constructions. The use of fiber composites in construction is not only an invention of modern technology because fiber composite constructions also exist in nature. Woods and bones are examples of fiber composite constructions in nature.

Acknowledgement: The authors gratefully acknowledge the financial support of the work by the Deutsche Forschungsgemeinschaft (DFG), on the one hand, under the contract numbers HE 2574/31-1 and LA 1471/17-1 and, on the other hand, within the project MAPEX Bremen "Material, Process, Excellence" under the contract number CO 1043 12-1.

Michael Koerdt, Axel Siegfried Herrmann, Faserinstitut Bremen, Bremen, Germany, e-mail: mkoerdt@faserinstitut.de
Martina Hübner, Faserinstitut Bremen, Bremen, Germany
Maryam Kahali Moghaddam, Faserinstitut Bremen, Bremen, Germany
Walter Lang, Faserinstitut Bremen, Bremen, Germany

https://doi.org/10.1515/9783110584455-012

The principle of load-carrying by high strength fibers, which are surrounded with a rigid load transfer matrix material has been developed by nature as the best suited lightweight construction principle in a millennia-long evolution [1].

Fiber-reinforced polymer composites are generally used for a wide range of applications. For example, fiber-reinforced polymer composites are used in the automotive industry for cover parts, in the aviation industry for primary structures of airplanes, for rotor blades of wind turbines, in the building sector for the reinforcement of bridges, for sports equipment and for long booms of agriculture machinery and arms of robots.

Due to the increasing demand for fiber-reinforced polymer composite materials that continuously replace metals as the classical construction material in many areas, the interest and the demand for cost-efficient processes and for smart structures have risen sharply. For many years, the idea of integrating sensors and actuators into structural components of fiber-reinforced polymer composites exists, which enable a continuous monitoring of the component over its entire life cycle from manufacturing to qualification and in service. The integration of sensors could be an enabling technology to reach higher resource efficiency.

The laminate is a common form in which fiber-reinforced polymer composites are used in structural applications. Laminates are particularly suitable for the integration of actuators and sensors since they have a layered structure and their production takes place at moderate temperatures, often below 200 °C, which are tolerable for many actuator and sensor materials.

Many different types of actuators and sensors and combinations of these have already been embedded and evaluated in laminates of fiber-reinforced polymer composites, on the one hand for the online monitoring of the production,and on the other hand, to simultaneously create functional materials for structural health monitoring. The goal is often to use the sensors both for online process monitoring as well as for structural health monitoring.

Integrated sensor systems can be efficiently used for online process monitoring of fiber-reinforced polymer composites. For example, information about the solidification process of the polymer matrix could be available using in-situ online process monitoring. The component development process can also be carried out more efficiently with real-time process data. The process times for component production and the effort for component quality control can be reduced. Further consequences are that rejects can be reduced and the quality of the component can be increased. Structural health monitoring techniques will enhance the prediction of structural failures, for example, in transportation structures and in buildings. In these applications it is essential to integrate durable, reliable and small transducers, which do not weaken the structure.

In summary, an even higher added-value rate will be generated by integrating smart systems into lightweight structures. This publication presents technologies and systems for the online process and structural health monitoring of fiber-reinforced lightweight structures. Wired and wireless sensor and actuator systems as well as systems for energy harvesting, energy supply and communication devices can be integrated into the structure and material of lightweight systems. A major challenge is the reliable integration of such systems without increasing the weight and affecting the structural integrity of the fiber-reinforced polymer composite structure. One vision is to meet this challenge with functional elements that have nearly the same size as or a smaller size than the reinforcement fiber of the composite material.

In this regard, miniaturized sensors with a fibrous shape, such as optical fiber sensors [2] and miniaturized sensors which have a laminar shape like a part of a strip or a foil, such as film sensors [3], received particular attention because they can be embedded in the fiber-reinforced polymer composite without substantially disturbing the structural integrity due to their shape and their dimensions. These sensor shapes cause the least possible disruption of the laminate structure and minimize unavoidable structural weak points arising from the embedding of the sensors per se and pores, resin-rich zones and undulations of the reinforcement fibers, which can be increased due to the embedding of sensors. The reliable adhesion of the polymer matrix with the sensor material is important in order to achieve long-term stability.

In the case of the optical fiber sensors, the feed line is also an optical fiber. Electrical sensors like dielectrical film sensors usually need electrical conductive wires for electrical contacting if antenna solutions for radio-based signal transmission are excluded. Even in the case of miniaturized dimensions in the decimicrometer range sensors and their feed lines embedded inside the material are foreign objects in the composite because they are typically composed of other materials than the host composite material. The integration of sensors into reinforced polymer composites always leads to unavoidable disturbances of the reinforcement fibers and to additional residual stresses due to material inhomogeneities. As a result, the mechanical strength and stiffness of the structure decrease [4]. Before materials functionalized with sensors can be used industrially, the influence of the sensor on the structural integrity has to be investigated [5]. One of the critical failure modes in fiber-reinforced polymer composite laminates is the interlaminar shear failure, which can be initiated in the short beam shear test [6]. The short beam test method can be used for the quality control of fiber-reinforced polymer composites [7]. The short beam test method can also be used as a destructive test method for the comparison of differently sensorized fiber-reinforced polymer composites [8]. The aim of the short beam test method is to determine the apparent interlaminar shear strength (ILSS) [9].

In this work, the influence of embedded sensors and sensor dummies of different shape, dimension, design and material on the structural integrity of the composite was compared. Sensors with a fibrous shape, miniaturized film sensor dummies and pressure sensors were considered. Related optical fiber sensors [10], miniaturized, dielectric film sensors [11] and pressure sensors [12] have already been used and demonstrated for process monitoring. In this publication, results which were published in the conference proceedings paper [13] are presented in an English version.

2 Materials and methods

2.1 Fabrication of the test specimens

Two carbon fiber unidirectional plates, Laminate 1 (L1) and Laminate 2 (L2), with optical fibers, electrical wires, sensors and dummy sensors were fabricated with the vacuum infusion method. The dimensions of the fabricated panels were 330 mm × 840 mm × 2 mm. Hexcel®HexForce ®G1157 carbon fiber plies and Hexcel Hex-Flow ®RTM6 resin were used to make the panels. Eight plies of G1157 were used for each panel. The unidirectionally stacked plies on an aluminum base plate were infused at 120 °C within a furnace. The resin was preheated at 80 °C and the pressure was 100 mbar at the suction side of the vacuum assembly. The final stage of curing took place for 90 min at 180 °C. The demolded panels were separated with a circular saw into parts with dimensions of 100 mm × 100 mm. The test specimens were cut out of these parts with a wafer saw. The nominal size of the test specimen is 20 mm × 10 mm according to the standard DIN EN 2563: 1997-03 [14]. All of the specimens were labeled such that the assignment of the sample to the origin position in the panel is documented.

In addition to the reference specimens (REFE) without embedments, test specimens were made with optical fibers, various metallic wires, various films and pressure sensors. The embedments were aligned with their long side always parallel to the carbon fibers and embedded between the fourth and fifth G1157 layer. The number of embedded objects per test specimen is indicated in the sample designation by the first number. The designations of the samples in which reference test specimens have been tested thus start with 0-REFE-.

A conventional single-mode optical glass (G) fiber used in telecommunications was embedded in the optical fiber specimens. The polymer coating of the glass fiber, the so-called primary buffer coating, was mechanically removed prior to embedding with a standard glass fiber stripping tool so that a glass fiber with a diameter of 125 μm (G125) was existent for embedding. The glass fiber without the primary buffer coating was cleaned with isopropanol. The electrical conductive wires used were conventional copper (C) wires with a diameter of approximately 250 μm (C250) and polyimidecoated wires with a diameter of 200 μm (C200) from MWS Wire Industries. The wires were cleaned with isopropanol before embedding. Both types of wires were embedded in pairs. Test specimens with only one wire were only prepared with the wire C250. Examples of cross sections of these specimens are shown in Figure 1.

Eleven different designs of polyimide film (F) objects were embedded. The film objects are subdivided into the two categories film sensor and film substrate. Films used were 5 μm thick and perforated. The films of the sensors have a sputtered and structured 5 μm thick layer made of tantalum, which was electrically isolated with a sputtered 5 μm thick layer made of polyimide. The perforated film sensors and film substrates were manufactured by IMSAS. Figure 2(a) shows a film sensor with a 3 mm

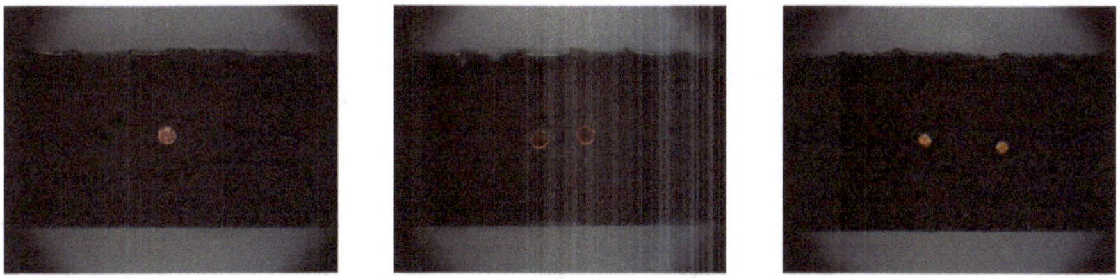

(a) (b) (c)

Figure 1: Cross-sections of specimens with embedded copper wires: (a) 1-C250, (b) 2-C250, (c) 2-C200.

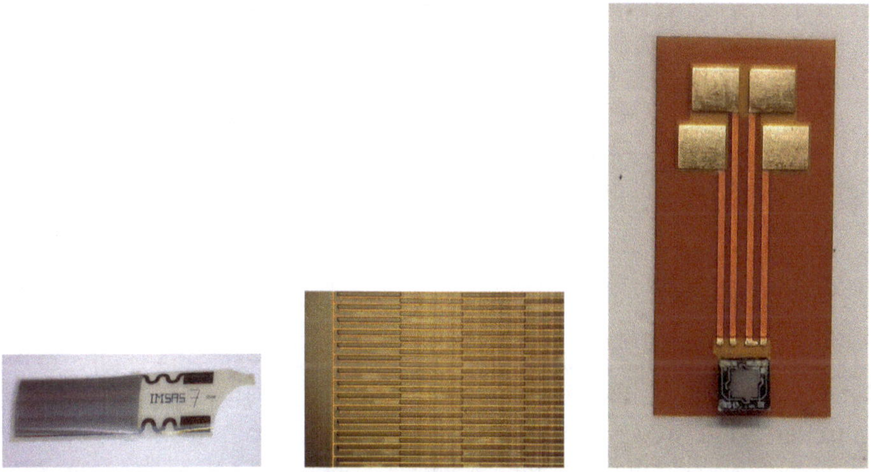

(a) (b) (c)

Figure 2: (a) Film sensor: The right part of the film sensor is not perforated and contains the contact pads for the electrical wires; (b) Perforation of a film sensor in the area of the comb electrode structure; (c) Pressure sensor on a Kapton board.

wide and 6 mm long sensor surface. Figure 2(b) shows an example of the perforation of a film sensor. The perforated film sensors were categorized according to two criteria: the design of the perforation and the dimension of the sensor surface. The characteristic values of the eight different film sensors are listed in Table 1.

The perforated film substrates differ in their proportion of the perforation area. There are three groups with a proportion of the perforation area of 0 % (F00%), 20 % (F20%) and 40 % (F40%), respectively. All film substrates of the three groups had a length of 6.65 mm and a width of 6.40 mm.

The pressure (P) sensors used for embedding were from the first sensor. They consist of a composite of silicon and the borosilicate glass Pyrex. The composite of silicon and Pyrex has a length and width of 1.8 mm and a thickness of 1.2 mm. The pressure sensors were each mounted on a board with a width of 4 mm and a length of 12 mm.

Table 1: Characteristic values of the eight different film sensors.

Name	F051	F052	F053	F054	F071	F072	F101	F102
Length of the sensor area/mm	9	6	9	6	9	6	9	6
Width of the sensor area/mm	2	3	2	3	2	3	2	3
Length of the perforation hole/μm	850	1350	50	50	50	50	50	50
Width of the perforation hole/μm	5	5	5	5	7	7	10	10
Distance between the perforation holes/μm	300	290	50	50	50	50	50	50
Distance between the electrodes/μm	5	5	5	5	7	7	10	10

The board consisted of a flexible Kapton film with a thickness of 80 μm (P080) or of a rigid glass fiber reinforced plastic with a thickness of 200 μm (P200). Figure 2(c) shows a pressure sensor on a Kapton board. The optical fibers and electrical wires were stretched at the intended positions over the corresponding G1157 layer and fixed by the vacuum sealing tape and fixed with an adhesive tape on the aluminium base plate during the lay-up of the laminate.

2.2 Test methods

The apparent interlaminar shear strength (ILSS) can be determined according to the standards DIN EN 2563: 1997-03 [14], DIN EN 2377: 1989-10 [15], DIN EN ISO 14130: 1998-02 [16] and ASTM D 2344/D 2344M: 2016 [17]. The mechanical tests were carried out in accordance with the standard DIN EN 2563: 1997-03. The ILSS was determined from the measured data according to [14] and [17].

The apparent interlaminar shear or short beam strength τ was calculated as follows:

$$\tau = \frac{3P_R}{4bh}, \quad \tau = \begin{cases} \mathrm{ILSS}_{\mathrm{ini}}, & P_R: \text{load at the moment of first failure} \\ \mathrm{ILSS}_{\mathrm{max}}, & P_R: \text{maximumloadobservedduringthetest.} \end{cases} \tag{1}$$

b is the width of the specimen, h is the thickness of the specimen and according to [14] P_R is the load at the moment of first failure and according to [17] P_R is maximum load observed during the test. In this publication τ according to [14] corresponds to $\mathrm{ILSS}_{\mathrm{ini}}$ (ini for initial) and τ according to [17] corresponds to $\mathrm{ILSS}_{\mathrm{max}}$ (max for maximum). The apparent interlaminar shear or short beam strength τ_i of each specimen i of a sample was calculated. If a specimen fails in shear approximately at the neutral fiber axis, the interlaminar shear strength is calculated in accordance with the formula (1) [14].

The sample mean of the apparent interlaminar shear or short beam strength $\bar{\tau}$ ($\overline{\mathrm{ILSS}}_{\mathrm{ini}}$ and $\overline{\mathrm{ILSS}}_{\mathrm{max}}$ of the sample) was calculated. The sample standard deviation s

was calculated as follows:

$$s = \sqrt{\frac{1}{n-1}\sum_{i=1}^{n}(\tau_i - \bar{\tau})^2} \quad (2)$$

where n is the number of specimens. It was not always possible to test five specimens. The number (N) of specimens of each sample is indicated in the sample designation by the last number directly behind the capital N. The sample tests were carried out in three separate measurements (M). The respective measurements M1, M2, M3 are indicated in the sample designation. The specimens were investigated with the X-ray microscope ZEISS Xradia 520 Versa before the destructive test. Xradia 520 Versa enables a spatial resolution of 0.7 µm and voxel sizes below 70 nm.

3 Results

The investigations with the X-ray microscope show that the embedding of objects with a fibrous and a laminar shape neither results in increased pore formation, areas with matrix accumulation, fiber waviness nor increased delaminations in the zone between the embedded object and the structural material. X-ray microscope images of an embedded optical fiber are shown exemplary in Figure 3. Fiber undulations and resin rich pockets were detectable in the zone of the embedding of the pressure sensors, in particular at the edges of the sensor. However, an increased pore formation was not detected. The thickness of each specimen was measured according to the standard. The average thickness of all measured test specimens made from the laminate L1 is 2.2 mm, slightly lower than the average thickness of the test specimens from the laminate L2, which is 2.3 mm. The specimens with the embedded pressure sensors were not taken into account for the determination of the average thickness of the specimens of the laminate L2, since these specimens were significantly thicker. The average thickness of the specimens with an embedded pressure sensor is 2.8 mm. The thickness of almost all specimens of the laminate L2 are above of 2 ± 0.2 mm recommended in the standards for the specimen thickness. In addition, the length-to-thickness ratio of the specimens is not in the recommended range between 5 and 6 [18].

Figure 4(a) shows representative load-displacement curves of various test specimen assemblies. The slope of the curves of the specimens with pressure sensor was smaller than the slope of the other specimens, similar as shown in Figure 4(b). During the mechanical test, one of the pressure sensor specimens slipped over the supports due to its uneven surface so that the measurement results could not be used in the evaluation. The maximum load at the moment of first failure and the maximum load observed during the test were determined from the measured curves. The moment of the first failure in the measurement curves of the reference test specimen and the test

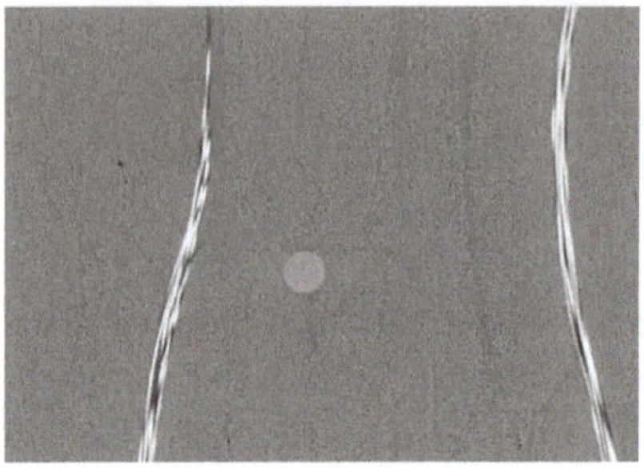

(a)

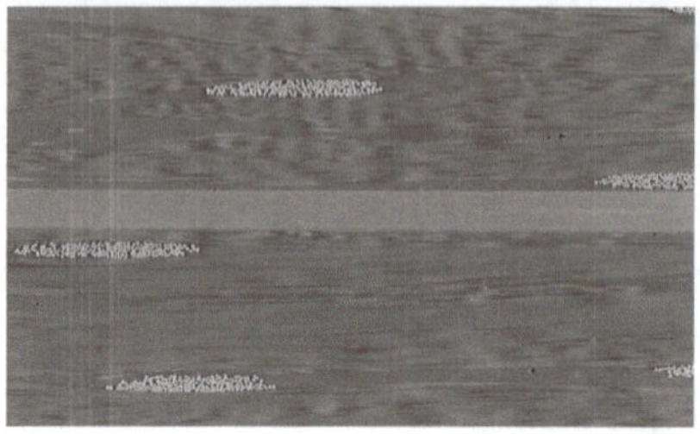

(b)

Figure 3: X-ray micrographs of a test specimen configuration with an embedded optical fiber 1-G125 (voxel size 2.77 μm); (a) The optical fiber is visible in cross-section as a circular object in the center of the image. (b) The optical fiber is shown in longitudinal section.

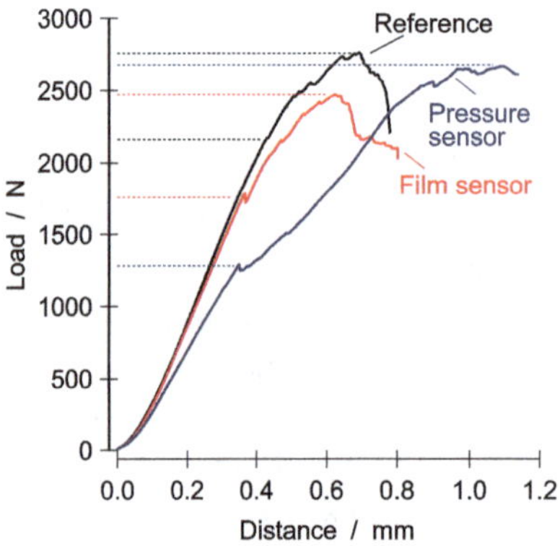

(a)

(b)

Figure 4: (a) Typical load-displacement curves of the short beam tests; in the diagram the forces identified for the calculation of the ILSS are marked. The respective smaller value corresponds to the maximum force at the moment of the first failure used for the calculation of $ILSS_{ini}$. The respective larger value corresponds to the maximum load observed during the test used for the calculation of $ILSS_{max}$. (b) Cross-section of a specimen after loading.

specimen assemblies with optical fibers and electrical wires can not always be clearly defined. The curve of a reference specimen shown in Figure 4(a) is representative for this behavior. However, the moment of the first failure in the test curves of the test specimens with embedded foils and pressure sensors could always be clearly identified. The first failure is usually associated with the beginning of a nonlinear curve progression and an acoustic emission event.

Figure 4(b) shows the cross section of the longitudinal side of a specimen after loading. All tested specimens show the failure mode multiple or single shear as it is shown in Figure 4(b). Failures such as matrix cracks and indents in the contact zone

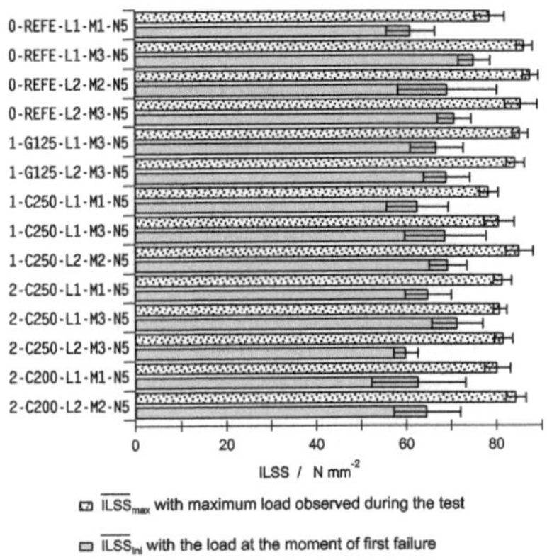

(a)

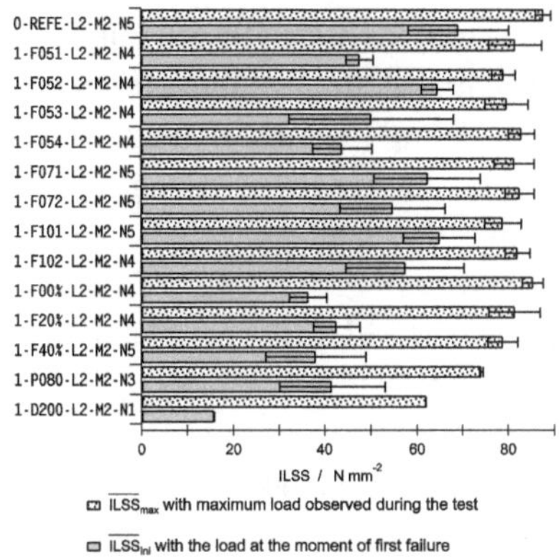

(b)

Figure 5: Mean apparent interlaminar shear strengths and their standard deviation of various samples of different test series: (a) Results of the samples of the test specimens without or with fibrous-shaped embedments, (b) Results of the samples of the test specimens with laminar-shaped embedments and pressure sensors and of the corresponding reference sample 0-REFE-L2-M2-N5.

of the loading nose are also visible. Figure 5 shows the test results of the three-point bending tests. The ILSS obtained from the experiments is used as a parameter to compare the structural integrity of various test specimen configurations and is shown in Figure 5.

4 Discussion

Figure 4(b) makes it clear that different types of failures are superimposed. It is critical to point out that the shear strength determined from the tests cannot be readily attributed to a shear property [9]. The calculated $ILSS_{max}$ therefore represents the ultimate failure through a combination of different failure modes [19]. Other types of failure modes due to superimposed compressive stress or tensile stress [20] may occur. It cannot be clearly stat-ed on which type of failure mode $ILSS_{ini}$ is based. The deviations from linearity of the load-displacement curve starting from the first failure (see Figure 4(a)) may be an indication of viscoelastic effects or matrix failures [21]. It can be assumed that pure shear failure occurs only in lower quality specimens and the real shear strength of higher quality samples cannot be determined in the short beam shear test [22]. Initiating damage in the form of vertical cracks appears to be necessary to induce shear failure [20]. Other types of failure modes can occur through embedding. In particular, in the specimens with the pressure sensors further failure modes can occur because of the increased fiber undulations and resin rich areas in these specimens.

The samples with the pressure sensors have the lowest $\overline{ILSS}_{max}$ values of samples. The undulations of the reinforcement fibers caused by the volume-shaped pressure sensor and the resin rich pockets at the edges of the pressure sensor can be held responsible for the lower strengths. Undulations and waviness of the reinforcement fibers and thickened specimens can be avoided through appropriately sized cut-outs for the sensor in the plies [23]. The results of the samples 0-REFE-L1-C1-N5, 1-C250-L1-M1-N5, 2-C250-L1-M1-N5 and 2-M200-L1-M1-N5 of the measurement M1, shown in Figure 5(a), manifest that the $\overline{ILSS}_{max}$ and $\overline{ILSS}_{ini}$ of the samples do not differ significantly. The $\overline{ILSS}_{max}$ of the samples are between 78.4 N/mm^2 and 80.3 N/mm^2. The $\overline{ILSS}_{ini}$ of the samples are between 61.0 N/mm^2 and 64.9 N/mm^2. It is worth pointing out that the sample 0-REFE-L1-M1-N5, which specimens were without embedments, had the lowest strength.

The $\overline{ILSS}_{max}$ of the samples 0-REFE-L2-M2-N5, 1-C250-L2-M2-N5 and 2-C200-L2-M2-N5 of the measurement M2 are between 84.3 N/mm2 and 87.6 N/mm2. The $\overline{ILSS}_{ini}$ of the samples are between 64.6 N/mm^2 and 69.3 N/mm^2. Thus, the values of the measurement M2 are consistently higher than the corresponding values of the measurement M1. Since only specimens from the laminate L1 were tested in the measurement M1 and only specimens from the L2 laminate in the measurement M2, the differences between the $\overline{ILSS}$ values of the samples of the measurements M1 and M2 may result from differences in the properties of the laminates L1 and L2. Fiber-reinforced polymer composite materials often display differences in their mechanical properties of between 5 and 10 % between two batches [24]. Differences in laminate thicknesses of around 0.1 mm were found between the laminates L1 and L2. Influences due to manual fabrication and embedding as well as changing environmental influences are not completely avoidable. If one considers the $\overline{ILSS}$ values of the measurement M3, in which specimens of the laminate L1 and the laminate L2 were tested, the $\overline{ILSS}$ values displays that no significant differences between the corresponding samples with the same specimen configuration can be recognized. The results of the measurement M3 with the specimens from laminate L1 point the higher $\overline{ILSS}$ values with one exception: $\overline{ILSS}_{max}$ value of the sample 2-C250-L2-M3-N5.

Figure 5(b) displays the results of the samples with the specimens that have films and pressure sensors embedded and the corresponding reference sample 0-REFE-L2-M2-N5. All samples were carried out within the measurement M2. $\overline{ILSS}$ of the sample with the specimens without embedments is the largest of all samples, which meets the expectations. Although the differences in the $\overline{ILSS}$ values of samples with the specimens with the film embedments are in some cases greater than their simple standard deviations, no significant influences can be attributed to the different film designs due to the small number of specimens.

Compared to the samples with the specimens with the fibrous-shaped embedments, the samples with the films only result in lower $\overline{ILSS}_{ini}$ values. In addition, the standard deviations of the $\overline{ILSS}_{ini}$ of samples with the films are generally significantly greater than those with the fibrous-shaped embedments, which cannot be explained only by the smaller number of specimens per sample. It is worth mentioning that the samples 1-F00%-L2-M2-N4, 1-F20%-L2-M2-N4 and 1-F40%-L2-M2-N4 with the 6.40 mm wide perforated films have similarly relatively low $\overline{ILSS}_{ini}$ values. An influence of the different proportions of the perforation is not recognizable. The relatively low $\overline{ILSS}_{ini}$ values can be explained by the relatively large film width of 6.40 mm compared to the specimen width of 10 mm. In particular, the nonperforated contacting area (see Figure 2(a)) of the sensor substrate film could be responsible for the relatively low $\overline{ILSS}_{ini}$. The results generally show higher $\overline{ILSS}_{ini}$ for the 2 mm wide films compared to the corresponding 3 mm wide films. The same behavior was observed in the investigations presented in [25]. One exception to this is the sample pair 1-F051-L2-M2-N4 and 1-F052-L2-M2-N4. In this case, the sample 1-F052-L2-M2-N4 with the 3 mm wide film has a higher $\overline{ILSS}_{ini}$. However, in contrast to the other sample pairs not only the film dimensions but also the perforations differ in this case (see Table 1). It is noted that in [25] the films used, were perforated over the entire surface, unlike the films used in this work.

In summary, it can be stated that with the present results it is not possible to assess which film configuration has the least influence on the structural integrity. All embedded films cover a relatively large area, therefore the results cannot be transferred directly to large outer skin components. Further approaches for the improvement of the structural integrity are presented in [26].

5 Conclusion and outlook

In this work, the effects of embedding different films, metal wires, optical fibers and pressure sensors on the structural integrity of fiber-reinforced polymer composite materials were investigated. The focus was on mechanical experiments according to the short beam shear test method. The samples with the pressure sensors have the lowest $\overline{ILSS}_{max}$ values of samples. The usability of the values is limited due to the small number of specimen used. In the results of the samples with the metal wires and optical fibers, an influence of the embedments on the $\overline{ILSS}$ is not identifiable. Only with the $\overline{ILSS}_{ini}$ values of the samples with the films, a weakening of the structural integrity of the test specimen with films is identifiable. However, which design of the film has the least influence on the structural integrity cannot be assessed with the present results, since the scattering of the measured values is too large. With a larger number of specimens per sample, the results should be become more meaningful. X-μCT studies showed that the amount of pres does not increase if embedments are present. In order to gain further insights into the failure behavior of fiber-reinforced polymer composite materials with embedded sensors, extended X-μCT investigations are planned. Furthermore, fatigue tests are planned, which better simulate the real operating conditions than in quasistatic tests.

Bibliography

[1] Schürmann, H.: Konstruieren mit Faser-Kunststoff-Verbunden, 2., bearbeitete und erweiterte Auflage. Springer, Berlin Heidelberg (2007).
[2] Di Sante, R.: Fibre Optic Sensors for Structural Health Monitoring of Aircraft Composite Structures: Recent Advances and Applications. Sensors 15, 18666 (2015).
[3] Boll, D., Schubert, K., Brauner, C., Lang, W.: Miniaturized flexible interdigital sensor for in situ dielectric cure monitoring of composite materials. IEEE Sens. J. 14(7), 2193–2197 (2014).
[4] Kim, K.-S., Breslauer, M., Springer, G. S.: The effect of embedded sensors on the strength of composite lami-nates. J. Reinf. Plast. Compos. 2, 949–958 (1992).
[5] Warkentin, D. J., Crawley, E. F.: Embedded electronics for intelligent structures. In: Proceedings of the 32nd Structures, Structural Dynamics, and Materials Conference, Structures, Structural Dynamics, and Materials and Co-located Conferences, Baltimore, MD, U.S.A., pp. 1322-–1331 (1991).
[6] Mallick, P. K.: Fiber-reinforced composites: materials, manufacturing, and design, 3rd edn. CRC Press Taylor & Francis Group, Boca Raton (2008).
[7] Abdallah, M. G.: Review of the state-of-the-art of advanced composite interlaminar shear test methods. In: SEM Spring Conference on Experimental Mechanics, Proceedings (A86-48676), New Orleans, LA, June 8-–13, 1986, vol. 301-–313, pp. 24–35. Society for Experimental Mechanics, Inc., Bethel, CT (1986).
[8] Partridge, I. K., Maistros, G. M.: Dielectric cure monitoring for process control. Editors-in-Chief: Kelly, A., Zweben, C. In: Carlsson, L., Crane, R. L., Uchino, K. (eds.) Comprehensive Composite Materials, Test Methods, Nondestructive Evaluation, and Smart Materials, vol. 5, pp. 413–426. Elsevier Science Ltd. (2000).
[9] Schaaf, K. L.: Composite materials with integrated embedded sensing networks. Dissertation, University of California (2008).
[10] Gaitzsch, R. R., Koerdt, M., Brauner, C., Kroll, L., Herrmann, A. S.: Signal evaluation of fibre optical sensors embedded between unidirectional thermoplastic prepreg tapes in a hot-press consolidation for online process monitoring. In: ICCM20 – 20th International Conference on Composite Materials, July 19-–24, Copenhagen, Denmark (2015).
[11] Kahali Moghaddam, M., Breede, A., Chaloupka, A., Bödecker, A., Habben, C., Meyer, E.-M., Brauner, C., Lang, W.: Design, fabrication and embedding of microscale interdigital sensors for real-time cure monitoring during composite manufacturing. Sens. Actuators A, Phys. 243, 123 (2016).
[12] Kahali Moghaddam, M., Breede, A., Brauner, C., Lang, W.: Embedding piezoresistive pressure sensors to obtain online pressure profiles inside fiber composite laminates. Sensors 15, 7499 (2015).
[13] Koerdt, M., Hübner, M., Kahali Moghaddam, M., Lang, W., Herrmann, A. S.: Integration von minimalinvasiven Sensoren in Faserkunststoffverbunde zur Herstellungsprozesskontrolle und Zustandsüberwachung. In: Wiedemann, M., Melz, T. (eds.) Smarte Strukturen und Systeme, Tagungsband des 4SMARTS-Symposiums, June 21-–22, 2017, pp. 315-–326. Shaker Verlag Aachen, Braunschweig, (2017).
[14] Aerospace series – Carbon fibre reinforced plastics – Unidirectional laminates – Determination of the apparent interlaminar shear strength, German version: Luft- und Raumfahrt – Kohlenstoffaserverstärkte Kunststoffe – Unidirektionale Laminate, Bestimmung der scheinbaren interlaminaren Scherfestigkeit. Deutsche Fassung EN 2563 (1997).
[15] Aerospace series – Glass fibre reinforced plastics, Test method, Determination of the apparent interlaminar shear strength, German version: Luft- und Raumfahrt, Glasfaserverstärkte Kunststoffe, Prüfverfahren zur Bestimmung der scheinbaren interlaminaren Scherfestigkeit. Deutsche Fassung EN 2377 (1989).

[16] Fibre-reinforced plastic composites – Determination of the apparent interlaminar shear strength by short beam method, German version: Faserverstärkte Kunststoffe – Bestimmung der scheinbaren interlaminaren Scherfestigkeit nach dem Dreipunktverfahren mit kurzem Balken (ISO 14130: 1997), Deutsche Fassung EN ISO 14130 (1997).

[17] Standard test method for short-beam strength of polymer matrix composite materials and their laminates. ASTM Desig-nation: D 2344/D 2344M – 00 (2016).

[18] Chaterjee, S., Adams, D., Oplinger, D. W.: Shear test methods. In: Test methods for composites, a status report, Volume III, DOT/FAA/CT-93/17, U.S. Department of Transportation, Federal Aviation Administration, Final Report (1993).

[19] Fan, Z., Santare, M. H., Advani, S. G.: Interlaminar shear strength of glass fiber reinforced epoxy composites enhanced with multi-walled carbon nanotubes. Composites, Part A 39, 540 (2008).

[20] Whitney, J. M., Browning, C. E.: On short-beam shear tests for composite materials. Exp. Mech. 25(3), 294–-300 (1985).

[21] Carlsson, L. A., Gillespie, J. W., Trethewey, B. R.: Mode II interlaminar fracture of graphite/epoxy and graphite/PEEK. J. Reinf. Plast. Compos. 5, 170–187 (1986).

[22] Stinchcomb, W. W., Henneke, E. G., Price, H. L.: Use of the short-beam shear test for quality control of graphite-polymide laminates. In: Reproducibility and Accuracy of Mechanical Tests. A symposium sponsored by ASTM Committee E-28 on Mechanical Testing, American Society for Testing and Materials, St. Louis, Mo., 5 May 1976. ASTM Special Technical Publication, vol. 626, pp. 96-–109. ASTM, Philadelphia, Pa. (1977).

[23] Wierach, P.: Nano-micro-macro. In: Wiedemann, M., Sinapius, M. (eds.) Adaptive, tolerant and efficient composite structures, pp. 17–-27. Springer, Berlin Heidelberg (2013).

[24] Davidson, R., Roberts, S.: Optical fiber sensor compatibility and integration with composite materials. Editors-in-Chief: Kelly, A., Zweben, C. In: Carlsson, L., Crane, R. L., Uchino, K. (ed.) Comprehensive Composite Materials, Test Methods, Nondestructive Evaluation, and Smart Materials, vol. 5, pp. 591–606. Elsevier Science Ltd. (2000).

[25] Kahali Moghaddam, M., Boll, D., Lang, W.: Embedding rigid and flexible inlays in carbon fiber reinforced plas-tics. In: Proceedings of the 2014 IEEE/ASME International Conference on Advanced Intelligent Mechatronics (AIM), Besançon, France, July 8–-11, pp. 1387-–1392 (2014).

[26] Kahali Moghaddam, M., Hellmann, J., Lang, W.: Plasticisation of epoxy resin transfer molding substrate for fabrication of interdigital capacitive sensors. In: 30th Eurosensors Conference, EUROSENSORS 2016. Procedia Engineering, vol. 168, pp. 1110–-1113 (2016).

Silvia Titotto

The role of biologically inspired design to 4D printing development

Abstract: Additive manufacturing allows the assembly of objects layer-by-layer via ejection of proper material. It leads to an accurate programming of the properties as well as dimensions of intricate printed objects. Whereas 3D printing manufactures motionless structures, the novel 4D printing empowers the fabrication of animated structures. In this kind of additive manufacturing, the printed objects change their shape over time via passive energy while in contact with certain external stimuli. Therefore, it has deep relations with the self-assembly concept that is a process frequently experienced in nature in that the disordered parts fabricate an organized structure via local association. The reason for that is those beings are prone to reorder their shape according to specific challenges imposed by the environment itself and by customized needs for growth and survival along their life span. Technological innovative solutions can be created from the observation of living beings' biomechanics and biological systems that are responsive to environmental stimuli. Avenues for that are the hybrid technologies that enable new functionalities where shape adaptation and reaction to stimuli are key requirements. This may happen via the partnership between deployment mechanisms and sustainable materials. This field totally integrates the digital creation process and manufacturing with materials design within the context of self-assembly. Bioinspired self-assembly has already enabled advances in applications in fields such as material science, medicine, software, robotics, transportation, manufacturing, infrastructure and aerospace. It is also expected to impact even the food sector and the creative industry in the next few years.

Keywords: Additive manufacturing, 4D printing, bioinspired technologies, deployable structures, sustainability

Acknowledgement: The research was supported by the Brazilian National Council for Scientific and Technological Development (CNPq) research group "4D Printing & Biomimetics" (4DB) at the Center of Engineering, Modeling and Applied Social Sciences (CECS) of the Federal University of ABC (UFABC). It was partially funded by CAPES (for the BRAGFOST conference in 2017) and the Alexander von Humboldt Foundation (for the BRAGFOST conference in 2017 and for the Follow-up Program CONNECT at Fraunhofer Institute for Mechanics of Materials in 2018).

Silvia Titotto, Federal University of ABC (UFABC), Santo André, Brazil, e-mail: silvia.titotto@ufabc.edu.br

https://doi.org/10.1515/9783110584455-013

1 Introduction

Very recent researches have pointed out that the creation of hybrid technologies via additive manufacturing that enable new functionalities for shape adaptation and reaction to stimuli may be the shortest avenue to reach solutions to kinetic objects that demand a responsive behavior. One of the most well-succeeded shape alteration methods is via self-assembly. It is a process frequently experienced in nature in that the disordered parts fabricate an organized structure via local association.

The chapter aims to discuss possibilities of additive manufacturing of kinetic objects inspired in biological morphological changes as reactions to environmental stimuli. 3D printing is a kind of additive manufacturing that allows the fabrication of objects using layer-by-layer material deposition, leading to accurate programming of some geometrical properties as well as the dimensions of intricate printed objects.

Whereas 3D printing manufactures motionless objects, the prominent technology of 4D printing permits the fabrication of kinetic structures that alter their form over time while kept in contact with certain external stimuli. It also agrees with sustainability principles as it demands very low to none energy expenditure.

First, this chapter introduces concepts related to the three-dimensional additive manufacturing. In sequence, it presents the contextualization of the biologically inspirational factor for the surge of 4D printing. Then it examines some mechanical mechanisms that are used to design shape alteration such as deployability and sustainable possibilities of materials for 4D printing. From that, the specific case of bioinspired polymer robots is introduced as an example of dual relationship to nature via bioinspiration, as their form and function merge along with movement. Finally, the chapter mentions the expansion of biologically inspired solutions from more traditional fields such as engineering to surprising examples in the food and creative industries. Then the chapter is concluded via final considerations.

2 The additive manufacturing background and main developments in 3D

Although additive manufacturing technologies have been largely accessed by the public in the last decade, the concept of 3D printing was laid out in a New Scientist article in 1974 [1], the first additive methods for fabricating three-dimensional models with photo-hardening thermoset polymer were created in 1981 [2].

Charles Hull initially came up with the idea to use UV light in order to solidify tabletop coatings in 1983 although the first patents for both stereolithography processes and fabrication system were filed in 1984 [3]: in July 1984, the French inventors

Olivier de Witte, Jean Claude André, and Alain Le Méhauté used their patent for stereolithography [4] and a few weeks later Chuck Hull deposited his own patent for the stereolithography process. In 1986, commercial rapid prototyping became available via his company 3D Systems in California [5].

The major characteristic of additive manufacturing (AM) techniques link to the layer-by-layer assembly pattern, that literally casts building material throughout tiers that must have been previously computer designed.

It results in a great design spectrum, versatility and precision for construction of complex objects, ranging from the overall geometry of a part to the topology of its internal architecture to the spatial distribution of its material composition [6].

One of the main advantages of this kind of technology is that it also reduces the total number of process steps if in comparison to other usual prototyping systems, which lead to costly processes. This is especially significant if custom designs are required.

In standard 3D printing systems, materials are expected to be stable and fabricated models are designed and printed as static or dead objects. Technologies to create inert 3D objects kept progressing mainly strengthened by academia motivation and industry interests, a number of researchers started to move forward to assess how more impact would derive from active structures with responsiveness to natural and environmental stimuli via movement along time [7, 8].

Recently, a broad diversity of soft and advanced materials of contrasting programmable functionalities has been established for additive manufacturing to create intricate designs using a smart and moldless technique. But several of these materials had only static structures as their printing outcome and that restricted applications in which time evolving form alteration is essential.

3 Biologically inspired factor for the surge of 4D printing

Although a number of authors had already referred to correlation between advancements in additive manufacturing and biologically inspired engineering [9, 10], it was via predicting connections among several self-evolving and self-assembly systems observed in the natural biological environments and additive manufacturing projects that Skylar Tibbits pioneered the field. He performed a variety of conceptual tests at the Massachusetts Institute of Technology (MIT) since 2011 (Tibbits, 2017; Tibbits, 2014) and introduced the concept of 4D printing to the broad public in his TED talk in the year of 2013.

In terms of Physics, 4D, meaning the 4 common dimensions, usually alludes to three-dimensional space (3D) and this builds on the dimension of time to the other three dimensions of length, width and depth.

The fourth dimension in 4D printing emerges from the capacity of the printed objects to transform their functionality and/or shape over time while being kept exposed to certain environmental stimuli as well as during their post-printed existence [11]. It means the fourth dimension is also understood as time because the printed structure has the ability to change its shape, functionality and properties across time when it is in contact with environmental stimuli, for example, temperature, ion concentration, pH, magnetic and electric fields, light or even hydration/dehydration [12, 13].

The pioneers of 4D printing unfolded the field by associating it to specific stimuli throughout passive energy. It means that the structure was not to be placed after active sources of energy to calibrate its flown to the system [11]. Recently, some research groups began to consider the probability of using diversified energy sources. It might include active energy to be applied locally [11].

In this case, to resist responsiveness and stimuli variation, one of the most important factors is the capability to return on their own to the primary arranged status. On the other hand, in a few special cases a partial reversion cycle may really change the reaction and the properties of the material. Anyway, the response and reversion processes are usually expected to not happen in the same amount of time (Pei, 2014) as it can be observed in many examples in botany and also invertebrate sea life.

4 Mechanical mechanisms for shape alteration in 4D printing

Although deployable structures are most studied for aerospace applications, there are recent indirect indications that interesting new paths for technological innovations in deployable type devices can be opened to 4D printing via the study the structural morphology of strands of spider webs or membranes supported by branched structures, such as insect and chiroptera wings, which are structures subject to large displacements and deformations.

Such biological structures have structural synthesis, lightness, tenacity, flexibility, adaptability and portability, which are quite desirable characteristics for applications with space and mass limitations, in addition to being exposed to random dynamic excitations and hostile environments.

Deployable mechanisms are mainly necessary to create weight-efficient spatial structures that can be either large or small. They follow deployment principles that ensure design effectiveness such as design and functionality, behavior and control [14].

Up to date, the most used deployment systems are basically put together in three main clusters: the pantographic structures, the inflatable systems and the elastically deformed systems. Their mostly rigid structures are more easily visualized in 3D final configurations in conventional applications but the key concept is understanding the role of the deployment as a kind of shape alteration across the time dimension.

(a) Pantographic systems
Also known as manifold structures, they are usually built of rigid parts to be connected via hinges. In most cases, the hinges are spring driven and they benefit from a locking mechanism which fixes the hinge after achieving their final configuration [15].

(b) Inflatable systems
Inflatables are usually membranes made of a slim and light material but they need a system to inflate them so they are taken into account for the weight calculation. When an inflatable structure is required to keep stiffness during a long time, the system must maintain the pressure along the whole time or the overall object must be tensioned post-deployment when it reaches the final configuration so that it keeps rigid [15].

(c) Elastically deformed systems
Masts are usually stowed by twisting and the boom is under compression, so most of the times the deployment is controlled by a nut under a rotating state. The functionalities of hinge and hinge-driving spring are almost always related to the structure to prevent weight load and to reduce the system intrication. This increases the system reliability. Other advantages of this concept are the lack of friction that might happen as thermo-mechanical effects [15].

The application of biomimetic concepts to the deployable class structures can optimize their performance in several applications from the conceptualization to the final discard.

5 Sustainable possibilities of materials for 4D printing

The blossoming of 4D printing has brought to light many possibilities for the development of smart printing by the use of materials that are designed to forecast alteration processes.

Shape memory polymers are already reasonably well known for their application in 4D printing [13]. A late innovation that is expected to revolutionize the deployables design is the improvement of programming for shape memory materials that could be applied to the design of deployment mechanisms [16].

Shape memory devices are prone to eliminate the need for traditionally complex mechanical deployment systems as well as systems that independently control deployment. Shape memory mechanisms may also reveal very concise designs that include less nonstructural parts. This way they are expected to be considerably lighter in weight.

Elastic Memory Composites (EMC) belong to the family of shape memory materials. The greatest advantages of EMC materials in comparison to shape memory alloys and shape memory ceramics are that they are a lot less dense, they have higher damping and also higher strain capacities. Hence EMC materials present desirable characteristics for deployable spatial objects and structures, and for this reason they have been so much investigated within the additive manufacturing research field.

Apart from these polymers with shape memory characteristics, hydrogels are another type of polymers that have been seen as very important to the development and applicability of 4D printing [17]. In fact, to date, smart hydrogels are the favorite materials for the sensitivity and responsiveness to environmental stimuli, but nonresponsive hydrogels are also relevant to researchers as they may expand the range of materials.

Another reason for that is obtaining programmed and complex shape alterations by dehydration/rehydration via adjusting internal structures of cross-linking and density or via their combination with other kinds of materials. In reality for most smart hydrogels, the stimuli produce swelling alterations. This advantage has been used to generate shape transformations in hydrogel sheets built under traditional techniques and it seems to be the reason for being chosen as a kind of favorite for 4D printing [18, 19].

6 Bioinspired robots

So that multifunctionality of organisms is achieved, soft biomimetic robots are expected to match the feeling and the programming under passive as well as active mechanics.

Therefore, 4D printed actuators are relevant to the relatively novice field of soft robotics. Soft actuators must be strong and flexible besides mechanically stable so that after a number of cycles of stimuli application, they keep the same form alteration amplitude beyond their mechanical properties over time and they do not risk disintegrating.

Hydrogels are hydrophilic and stimuli-responsive, on the other hand, the conventional soft actuators made of silicone are not [20]. Several of the soft robots built via 4D printing technology react to ions concentration and magnetic field, but the most successful parameter to date has been temperature. They convert energy triggered by the external stimuli to promote specific duties, such as those that can be performed by artificial muscles, walking robots and activating grippers.

Soft robots have been more known for being able to reach remote areas of a body during a surgery intervention for instance, as the robot is used to traverse an orifice to perforate a chosen organ, after that it self-deploys or self-locks to perform a specific endeavor. But several other applications are being investigated very recently so that fields such as construction and precision manufacturing are benefited.

7 From bioinspired engineering to furniture design, architecture and installation artworks

Several biological organisms have tissue and microstructural compositions whose dynamic morphologies can alter form on demand to alterations to their own environments. In the biologically natural world, such kinds of changes are carried out by retracting leaves (Mimosa pudica L.), by tendrils (Macfadyena), by the loss of pigmentation of the skeletal flowers (Diphylleia grayi), etc.

In the last years, one of the research trends at the Institute for Biologically Inspired Engineering at Harvard University has been to mimic biological structures via hydrogel composites or multimaterials printed in 4D, which alters in a dynamical way as an answer to alterations in temperature or humidity [21, 22].

The Self-assembly Lab at the International Design Center of the Massachusetts Institute of Technology (MIT) has pioneered in examining the self-assembly phenomenon for 4D printing. Skylar Tibbits, who is an architect and his colleagues departed from the concept that those systems are scale-independent and they can be used for self-constructing and manufacturing systems, from a simple set of responsive building blocks, energy and interactions that can be designed in a wide range of materials and machining processes. This can happen in the case of additive manufacturing, where programmable materials get involved in environmental stimuli, and also via the use of vernacular techniques whose results are generated randomly. For instance, it is the case of agitation of particles that Tibbits experimented and exhibited.

Since the beginning, architecture has been widely in touch with additive manufacturing research from both conception and implementation approaches. It is an engaging field for designing novel shape-morphing solutions in the near future. The usage of 4D printing into built structures permits the architecture to respond to environmental conditions (such as light, humidity, etc.) as an interface between the synthetic matter and climate conditions [23].

To date, three major directions have been recognized as applications for 4D printed structures to building design. Khoo and Shin have chosen heat-sensitive stretchable hydrogel skeletons for their experimentations towards a soft and evolving architecture implementation [23]:

(a) improving glazing and window performance via a thermosensitive hydrogel between two glasses sheets. The thermally sensitive polymer becomes opaque under heat and works as a thermotropic shading [23].

(b) to create new architectural skeletons for existing building components for temperature regulation and energy consumption reduction [23].
(c) to use printed polymers as a responsive cover for building façades under the role of structure, sensor and actuator [23].

Those insights may be perceived via immersive installation artworks that respond and that may even evolve according to the movements of people around them. Some of that kind of bioinspired responsive installation artworks are powered by protocells as some of the artworks conceived by the architect Philip Beesley. The translation of the concept of "metabolic architectures" can be understood as artificial four-dimensional living structures that may be even capable of breathing, regenerating and growing, instead of inanimate, fixed objects displayed at a white cube of a museum architecture. Therefore revision in depth of the concepts of naturalness and artificiality urges more investigation and one way of doing so is via experimental disruptive bioinspired design.

8 Final considerations

This chapter discussed technological innovative solutions from observations of biological structures and systems that respond to environmental stimuli. This research topic was built on the vision to seamlessly integrate materials design with digital fabrication processes within the context of self-assembly manufacturing.

Bioinspired self-assembly undertakes enabling more and more advances in a variety of applications in very diverse fields such as material science, biology, computation, medicine, manufacturing, robotics, infrastructure, transportation, aerospace and construction industries within the next decade, as well as the creation of new categories in the creative industry such as living artworks that alter shape and behavior according to their surrounding environment.

As per definition 4D printing demands alteration of form and/or function of the printed object over time, improvements in material properties are still demanded for each application to be sustainable and to continue to be efficient over a number of cycles. This way, self-healing, biodegradation, speed response, actuation capability, besides improved toughness should become mainstream characteristics for self-assembled 4D printed objects in the future.

The potentiality of 4D printing was approved for tissue engineering and soft actuators but in order to kick off printed devices to many if not all the other fields of engineering and creative industries, more technology integration of additive manufacturing, programmable materials and parametric design is highly needed.

Interdisciplinarity and transdisciplinarity are ways of giving continuous support and input to researchers from different fields to work together and to investigate several design possibilities based on people's lifestyles in different contexts.

Everyday more researchers seem to be interested in investigating 4D printing for innovative applications and their access to a wider range of materials encourages the prospect of performing kinetic, adaptive and performative accomplishments via unhackneyed design possibilities. Real-time decision-making programmable 4D printing models may emerge as potential future development, revealing 5D printing as a frontier technology.

Declaration of interest: The author wishes to confirm that there are no known conflicts of interest associated with this publication.

Bibliography

[1] Jones, D.: Ariadne column. New Sci. October 3rd, 80 (1974).

[2] Kodama, H.: Automatic method for fabricating a three dimensional plastic model with photo hardening polymer. Rev. Sci. Instrum. 52, 1770 (1981). https://doi.org/10.1063/1.1136492.

[3] Mendoza, A. J.: Legal and Social Implications of the 3D Printing Revolution (2015). CMC Senior Theses. Paper 1032 http://scholarship.claremont.edu/cmc_theses/1032..

[4] Andre, J. C.: Disdpositif pour realiser un modele de piece industrielle – Notice FR2567668. Institut National De La Propriété Industrielle – Demande de Brevet dÎnvention (1984). https://bases-brevets.inpi.fr/fr/document/FR2567668/publications.html, Accessed 06 Oct 2017..

[5] Ponsford, M.: The night I invented 3D printing. Cable News Network (2014). https://edition.cnn.com/2014/02/13/tech/innovation/the-night-i-invented-3d-printing-chuck-hall/ Accessed 06 Oct 2017..

[6] Seepersad, C. C.: Challenges and Opportunities in Design for Additive Manufacturing. 3D Printing Addit. Manuf. 1(1), 10–13 (2014). https://doi.org/10.1089/3dp.2013.0006.

[7] Raviv, D., Skylar, T., et al.: Active Printed Materials for Complex Self-Evolving Deformations. Sci. Rep. 4, 7422 (2014).

[8] Raviv, D., Zhao, W., McKnelly, C., Papadopoulou, A., Kadambi, A., Shi, B., Hirsch, S., Dikovsky, D., Zyracki, M., Olguin, C., Raskar, R., Tibbits, S.: Active Printed Materials for Complex Self-Evolving Deformations. Sci. Rep. 4, 7422 (2014). https://doi.org/10.1038/srep07422.

[9] Fratzl, P., Burgert, I.: Actuation systems in plants as prototypes for bioinspired devices. Philos. Trans. R. Soc. Lond. A 6, 1541–1557 (2009).

[10] Ionov, L.: Bioinspired micro origami by self-folding polymer films. Macromol. Chem. Phys. 214, 1178_1183 (2012).

[11] Ge, Q., Sakhaei, A. H., Lee, H., Dunn, C. K., Fang, N. X., Dunn, M. L.: Multimaterial 4D Printing with Tailorable Shape Memory Polymers. Sci. Rep. 6, 31110 (2016). https://doi.org/10.1038/srep31110.

[12] Momeni, F., Mehdi Hassani, S. M. N., Liu, X., Ni, J.: A review of 4D printing. Mater. Des. 122, 42–79 (2017). https://doi.org/10.1016/j.matdes.2017.02.068.

[13] Ge, Q., Qi, H. J., Dunn, M. L.: Active materials by four-dimension printing. Appl. Phys. Lett. 103, 131901 (2013).

[14] Fetchko, P., et al.: Deployment Optimization of a Boom for FalconSAT-3 using Elastic Memory Composite Material. In: 18th Annual AIAA/ USU Conference on Small Satellites. SSC04-VIII-4 (2004).
[15] Straubel, M., Block, J., Sinapius, M., Hühne, C.: Deployable Composite Booms for Various Gossamer Space Structures. In: 52nd AIAA/ASME/ASCE/AHS/ASC Structures, Structural Dynamics and Materials Conference (2011).
[16] Liu, Y., et al.: Smart Mater. Struct. 23, 023001 (2014). https://doi.org/10.1088/0964-1726/23/2/023001.
[17] Shin, D. G., Kim, T. H., Kim, D. E.: Int. J. Pr. Eng. Man-GT 4, 349 (2017).
[18] Ding, Z., Yuan, C., Peng, X., Wang, T., Qi, H. J., Dunn, M. L.: Sci. Adv., 3, e1602890 (2017).
[19] Thérien-Aubin, H., Wu, Z. L., Nie, Z., Kumacheva, E.: J. Am. Chem. Soc. 135, 4834 (2013).
[20] Wang, N., Li, Y., Zhang, Y., Liao, Y., Liu, W.: Langmuir 30, 11823 (2014).
[21] Gladman, A. S., Matsumoto, E. A., Nuzzo, R. G., Mahadevan, L., Lewis, J. A.: Biomimetic 4D printing. Nat. Mater. 15(4), 413–418 (2016). https://doi.org/10.1038/nmat4544.
[22] Gladman, A. S., Matsumoto, E. A., Nuzzo, R. G., Mahadevan, L., Lewis, J. A.: Biomimetic 4D printing – Supplementary Information. Nat. Mater. 15(4), 413–418 (2016). https://doi.org/10.1038/nmat4544.
[23] Markstedt, K., Mantas, A., Tournier, I., Martinez Avila, H., Hägg, D., Gatenholm, P.: Biomacromolecules 16, 1489 (2015).

Elaine Yoshiko Matsubara, Maria Isabel Felisberti, and
José Mauricio Rosolen

Influence of different carbon nanotubes types in dynamic-mechanical properties of lightweight carbon felt/CNTs composites

Abstract: The improvement of the quality of the interface between carbon nanotubes (CNTs) and polymers is an important issue that has a lot of challengers to be solved. The use of CNTs in the preparation of a hybrid composite of carbon felt/CNTs/Polymer provides a means of controlling all the special and important properties of these nanomaterials. This happens because the dispersion of CNTs in polymers (as thermoplastics, silicone, and epoxy-resins) is not an easy task mainly due to the van der Walls interactions between CNTs themselves and their weak interaction with polymers in the pristine form. The use of carbon felt/CNTs as a hybrid composite matrix for the improvement and development of polymer composites and its mechanical properties have never been studied before in this hybrid configuration, where the composite is prepared by the Chemical Vapor Deposition method (CVD) and the CNTs cover all the surface of the fiber. The effect of the CNTs type (e. g., SWCNT, MWCNT, Cup-Stacked CNTs, Helical CNTs, DWCNT and others) in the properties of the polymer nanocomposite is not studied yet. In this research report, we present the study of the carbon felt/CNTs/Polymer Hybrid Composite on dynamic conditions and their mechanical behavior using two different types and less studied CNTs: CSCNTs and helical CNTs.

Keywords: Carbon nanotubes composites, carbon felt, lightweight composites, mechanical reinforcement

1 Introduction

Owing to their unique mechanical properties, Carbon Nanotubes (CNTs) are very well-rated materials to the improvement of mechanical features of several polymers and

Acknowledgement: The authors would like to thank FAPESP (06/07253-8 and 2017/04759-2) and CAPES (PROEX 1869 – PNPD and PROEX 0487 – Master fellowship) for the financial support. The electron microscopy work has been performed with the TEM-MSC JEOL 2100 or SEM-FEG JEOL 6330 microscopes of the Brazilian Nanotechnology National Laboratory LNNano – Campinas.

Elaine Yoshiko Matsubara, José Mauricio Rosolen, Chemistry Department, FFCLRP – University of São Paulo, Ribeirão Preto, SP, Brazil, e-mail: elainematsubara@yahoo.com
Maria Isabel Felisberti, Chemistry Institute, IQ – University of Campinas, Campinas, SP, Brazil

https://doi.org/10.1515/9783110584455-014

lightweight composites [1–6], mainly in the recent scenario in which the costs to produce or buy them are in constant decreasing. Nevertheless, there are a lot of problems to study and elucidate in order to be possible that the use of CNTs shows their full potential. he effective processing of CNTs and polymers to fabricate new advanced and lightweight composites materials are still a great challenge. There are several mixing methods described in the literature, from the simple mechanical mixing to the in situ polymerization technique, where the polymer is prepared in the presence of CNTs, producing composites with higher CNTs loading. Although all these techniques may contribute to produce composites with higher elastic modulus than the original polymers, the final material is still not so good as expected, mainly taking into account the intrinsic (and excellent) properties of CNTs. The final composite prepared with CNTs have properties derived from CNT-CNT and CNT-polymer interfaces and all the electronic interactions between the materials [7–9]. In the particular case of polymeric composites and reinforcement fillers, the final product must be versatile and lightweight as a traditional polymer, but with enhanced properties, as better stiffness and higher strength [10]. Besides that, the addition of CNTs can provide other properties to the composite, as electrical conductivity [11, 12]. The composites constituted by polymers and CNTs are classified as fiber reinforced composites (FRC). This type of composite tends to be very efficient on the improvement of the properties because of the micrometric fibers of carbon felt and the nanometric ones, due to the presence of CNTs. They are about the same size of the defects along the polymeric matrix and this leads to a superior mechanical strength. However, this class of composites is very anisotropic and is strongly dependent on the direction and alignment of the fibers. The strength tends to be at its most in the parallel direction of the fibers. Thereby, according to the FRC micromechanics theory, the final values of the mechanical parameters of the composite, e. g., the Young's modulus, are the result of the perfect interfacial bonding and the uniform distribution and alignment of the fibers along the direction of stress loading [13, 14]. Therefore, deviations of this ideal behavior are extensively found. CNTs are easily agglomerated and bundle together and entangle, causing defects sites in the composite structures and deviating from the ideal aligned behavior. Moreover, Multiwalled Carbon Nanotubes (MWCNT) are composed by several concentric graphene shells, and these shells are connected by weak van der Waals forces, that makes only the outer shells taking part into load transfer for the polymeric matrix [15]. In this context, it becomes clear that a good CNT-polymer interface and the kind of CNTs (MWCNTs, SWCNTs, HCNTs or CSCNTs) can properly transfer the load superior properties of the CNT [4, 16] to the polymeric material and provide enhanced mechanical properties, like better elastic modulus, while the dispersion of CNTs is ascribed to be very important to obtain a uniform stress distribution [8, 17]. A homogeneous dispersion of CNTs is a crucial step [18]. The aggregation of CNTs (Figure 1) is frequently reported, and it occurs due to the van der Waals forces between CNTs. Thereby, the homogeneous distribution of the CNTs remains an important issue to be solved. As alternative could be to use the CNTs' functionalization or/and an addition of surfactants in the polymer solution to obtain a better dispersion. Unluckily, the chemical reactions and the chemical reagents used in both cases introducing some byproducts which could be inappropriate to the polymeric composite material [19]. In this work, in order to facilitate the dispersion of CNTs on the polymer matrices, it was studied the incorporation of CNTs directly onto the surface of the carbon felt. Carbon felt is a micrometric material used in various aerospace applications due to thermal, mechanical and electromagnetic conditions often required for these applications [20, 21]. The use of CNTs in the interface between the polymer and carbon felt could improve the performance of these materials since this combination modify the properties of both, carbon felt and carbon nanotubes [22–24]. In this scenario, this research work presents and discuss the results obtained in the study of dynamic-mechanical properties of the carbon felt/CNT/silicone hybrid composite sinthetized with two different types of CNTs.

Figure 1: Illustration of a (left) CNTs aggregation and (right) CNTs well-dispersed solution.

This approach uses Chemical Vapor Deposition (CVD) method to grow CNTs on the surface of the carbon felt (Figure 2) or in other micrometric particle, being possible to control the length, quantity and the dispersion of the nanomaterial [25]. The pre-growing of CNTs into the carbon felt is followed by the incorporation of the polymer. The mechanical behavior of the CNTs network on the surface of micrometric materials is a relevant subject that deserves investigation since it can also be a valid alternative to the mechanical reinforcement of several materials, as instant polymers, cement [26] and other viscous substances where the dispersion of CNTs could be problematic. This class of composite materials can be easily dispersed in substances with high viscosity because the CNTs are supported on the micrometric material (carbon felt fiber, for exemple). Moreover, the mechanical behavior of CNTs micrometric composites has been little investigated and there is a lack of specialized literature about this subject.

Figure 2: Optical microscope images (A) and (B) of carbon felt produced by the carbonization of polyacrylonitrile (PAN); Scanning electron microscope images of: (C) the fibers of carbon felt, (D) one fiber of carbon felt, (E) one fiber of carbon felt covered by CNTs and (F) CNTs on the surface of carbon felt.

2 Experimental data

2.1 Pregrowing of CNTs on the surface of carbon felt – production of the carbon felt/CNTs hybrid composite

The hybrid composite was obtained by growing CNTs via chemical vapor deposition (CVD) synthesis using alcohol as carbon source, and cobalt and manganese as a catalyst by chemical impregnation deposition of them on the carbon felt. The CVD synthesis was conducted at 650 ° C during 30 min using nitrogen as an inert carrier gas. To remove the catalyst particles after CNTs growth, the hybrid composite carbon felt/CNTs was submitted to a purification process by the immersion in concentrated hydrochloric acid during 15 minutes at room temperature (25 ° C), washed with deionized water until pH 7, and dried under vacuum at 100 ° C.

2.2 Incorporation of silicone in the carbon felt/CNTs hybrid composite

For this research work, carbon felt was synthetized with two different CNTs types: cup-stacked (CSCNT) or Helical (HCNT) carbon nanotubes (Figure 3). A solution of commer-

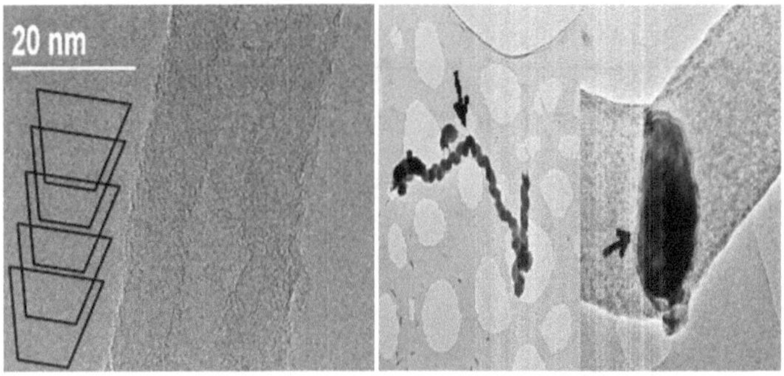

Figure 3: High resolution transmission electron microscope images of (left) cup-stacked carbon Nanotubes (CSCNT) and (right) Helical Carbon Nanotubes (HCNT).

cial silicone (Silastic, Figure 4) and dichloromethane was prepared and homogenized by magnetic stirring for 20 minutes at room temperature (27 °C) as shown in Figure 5. Samples of carbon felt/CNTs with dimensions about 20×2×5 mm were immersed in this solution during 2 minutes. The number of dips ranged between one and three times; samples were dried at 100 °C for 30 minutes between immersions to evaporation of the solvent. The cure was carried out at room temperature during 2 days. Table 1 contains the initial quantities (in grams) of carbon felt, CNTs and silicone incorporated into the hybrid composite for each studied sample. The first column has the identified number of immersions held therein.

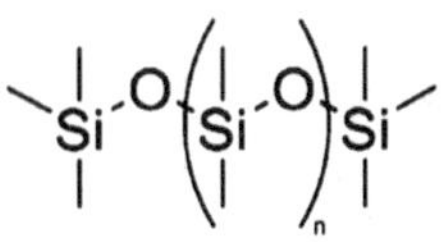

Figure 4: Structure of silastic (silicone).

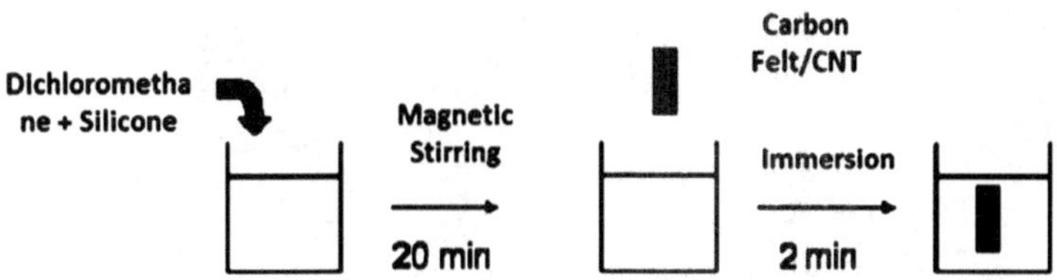

Figure 5: Scheme of incorporation of silicone in the carbon felt/NTCs hybrid composite.

It is possible to observe that the samples preparation versus the number of immersions and CNTs type is well correlated. With one or two immersions, mass embedded silicone is always higher for the felt with CSCNT. Although the mass of the CNTs is greater

Table 1: Parameters of the samples.

Sample/Number of immersions	NTC	Felt Mass (mg)	CNT Mass (mg)	Silicone Mass (mg)	Total Mass (mg)	Silicone (%)
1A/1	–	18:6	–	42.8	61.4	69.7
1B/1	HCNT	15.6	29.2	45.2	91.7	49.3
1C/1	CSCNT	12.6	3.83	40.6	57.7	70.4
2A/2	–	18.6	–	61	79.6	76.6
2B/2	HCNT	16.5	18	79.9	116.7	68.5
2C/2	CSCNT	16.3	4.98	89.3	111.5	80
3B/3	HCNT	12.6	17.86	114.6	139.6	82.1
3C/3	CSCNT	13.7	3.79	83.6	101.9	82

in the sample containing helical CNTs, these tubes do not completely cover carbon felt and are much larger than the CSCNT and have a diameter higher than 100 nm while the CSCNT have diameters in the range of 20 to 60 nm. In the carbon felt/CNTs helical composite, the CNTs are larger and are in fewer quantities. In samples immersed three times in the dispersion, it is observed that the amount of silicone incorporated does not dependent on the CNTs type. Figure 6 shows the images obtained by scanning electron microscopy for the carbon felt/CNTs/silicone hybrid composite. The images show the carbon felt/carbon nanotube composite fiber totally covered with silicone after three immersions in the dispersion.

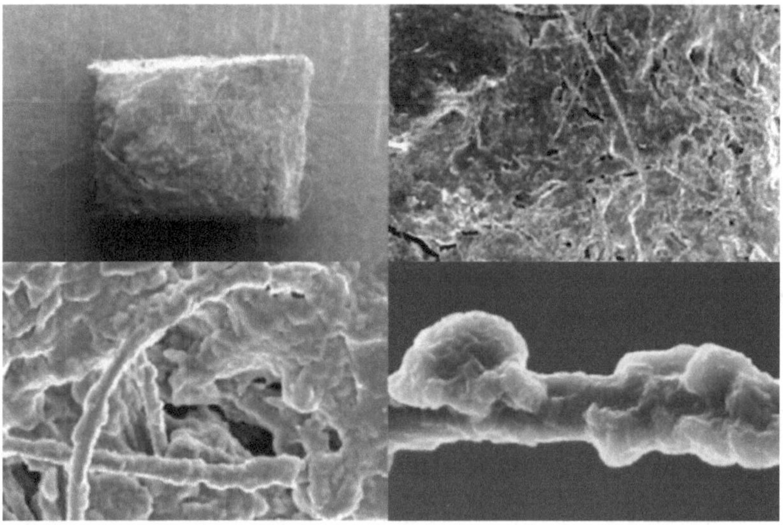

Figure 6: Scanning electron microscope images of carbon felt/CNTs/silicone hybrid composite.

2.3 Dynamic-mechanical characterization of carbon felt/CNTs hybrid composite

The dynamic mechanical characterization of the hybrid composites was conducted using samples of 20 × 2 × 5 mm in tensile mode at a frequency of 1 Hz. Samples were heated from –120 to 300 °C at a rate of 2 °C/min. Figure 7 shows the storage and loss module for composites prepared with different loads. In the case of carbon felt/silicone, carbon felt/HCNT/silicone and carbon felt/CSCNT/silicone, load on the samples were 72, 17 and 5.8 %, respectively. The major change was observed when the composite was prepared with CSCNT: an increase of 122 % in the positive temperature range with only 5.8 % of CSCNT. Moreover, the changes caused due to HNTC were smaller and depend on the temperature range. The HNTC makes the composite softer at lower temperatures and slightly enhanced at temperatures up to –22 °C. These results showed that the NTCs have a strong influence in the dynamic mechanical properties of the compounds. These changes are higher for the composite formed with CSCNT than for HCNT.

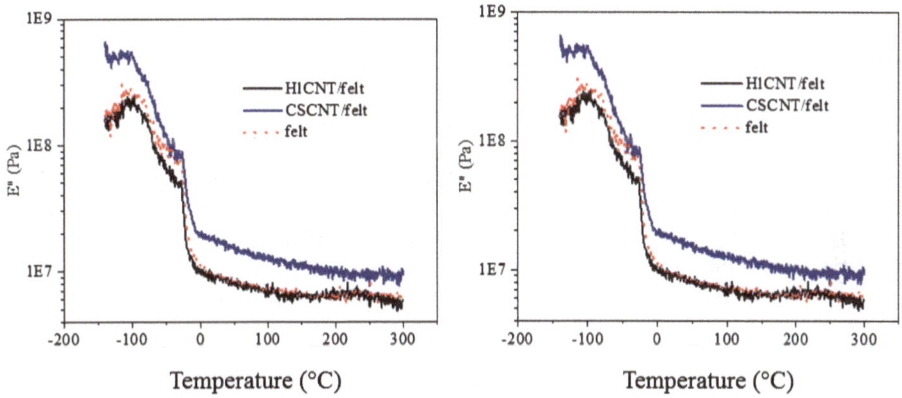

Figure 7: Storage modulus (E′) and loss modulus (E″) of carbon felt/CNTs/silicone hybrid composites where the CNTs are HCNT or CSCNT (Helical CNTs or cup-stacked CNTs).

3 Considerations

This research work shows that the use of carbon felt/CNTs hybrid composite can improve the mechanical properties of the carbon fiber. Storage Modulus (E′) and Loss Modulus (E″) increased 300 % and 250 %, respectively, when cup-stacked carbon nanotubes are grown directly on the surface of carbon fiber. Due to the strong and synergetic interaction between CNTs and carbon felt, the result hybrid composite is shown as an important, fast, easily and cheap alternative to improve the properties of carbon fibers. Beyond that, the polymer shows it is well dispersed on the surface of the fiber and the CNTs, due to the hydrophilic properties of CSCNTs without functionalization, which form a good interface between fiber-CNTs-polymer. This approach can be used with success to other combination of micrometric material or particles/CNTs/different polymers, as has been shown by other studies conducted by this research group.

Bibliography

[1] Assan, A. E.: Resistência dos materiais. Unicamp (2010).

[2] Ijima, S.: Helical microtubules of graphitic carbon. Nature 354(6348), 56–58 (1991).

[3] Lehman, J. H., Terrones, M., Mansfield, E., Hurst, K. E., Meunier, V.: Evaluating the characteristics of multiwall carbon nanotubes. Carbon 49(8), 2581–2602 (2011).

[4] Abu-Abdeen, M.: Investigation of the rheological, dynamic mechanical and tensile properties of single-walled carbon nanotubes reinforced poly(vinyl chloride). J. Appl. Polym. Sci. 124(4), 3192–3199 (2012).

[5] Bokobza, L.: Multiwall carbon nanotube elastomeric composites: a review. Polymers 48, 4907–4920 (2007).

[6] Lau, K., Chipara, M., Ling, H., Hui, D.: On the effective elastic moduli of carbon nanotubes for nanocomposite structures. Composites, Part B 35(2), 95–101 (2004).

[7] Sheikhnejad, O., Nakamoto, T., Kalteis, A., Rajabtabar, A., Major, Z.: Molecular Dynamic Simulation of Carbon Nanotube Reinforced Nanocomposites: The Effect of Interface Interaction on Mechanical Properties. MOJ Poly. Sci. 2(1), 00038 (2018). https://doi.org/10.15406/mojps.2018.02.00038.

[8] Eitan, A., Jiang, K., Dukes, D., Andrews, R., Schadler, L. S.: Surface Modification of Multiwalled Carbon Nanotubes: Toward the Tailoring of the Interface in Polymer Composites. Mater. Sci. Eng., Chem. Mater. 15(16), 3198–3201 (2003). https://doi.org/10.1021/cm020975d.

[9] Wong, M., Paramsothy, M., Xu, X. J., Ren, Y., Li, S., Liao, K.: Physical interactions at carbon nanotube-polymer interface. Polymer 44(25), 7757–7764 (2003). https://doi.org/10.1016/j.polymer.2003.10.011.

[10] Fu, S., Feng, X., Lauke, B., Mai, Y.: Effects of particle size, particle/matrix interface adhesion and particle loading on mechanical properties of particulate-polymer composites. Composites, Part B 39(6), 933–961 (2008).

[11] Zare, Y., Rhee, K. Y.: A simple model for electrical conductivity of polymer carbon nanotubes nanocomposites assuming the filler properties, interphase dimension, network level, interfacial tension and tunneling distance. Compos. Sci. Technol. 155(8), 252–260 (2018). https://doi.org/10.1016/j.compscitech.2017.10.007.

[12] Bauhofer, W., Kovacs, J. Z.: A review and analysis of electrical percolation in carbon nanotube polymer composites. Compos. Sci. Technol. 69(10), 1486–1498 (2009). https://doi.org/10.1016/j.compscitech.2008.06.018.

[13] Desai, A. V., Haque, M. A.: Mechanics of the interface for carbon nanotube-polymer composites. Thin-Walled Struct. 43(11), 1787–1803 (2005).

[14] Qian, D., Dickey, E. C.: In-situ transmission electron microscopy studies of polymer-carbon nanotube composite deformation. J. Microsc. 204(1), 39–45 (2001).

[15] Xie, X., Mai, Y., Zhou, X.: Dispersion and alignment of carbon nanotubes in polymer matrix: a review. Mater. Sci. Eng. 49, 89–112 (2005).

[16] Ruoff, R. S., Lorents, D. C.: Mechanical and thermal properties of carbon nanotubes. Carbon 33(7), 925–930 (1995).

[17] Andrews, R., Weisenberger, M. C.: Carbon nanotube polymer composites. Curr. Opin. Solid State Mater. Sci. 8(1), 31–37 (2004).
[18] Du, F., Scogna, R. C., Zhou, W., Brand, S., Fischer, J. E., Winey, K. I.: Nanotube Networks in Polymer Nanocomposites: Rheology and Electrical Conductivity. Macromolecules 37(24), 9048–9055 (2004). https://doi.org/10.1021/ma049164g.
[19] Sahoo, N. G., Rana, S., Cho, J. W., Li, L., Chan, S. H.: Polymer nanocomposites based on functionalized carbon nanotubes. Prog. Polym. Sci. 35(7), 837–867 (2010). https://doi.org/10.1016/j.progpolymsci.2010.03.002.
[20] Burkitt, B., Sivas, S. L., Thomaier, R.: Material and process innovations with filled silicone elastomers. In: International SAMPE Symposium and Exhibition (Proceedings) vol. 52 (2008). 9 pp.
[21] Dai, L., Zhang, Z., Zhao, Y., Liu, H., Xie, Z.: Effects of Polymeric Curing Agent Modified with Siloxanes on the Mechanical Properties of Silicone Rubber. J. Appl. Polym. Sci. 111, 1057–1062 (2009).
[22] Matsubara, E. Y., Rosolen, J. M., Silva, S., Ravi, P.: Composite electrode of carbon nanotubes and vitreous carbon for electron field emission. J. Appl. Phys. 104, 054303 (2008).
[23] Gonçales, V. R., Matsubara, E. Y., Rosolen, J. M., Torresi, S. I. C.: Micro/nanostructured carbon composite modified with a hybrid redox mediator and enzymes as a glucose biosensor. Carbon 49(3039), 3046 (2011).
[24] Gonçales, V. R., Colombo, R. N. P., Minadeo, M. A. O., Matsubara, E. Y., Rosolen, J. M., Torresi, S. I. C.: Three-dimensional graphene/carbon nanotubes hybrid composites for exploring interaction between glucose oxidase and carbon based electrodes. J. Electroanal. Chem. 775, 235–242 (2016).
[25] Rosolen, J. M., Matsubara, E. Y., Montoro, L. A., Lala, S. M., Marchesin, M. S., Tronto, S.: Carbon Nanotube/felt composite electrodes without polymers binders. J. Power Sources 162, 620–628 (2006).
[26] Shaughnessy, R., Clark, P. E.: The rheological behavior of fresh cement pasts. Cem. Concr. Res. 18, 327–341 (1988).

Leonardo G. Paterno, Ítalo Azevedo Costa, Maria José A. Sales, Michele A. Santos, Priscila R. Teixeira, and Priscilla P. Peregrino

Light-assisted synthesis of colloids and solid films of metallic nanoparticles

Abstract: Nanoparticles of noble metals are at the forefront of nanotechnology owing to their unusual electrochemical and optical properties. Different examples of biomedical, optoelectronic and sensor devices can be realized with these nanoparticles. However, they pose some health and environmental risks, mainly related to their synthesis, which uses toxic and environmentally unfriendly chemical reductants. Alternatively, light of different wavelengths can be used instead of chemicals to reduce metallic ions into metallic nanoparticles. The so-called light-assisted methods of synthesis are therefore much cleaner. Besides that, they allow for a better control of different synthesis features, such as spatial resolution, reaction selectivity and controllability and shape conversion via photoirradiation. Consequently, it permits one the careful tailoring of nanoparticle's properties including size, shape and surface chemistry and so on. Herein, it is provided an overview of light-assisted methods for the synthesis of colloids and solid films of noble metal nanoparticles, with emphasis on gold and silver nanoparticles.

Keywords: Nanoparticles, light assisted synthesis, biocompatible

1 Introduction

Nanoparticles (NP) with dimensions below 100 nm and made of noble metals (AgNP, AuNP, PtNP, PdNP, CuNP and others) are at the forefront of nanotechnology owing to their unusual electrochemical and optical properties. They show excellent electrocatalytic activity due to the enlarged surface to volume ratio that provides a plenty of uncoordinated atoms at the nanoparticle's surface. Indeed, they have been successfully employed to enhance the performance of fuel and photoelectrochemical cells [1, 2]. They can also be used as transducer elements in the active layer of electrochemical sensors. The binding event between the recognition element and the analyte can alter the physicochemical properties of the nanoparticles, such as conductivity, redox behavior and optical absorption, which can thus generate a detectable response signal [3]. Finally, they show strong localized surface plasmon resonance (LSPR). The LSPR is a consequence of the resonant oscillation of free electrons at the nanoparticles' sur-

Leonardo G. Paterno, Ítalo Azevedo Costa, Maria José A. Sales, Michele A. Santos, Priscila R. Teixeira, Priscilla P. Peregrino, University of Brasilia, Institute of Chemistry, Laboratory of Research on Polymers and Nanomaterials, 70900-910, Brasília-DF, Brazil, e-mail: lpaterno@unb.br

https://doi.org/10.1515/9783110584455-015

face in the presence of visible and near-infrared (NIR) light. Depending upon the size and shape, the LSPR phenomena can drive light scattering or its conversion into heat. These properties have been explored in cell and tissue imaging, photothermal therapies (tumor ablation), surface-enhanced Raman spectroscopy, immunosensors, to name a few [4–10]. Therefore, the research on noble metal nanoparticles has grown immensely since many potential applications can be realized in different fields, including energy conversion, monitoring of food quality, environment and home land security, medicine, electronics, home appliances and many others. Despite the recent success, the widespread use of metallic nanoparticles still faces the public skepticism, who is very concerned about the health hazards of fabrication, manipulation and disposal of these nanomaterials. Nonetheless, nanotechnology can address problems that previously established technologies cannot, while using a very tiny amount of nanomaterials. Consequently, researchers worldwide have put a great effort on the development of a sustainable production-supply chain, which has been found in high yield synthesis routes and environmentally friendly procedures, whenever is possible. Moreover, a considerable attention has been paid to understanding toxicity effects and safety risks posed by such nanomaterials [11, 12]. Nowadays, noble metal nanoparticles are mostly prepared by chemical reduction methods, in which metallic salts or complexes are reduced with a strong chemical reducing agent. Sodium citrate is a preferable reducing agent because of its low price, negligible toxicity and effectiveness. Additionally, citrate molecules when used in excess provide colloidal stability for nanoparticles in liquid suspensions (colloids). Other chemicals commonly used as reducing agents are sodium borohydride, hydrazine, hydroxylamine, ascorbic acid, glucose, etc. Despite the high yields and effectiveness of the chemical reduction method, the need for a chemical reducing agent counterbalances the pursuit for environmentally friendly routes. Moreover, noble metal nanoparticles have been playing a major role in biomedical applications, which restrict further the use of toxic reducing agents.

Fortunately, environmentally friendly alternatives for the synthesis of noble metal nanoparticles have been developed as well. Some natural extracts, nontoxic biocompatible polymers and electromagnetic radiation of different wavelengths have successfully replaced chemical reductants in this task [13–17]. In particular, light-assisted methods are very convenient for the synthesis of nanomaterials because they are cleaner and very versatile, since the extensive wavelength range that it can be used permits different reaction pathways to be established [15–17]. Besides that, the experimental setup is of low cost and can be implemented to provide a better control of the dynamics of (photo)reduction of metallic precursors, which in summary permits one the careful tailoring of nanoparticles' properties (size, shape, surface chemistry and so on) [15–17]. The next sections will provide an overview of actual light-assisted methods for the production of metal nanoparticles. Afterwards, some applications of AuNP and AgNP produced by these methods will be described and future perspectives will be highlighted.

2 Light-assisted synthesis of metallic nanoparticles

It has been accepted that in the conventional chemical method, nanoparticles can be formed via two distinct mechanisms, as schematically described in Figure 1: (a) nucleation and growth, (b) autocatalysis or seeded-growth [18]. In the mechanism (a), the metallic precursor (metallic ion) is reduced by a strong reducing agent, small metallic (zero valence) nuclei are then formed and finally they grow into nanoparticles. In the mechanism (b), a previously formed nanoparticle seed is introduced in the reaction mixture containing the metallic precursor and the reducing agent. The metallic precursor is then reduced at the seed's surface, which leads to nanoparticle growth. In light-assisted methods, the electromagnetic radiation of different wavelengths ignites several chemical reactions that ultimately lead to nucleation and growth of metallic nanoparticles in the same way as in the chemical approach [15–17]. However, they offer different advantages of fabrication. First, because depending upon the wavelength of the radiation source and composition of the reaction mixture, many reaction pathways are possible. A more extensive comprehension of the dynamics that regulate these processes should enable the careful tailoring of nanoparticles' properties. Second, and of technological importance, light assisted methods allow for high control of different synthesis features, such as reaction selectivity, spatial resolution and shape conversion via photoirradiation [15, 16]. As a general statement, for a photochemical reaction to occur, one specific species in the reaction mixture must absorb the specific wavelength provided by the light source. Therefore, the photochemical method permits the synthesis of nanoparticles in different transparent matrices, including micelles, synthetic and biological polymers, zeolites, etc. Also, the size and shape of as-formed nanoparticles is often changed by additional light irradiation doses. For example, spherical AgNPs can be controlled reshaped into Ag nanoprisms by additional light expositions [19]. Moreover, the light can be used to assist the synthesis and deposition of metallic nanoparticles with a predesigned spatial distribution, as in photographic and photolithography processes. Despite the problem of diffraction limits, individual nanoparticles and even bigger clusters can be assembled onto a predefined substrate, which permits 3-D writing and affords many optoelectronic applications. Most methods of synthesis of metallic nanoparticles do not offer so many possibilities as the light-assisted ones. Until now, most of procedures of light-assisted synthesis of metallic nanoparticles have employed gamma radiation, UV and/or visible light. They will be described in more detail hereinafter. The first attempt to use gamma rays to synthesize metallic nanoparticles was made by Fujita et al. in the 1960s [20]. They irradiated with 6°Co gamma ray source an aqueous solution of chloroauric acid and depending upon the applied radiation dose rate. They succeeded in producing AuNP of either spherical or polygonal shapes. AuNP of 80 µm could be prepared with plain $HAuCl_4$ solutions, while smaller particles were obtained when gelatin was mixed together. As shown in the scheme (a) of Figure 2, the gamma rays drive the water radioly-

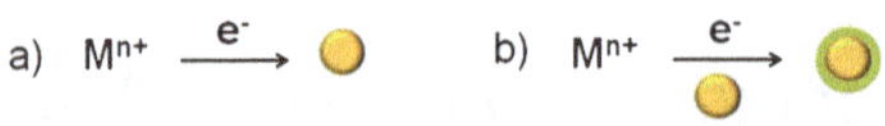

Figure 1: Mechanisms for the synthesis of metallic nanoparticles, via (a) nucleation and growth and (b) seeded-mediated.

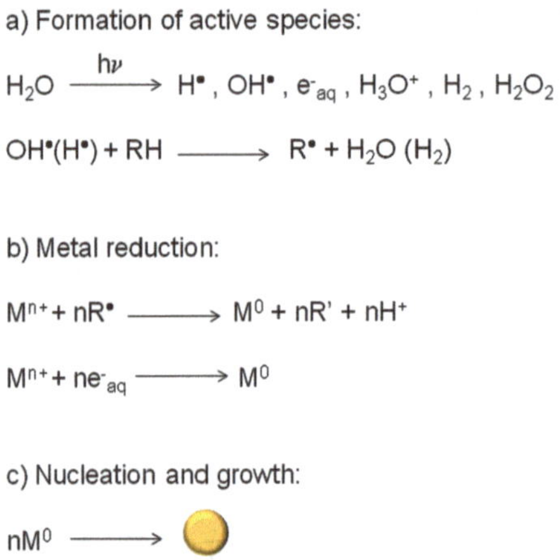

Figure 2: Mechanism for the gamma-ray assisted synthesis of metallic nanoparticles.

sis that generates energetic species such as hydrated electrons and hydroxyl radicals, which in turn can produce new radical species from the solvent (e. g., ethanol). These species are then responsible for the reduction of Au(III) ions to Au(0) and subsequent aggregation into AuNP. Other investigations that followed it employed the method to synthesize AuNP, AgNP and CuNP of controlled size and shapes. The main advantages of this method are: (i) the possibility to modulate the rate of the nucleation step and, therefore, the nanoparticles' size, by varying synthesis conditions (type of scavengers, presence of gases, solution pH); (ii) the reducing agent is uniformly generated in solution enabling the production of mono disperse samples. As a disadvantage, the method requires a strong energy input, which makes it expensive.

The synthesis assisted by UV irradiation is perhaps the most widespread among the light-assisted methods since it is of relative low cost, readily available and of easy implementation. As shown in Figure 3, the UV radiation can either generate a metal atom through the direct photoreduction of a metal source (metal salt or complex), or the reduction of metal ions by using photochemically generated intermediates such as excited molecules and radicals (photosensitization).

The direct photoreduction, Figure 3a, can be performed with several metal precursors. For example, for the direct photoreduction of Ag^+ in ethanol/water $AgClO_3$ solutions, excited $[Ag^+]^*$ induces electron transfer from the solvent (from ethanol) molecules and lead to Ag^0. The small Ag^0 nuclei then grow into AgNP. For metallic ions of higher valence states, the mechanism involves additional elementary steps [21, 22].

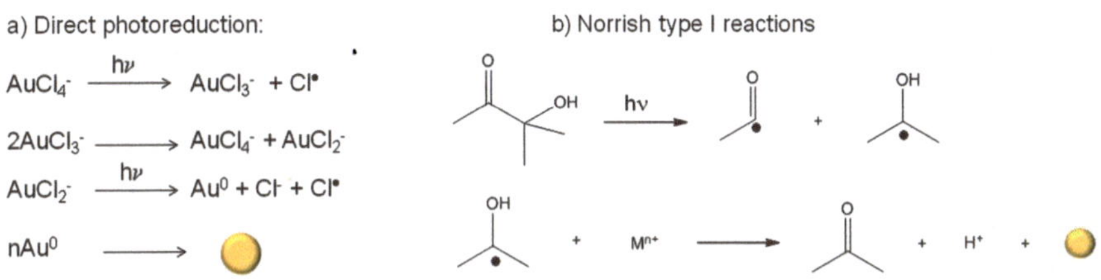

Figure 3: Mechanisms for UV-assisted synthesis of metallic nanoparticles via (a) direct photoreduction and (b) photosensitization (Norrish type I reactions).

In the direct photoreduction of Au^{3+} from tetrachloro aurate ($AuCl_4^-$), it initially forms the excited $[AuCl_4^-]^*$ species, which thus disproportionates and reduces into smaller valence ions until Au(0) [23]. Photoreduction of metal complexes is made possible by ligand radicals that are formed upon photoexcitation and reduce the metal ion. This mechanism has been observed with Cu and Au acetylacetonates as well as with Ag, Cu, Co and Fe acetylides [24, 25]. The solvent plays a key role, since excited metal ions or complex become highly oxidizing agents, and thus electron donating solvents may increase the reaction rate. In addition, the presence of polymers during synthesis can modulate the size and shape of nanoparticles, especially because they act more on the nucleation step [26].

UV assisted synthesis via photosensitization, Figure 3b, consists on the formation of intermediates from a photoactive reagent, like photoinitiators, which thus reduce the metal source to form zero valence metal species [27–29]. Generally, in the photosensitization processes the formation of nanoparticles is faster and more efficient than in direct photoreduction. Moreover, the excitation wavelength does not depend on the metal source but on the sensitizer, what makes it even more flexible. The synthesis can thus be performed in the presence of transparent matrixes, like polymers and biomacromolecules as well as surfactants. The size and shape of nanoparticles is dictated by the stoichiometric relation between the photoinitiator and the metal precursors as well as by the presence/absence of surfactant, polymer, whatsover. Also, the method can be implemented for seeded-growth of nanoparticles.

In the photosensitization, the intermediates are radicals that can be produced by either hydrogen abstraction or bond cleavage mechanisms, Figure 3b. Ketone derivatives and -alcohols are typical radical precursor/hydrogen donor couples. Many ketones such as acetone, acetophenone and benzophenone are excited into singlet state after UV absorption. The singlet excited state then decays to the triplet excited state via intersystem crossing, which thus abstract hydrogen from the hydrogen donor to generate strongly reducing ketyl radicals. The latter are responsible for the reduction of metal ions. On the other hand, aromatic ketones under UV excitation undergo the Norrish type I reaction to produce benzoyl radicals via C-C bond cleavage. These radicals then carry out the reduction of the metal source.

Gold nanorods (AuNRs) are a particular type of anisotropic AuNP that due to the rod aspect show two SPR bands. One is found in the visible and another one in the NIR and they are due to the different plasmon motion along the transversal and the longitudinal axes of the particle, respectively. The NIR absorption exhibited by AuNRs is of great interest for photothermal therapy (PTT) since it occurs in the biological wavelength window, which thus could permit the therapy of tumors located within bodily tissues [8, 9]. The PTT consists basically on the conversion of luminous energy input into heat. In more detail, in the resonance condition free electrons in the surface plasmon become photoexcitated by the incoming NIR energy and then return to equilibrium exchanging energy with lattice phonons in the solid nanoparticle.

Depending upon the rate in which this energy is exchanged with the lattice, many effects can result. For fast (~1 ps) to medium rates (~30 ps), in which occurs phase loss of the coherently excited electrons via electron-electron collisions, the temperature raises steeply up to 1000 K and it can occur loss of the capping agent, ablation, melting and even reshape of the nanoparticle. For slower rates (~100 ps), the energy is exchanged via phonon-phonon processes that leads to temperature rises on the order of a few tens of degrees and, therefore, it is useful and safe for tumor ablation.

The preparation of AuNRs of different aspect ratios (form 2 to 25) is made possible mainly by the seeded-growth mechanism, originally developed by Jana et al. and schematically illustrated in Figure 4 [30]. A previously prepared nanoparticle seed, made of either gold or silver, is introduced in the reaction mixture containing the metal precursor, a mild reducing agent and a combination of surfactants for shape controlling. The preferential attachment of surfactant molecules to the longitudinal faces of the growing particle induces particle growth to occur preferably through the nanoparticle's tip. Ag^+ and AgNP seeds can also be used to tune the size and aspect ratio of the AuNRs in a more complex mechanism. In the photochemical method, the reaction mixture has almost the same composition, except by the absence of the reducing agent that is replaced by UV light. The first successful photochemical synthesis of AuNR was accomplished by Kim et al. in 2002 [31], while many papers on the subject have appeared in the literature afterwards [32, 33]. They have produced AuNRs of different aspect ratios by irradiating reaction mixtures of $HAuCl_4$containing either AgNP seeds or Ag^+ (as AgBr) with 254-nm UV light (420 $\mu W/cm^2$) for about 30 h. It was concluded that both Ag^+ and AgNP seeds adsorb onto nascent AuNP at a preferably site making only the 100 facets available for the particle growth, which ultimately leads to the anisotropic shape.

Visible light can also be employed to carry out the formation of metal nanoparticles [19, 28]. Owing to the their LSPR characteristics, metallic nanoparticles can be used as photosensitizers. When they absorb light in the visible and NIR range they can drive an electron flow between an electron donor and the metallic ion, which consequently leads to the formation of nanoparticles. Compared to gamma irradiation and UV, plasmon-assisted nucleation is a relatively slow process, since the metallic

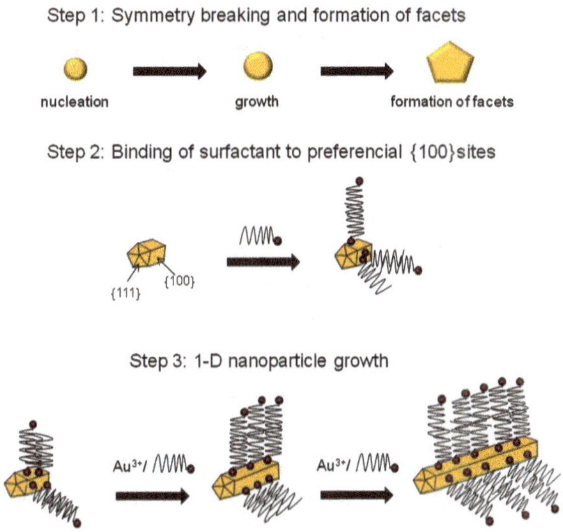

Figure 4: Schematic illustration of the formation of Au nanorods.

nanoparticles can storage photogenerated electrons. As such, there is found a delay in the reduction process (induction time) that can be useful to shape modulation. Indeed, many different bimetallic systems of different polygonal shapes have been achieved by this method (see Figure 5).

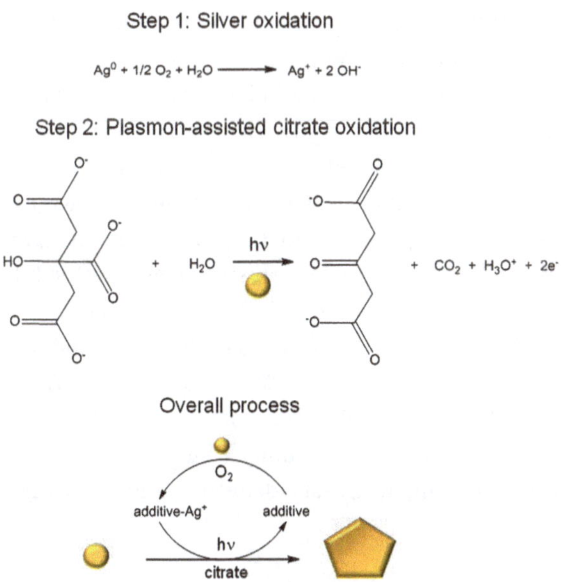

Figure 5: Schematic illustration of the formation of Au nanorods.

3 Biocompatible gold nanoparticles

AuNP are in the focus of nanomedicine because the LSPR effect can be harnessed in different applications as for instance imaging, cancer diagnosis and therapy as described earlier. The design of AuNP for nanomedicine should therefore account for biocompatibility and toxicity issues. Despite that AuNP are biologically friendly, surface functionalization is mandatory to prevent nanoparticles from elimination by the immunological system before reaching the target location and also to mitigate possible harmful effects. In this sense, AuNP for nanomedicine are usually functionalized with biopolymers. Besides the biocompatibility, biopolymers can also infer colloidal stability and provide anchoring sites for the immobilization of additional drugs in order to span the functionality of the nanoparticles.

The photochemically-assisted synthesis of AuNP colloids can be performed with the metal precursor in a solution already containing the biopolymer. This strategy is very advantageous since nanoparticles and surface functionalization are attained in a single step. Moreover, the metal precursor/biopolymer concentration ratio can be varied to control the nanoparticles' size. Teixeira et al. have applied this practical effect to photochemically synthesize biocompatible and nontoxic AuNP of different sizes and shapes. $HAuCl_4$ as the gold precursor was dissolved with the hyperbranched polyethylene imine (PEI) [34]. A schematic illustration of this process is provided in Figure 6. PEI is a biocompatible polymer and already tested as synthetic vector for DNA transfer in gene therapy. Because PEI's amino groups are capable of reducing AuCl4-ions solely at a very low rate, the reaction performed in the dark and under ambient light exhibit a lengthy induction step and it takes several hours (above 5 h) to produce sizeable amounts of AuNP (see Figure 6b). Nonetheless, the energy input provided by UV light almost eliminates the induction step, and larger amounts of AuNP-PEI nanocomposites are produced in much shorter periods (10 min. to 30 min.) (Figure 6b). As an additional feature, the PEI to AuCl4-molar ratio dictates the shape and size of formed AuNP. In the samples imaged by transmission electron microscopy reported in Figure 6, for a low $PEI/AuCl_4^-$ ratio, larger and prismatic particles are preferably formed because of the slow nucleation step (Figure 6c), while for higher $PEI/AuCl_4^-$ ratio, smaller (15 nm to 8 nm) and only spherical nanoparticles are formed (Figures 6d and 6e). PEI increases the nucleation rate by forming numerous ionic pairs with $AuCl_4^-$, which in turn will be nucleation sites for AuNP formation. AuNP-PEI samples prepared at different PEI/AuCl4-molar ratios show negligible toxicity effects, as confirmed by in vitro cell viability tests performed with both human connective tissue cell and human breast adenocarcinoma cell lines (MCF-7).

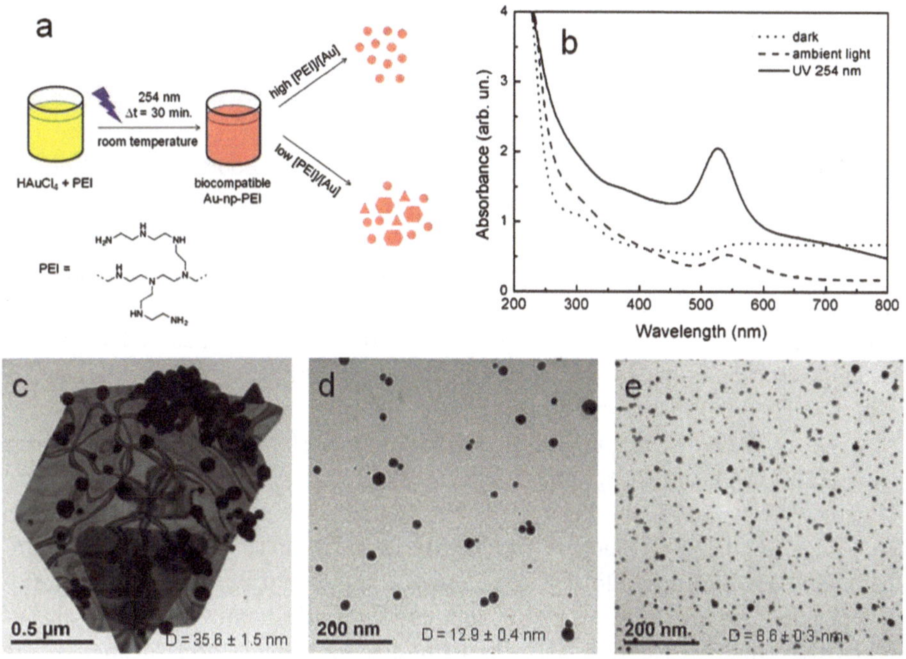

Figure 6: Formation of biocompatible gold nanoparticles by UV-assisted synthesis. (a) schematic of synthesis, (b) UV-vis spectra of AuNP samples produced in the dark, ambient and UV light, and transmission electron microscopy images of samples produced at (c) low, (d) medium and (e) high polymer to AuCl4-molar ratio.

4 Photocatalytic deposition of AgNP solid films

The photocatalytic deposition of AgNP and other noble metal nanoparticles has primarily envisaged the preparation of more efficient photocatalysts. TiO_2 is a ubiquitous semiconductor and certainly the most used photocatalyst to date, since it is of relative low cost, stable and innocuous. Nonetheless, TiO_2 electronic structure implies in a wide band gap (3.2 eV), therefore, limiting its use to UV light only, unless when it is sensitized for use in the visible and NIR ranges. Moreover, the wide band gap favors electron-hole recombination processes, which decreases the photocatalytic activity further. Metal nanoparticles adsorbed onto photocatalysts crystals suppress electron-hole recombination and improves the photocatalytic efficiency.

A general procedure for the photocatalytic deposition of noble metal nanoparticles is carried out in a suspension containing the dispersed photocatalyst (e. g., TiO_2 micro or nanoparticles), metal ions and hole scavengers. The photocatalyst once photoexcited produces holes and electrons, the latter being captured by the adsorbed metal ions. The photocatalytic cycle ends when the reduction of metal ions is accompanied by electron transfer from the hole scavenger to the photocatalyst semiconductor.

The pioneering works in this field date back the 1970s, when the main focus was to obtain kinetics parameters of the phototacatalytic reduction of different metal ions [35]. Many examples of the photocatalytic deposition were reported and the photocatalytic performance of the resulting metal-semicondutor photocatalyst for several model reactions was also evaluated [36, 37]. The physico-chemical effects on the scale-up of Ag photodeposition on TiO_2 nanoparticles were reported by Chen and Barteau [38]. Nowadays, there is a growing interest in using the photocatalytic deposition to produce micropatterns of metallic nanoparticles onto planar and curved substrates for the fabrication of optical and electronic devices. Some of these interesting applications will be briefly summarized below.

AgNP and AuNP have found an important use in surface enhanced Raman scattering (SERS) as plasmonic nanoparticles. The SERS effect was discovered in 1974 and ever since the research and advances in this topic have been enormous [39–41]. In practical terms, the SERS effect can enhance the Raman signal of a chemical target (analyte) up 1011, which thus allow for detection of very low concentrations or even a single molecule. Mostly of SERS spectra are obtained when the analyte molecules are adsorbed onto roughened metallic surfaces. In the electromagnetic mechanism (EM), the resonance of surface plasmons on the metallic surface with the incident light causes a strong increase of the local electromagnetic field. The enhanced field increases the induced dipole moment of the adsorbed molecule, which in turn enhances its Raman signal. In the chemical mechanism (CM), there must have a change on the molecular polarizability due to the interaction of the scattering molecule with the surface. This interaction can occur via conventional charge transfer complex, conventional chemical bonds with the surface or even through ionic pairing. The EM addresses most of SERS effect and, therefore, plasmonic nanoparticles are highly demanded for this application [40, 41]. Moreover, the SERS efficiency is improved when nanoparticles are partially aggregated or when they show truncated morphologies. The junctions between nanoparticles and sharpe edges are like SERS hotspots where large field enhancements occur. Nonetheless, the preparation of SERS active substrates with reproducible properties is still very difficult and, perhaps, it can be considered the main limitation of SERS applications [41].

One example of a SERS active substrate made by photocatalytic deposition of AgNP is shown in Figure 7. The AgNP were photocatalytic deposited using $AgNO_3$ solution (5 mmol.L^{-1}) and UV irradiation (254 nm, 16 W, 10 min.). As seen in Figure 7a, a glass slide dip coated with a TiO_2NP film is almost transparent while the one coated with AgNP produced by photocatalytic deposition is dark. The UV-vis absorption spectra provided in Figure 7b clearly show the formation of AgNP by the presence of a broad LSPR band centered at 580 nm. Finally, in Figure 7c is seen the Raman spectra of a model probe, methylene blue, which is more intense and better resolved when it is deposited onto the TiO_2 NP-AgNP SERS substrate.

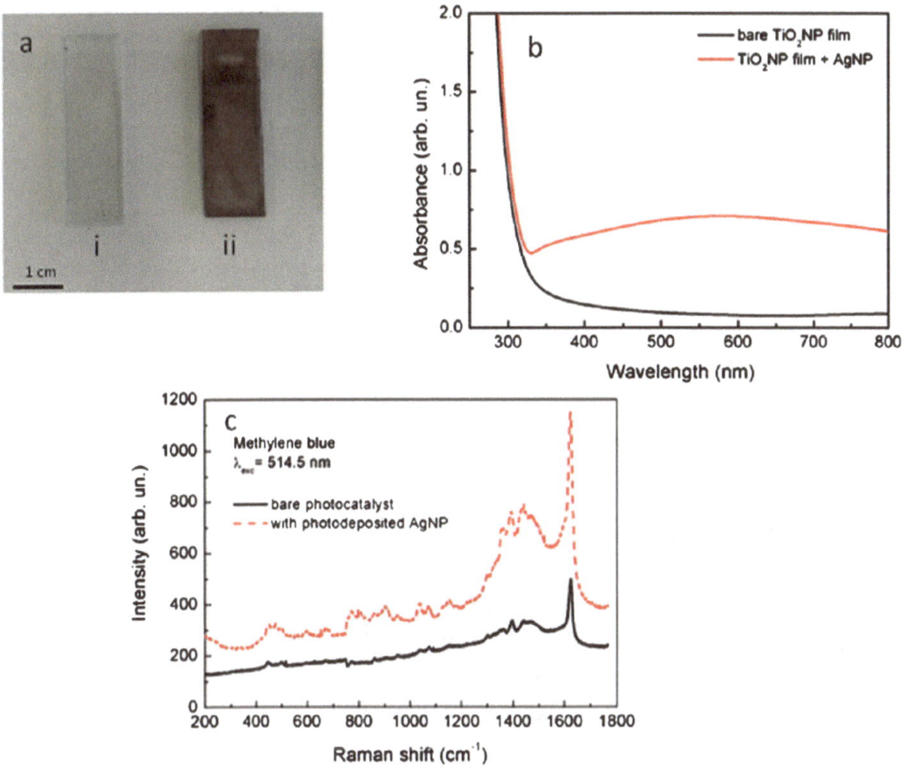

Figure 7: Digital photos (a) and UV-vis absorption spectra (b) of bare TiO_2NP photocatalyst substrate and after photocatalytic deposition of AgNP. In (c) it is shown the Raman spectra of methylene blue adsorbed onto bare and AgNP TiO_2 NP photocatalyst substrate.

The photocatalytic deposition of AgNP can also be employed in working electrodes for electrochemical sensors. Triclosan (5-chloro-2-(2,4-dichlorophenoxy)phenol) is an antibacterial and antifungal agent found in personal care products like toothpaste and soap. Nowadays, it is suspected of working as an endocrine disruptor chemical that can pose serious risk to human and aquatic animals' health [42]. Therefore, detection and monitoring of triclosan in natural and waste water resources is of great interest. Figure 8 displays cyclic voltammograms for the oxidation of triclosan (5-chloro-2-(2,4-dichlorophenoxy)phenol) performed with conducting glass electrodes (ITO, indium-doped tin oxide), bare and modified with photocatalytic deposited AgNP. The photocatalytic deposition of AgNP followed the same protocol used for fabrication of SERS active substrate shown in Figure 7. It is clear that the electrode containing AgNP is much more sensitive to triclosan than the bare electrode, since it provides a much higher anodic current at about +0.8 V. This current is associated to the electrochemical oxidation of triclosan.

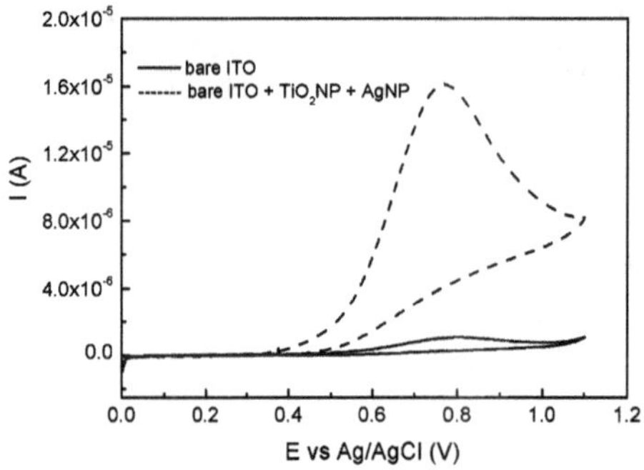

Figure 8: Cyclic voltammograms of triclosan registered with bare and AgNP modified ITO electrodes. Measurements performed in a three electrode electrochemical cell (Ag/AgCl reference electrode and Pt counter electrode) at 50 $mV.s^{-1}$ scan rate, phosphate buffer pH 9.

5 Final remarks

The light assisted methods are undoubtedly very promising for the synthesis of colloids and solid films of metallic nanoparticles. They can be employed to nucleate metallic nanoparticles from different metal sources, within a myriad of polymer and inorganic matrixes, and without using any hazardous or expensive chemical reducing agents. Moreover, they provide for spatial resolution, reaction selectivity and controllability, and shape conversion via photoirradiation of nanoparticles, something that no other available method alone is capable of providing. Since light of different wavelengths can be used, different mechanisms of metal ion reduction and nanoparticle formation may occur. The full comprehension of these mechanisms is used further to precisely tailoring of nanoparticles features like size, shape and surface chemistry. Metallic nanoparticles are demanded for several technological applications. Nonetheless, they may pose serious risks to human health as well as to the environment. The development of environmentally friendly synthetic routes and processing is a potential way of mitigating the eventual harms of nanomaterials supply chain. Therefore, light-assisted synthesis methods should be considered a viable way for producing nanomaterials with less negative effects to the environment and human health.

Bibliography

[1] Zhao, S., Yan, L., Luo, H., Mustain, W., Xua, H.: Recent progress and perspectives of bifunctional oxygen reduction/evolution catalyst development for regenerative anion exchange membrane fuel cells. Nano Energy 47, 172–198 (2018).

[2] Ueno, K., Oshikiri, T., Sun, Q., Shi, X., Misawa, H.: Solid-State Plasmonic Solar Cells Chem. Rev. 118, 2955–2993 (2018).

[3] Saha, K., Agasti, S. S., Kim, C., Li, X., Rotello, V. M.: Gold Nanoparticles in Chemical and Biological Sensing. Chem. Rev. 112, 2739–2779 (2012).

[4] Wang, W., Ramezani, M., Väkeväinen, A. I., Törmä, P., Rivas, J. G., Odom, T. W.: The rich photonic world of plasmonic nanoparticle arrays. Mater. Today 21, 303–314 (2018).

[5] Noguez, C.: Surface Plasmons on Metal Nanoparticles: The Influence of Shape and Physical Environment. J. Phys. Chem. C 111, 3806–3819 (2007).

[6] Abbas, A., Linman, M. J., Cheng, Q.: New trends in instrumental design for surface plasmon resonance-based biosensors. Biosens. Bioelectron. 26, 1815–1824 (2011).

[7] Shiohara, A., Wang, Y., Liz-Marzán, L. M.: Recent approaches toward creation of hot spots for SERS detection. J. Photochem. Photobiol., C, Photochem. Rev. 21, 2–25 (2014).

[8] Jain, P. K., Huang, X., El-sayed, I. H., El-sayed, M. A.: Noble Metals on the Nanoscale: Optical and Photothermal Properties and Some Applications in Imaging, Sensing, Biology, and Medicine. Acc. Chem. Res. 41, 1578–1586 (2008).

[9] Huang, X., El-Sayed, M. A.: Gold nanoparticles: Optical properties and implementations in cancer diagnosis and photothermal therapy. J. Adv. Res. 1, 13–28 (2010).

[10] Zhang, Y., He, S., Guo, W., Hu, Y., Huang, J., Mulcahy, J. R., Wei, W. D.: Surface-Plasmon-Driven Hot Electron Photochemistry. Chem. Rev. 118, 2927–2954 (2018).

[11] Schulte, P. A., Kuempel, E. D., Drew, N. M.: Characterizing risk assessments for the development of occupational exposure limits for engineered nanomaterials Regul. Toxicol. Pharmacol. 95, 207–219 (2018).

[12] Bar-Ilan, O., Albrecht, R. M., Fako, V. E., Furgeson, D. Y.: Toxicity Assessments of Multisized Gold and Silver Nanoparticles in Zebrafish Embryos. Small 5, 1897–1910 (2009).

[13] Alexandridis, P.: Gold Nanoparticle Synthesis, Morphology Control, and Stabilization Facilitated by Functional Polymers. Chem. Eng. Technol. 34, 15–28 (2011).

[14] Duan, H., Wang, D., Li, Y.: Green chemistry for nanoparticle synthesis. Chem. Soc. Rev. 44, 5778–5792 (2015).

[15] Sakamoto, M., Fujistuka, M., Majima, T.: Light as a construction tool of metal nanoparticles: Synthesis and mechanism. J. Photochem. Photobiol., C, Photochem. Rev. 10, 33–56 (2009).

[16] Sakamoto, M., Majima, T.: Photochemistry for the Synthesis of Noble Metal Nanoparticles. Bull. Chem. Soc. Jpn. 83, 1133–1154 (2010).

[17] Grzelczak, M., Liz-Marzan, L. M.: The relevance of light in the formation of colloidal metal nanoparticles. Chem. Soc. Rev. 43, 2089–2097 (2014).

[18] LaMer, V. K., Dinegar, R. H.: Theory, Production and Mechanism of Formation of Monodispersed Hydrosols. J. Am. Chem. Soc. 72, 4847– 4854 (1950).

[19] Pietrobon, B., Kitaev, V.: Photochemical Synthesis of Monodisperse Size- Controlled Silver Decahedral Nanoparticles and Their Remarkable Optical Properties. Chem. Mater. 20, 5186–5190 (2008).

[20] Fujita, H., Izawa, M., Yamazaki, H.: Gamma ray induced formation of gold sol from chloroauric acid solution. Nature 196, 666–667 (1962).

[21] Yonezawa, Y., Sato, T., Ohno, M., Hada, H.: Photochemical Formation of Colloidal Metals. J. Chem. Soc. Faraday Trans. 83, 1559–1567 (1987).

[22] Hada, H., Yonezawa, Y., Yoshida, A., Kurakake, A.: Photoreduction of Silver Ion in Aqueous and Alcoholic Solutions. J. Phys. Chem. 80, 2728–2731 (1976).

[23] Eustis, S., Hsu, H.-Y., El-Sayed, M. A.: Gold Nanoparticle Formation from Photochemical Reduction of Au3+ by Continuous Excitation in Colloidal Solutions. A Proposed Molecular Mechanism. J. Phys. Chem. B 109, 4811–4815 (2005).

[24] Roucoux, A., Schulz, J., Patin, H.: Reduced Transition Metal Colloids: A Novel Family of Reusable Catalysts? Chem. Rev. 102, 3757–3778 (2002).

[25] Eiwiov, B. G., Janata, E., Henglein, A.: Study of the formation of the dispersed phase in aqueous solutions of Cu(II). Int. J. Radiat. Appl. Instrum. C Radiat. Phys. 39, 123–126 (1992).

[26] Mallick, K., Witcomb, M. J., Scurrel, M. S.: Polymer stabilized silver nanoparticles: A photochemical synthesis route. J. Mater. Sci. 39, 4459–4463 (2004).
[27] Gupta, S., Prakash, R.: Photochemically mediated synthesis of a gold colloid by dithizone and its application in the amperometric sensing of thiocyanate. RSC Adv. 5, 81660–81667 (2015).
[28] Stamplecoskie, K. G., Scaiano, J. C.: Light Emitting Diode Irradiation Can Control the Morphology and Optical Properties of Silver Nanoparticles. J. Am. Chem. Soc. 132, 1825–1827 (2010).
[29] McGilvray, K. L., Decan, M. R., Wang, D., Scaiano, J. C.: Facile Photochemical Synthesis of Unprotected Aqueous Gold Nanoparticles. J. Am. Chem. Soc. 128, 15980–15981 (2006).
[30] Jana, N. R., Gearheart, L., Murphy, C. J.: Wet Chemical Synthesis of High Aspect Ratio Cylindrical Gold Nanorods. J. Phys. Chem. B 105, 4065– 4067 (2001).
[31] Kim, F., Song, J. H., Yang, P.: Photochemical Synthesis of Gold Nanorods. J. Am. Chem. Soc. 124, 14316–14317 (2002).
[32] Placido, T., Comparelli, R., Giannici, F., Cozzoli, P. D., Capitani, G., Striccoli, M., Agostiano, A., Curri, M. L.: Photochemical Synthesis of Water-Soluble Gold Nanorods: The Role of Silver in Assisting Anisotropic Growth. Chem. Mater. 21, 4192–4202 (2009).
[33] Ahmed, M., Narain, R.: Rapid Synthesis of Gold Nanorods Using a One-Step Photochemical Strategy. Langmuir 26, 18392–18399 (2010).
[34] Teixeira, P. R., Santos, M. S. C., Silva, A. L. G., Báo, S. N., Azevedo, R. B., Sales, M. J. A., Paterno, L. G.: Photochemically-assisted synthesis of non-toxic and biocompatible gold nanoparticles. Coll. Surf. B Bionterf. 148, 317–323 (2016).
[35] Tauster, S. J., Fung, S. C., Garten, R. L.: Strong metal-support interactions. Group 8 noble metals supported on titanium dioxide. J. Am. Chem. Soc. 100, 170–175 (1978).
[36] Bamwenda, G. R., Tsubota, S., Nakamura, T., Haruta, M.: The influence of the preparation methods on the catalytic activity of platinum and gold supported on TiO2 for CO oxidation. Catal. Lett. 44, 83–87 (1997).
[37] Li, D., McCann, J. T., Gratt, M., Xia, Y.: Photocatalytic deposition of gold nanoparticles on electrospun nanofibers of titania. Chem. Phys. Lett. 394, 387–391 (2004).
[38] Chan, S. C., Barteau, M. A.: Physico-Chemical Effects on the Scale-Up of Ag Photodeposition on TiO2 Nanoparticles. Top. Catal. 54, 378–389 (2011).
[39] Fleischmann, M., Hendra, P. J., McQuillan, A. J.: Raman spectra of pyridine adsorbed at a silver electrode. Chem. Phys. Lett. 26, 163–166 (1974).
[40] Lane, L. A., Qian, X., Nie, S.: SERS Nanoparticles in Medicine: From Label-Free Detection to Spectroscopic Tagging. Chem. Rev. 115, 10489–10529 (2015).
[41] Shiohara, A., Wang, Y., Liz-Marzán, L. M.: Recent approaches toward creation of hot spots for SERS detection. J. Photochem. Photobiol., C, Photochem. Rev. 21, 2–25 (2014).
[42] Petrie, B., Barden, R., Kasprzyk-Hordern, B.: A review on emerging contaminants in wastewaters and the environment: Current knowledge, understudied areas and recommendations for future monitoring. Water Res. 72, 3–27 (2015).

Mayara Bortolotti Rossini, Narasimha Rao Mekala,
Mauricio Chaves-Vargas, and Kai-Uwe Schröder

The influence of polymeric interlayers on damping behavior of a fiber metal laminate

Abstract: The influence of Polycaprolactone (PCL) layers on the dynamic response of the fiber metal laminate Glare 2A was proposed. The methodology was based on experimental and numerical analyses. An initial numerical evaluation of the natural frequency of the beams is performed to compare with the experimental results. The tests were performed to find the damping properties of the PCL viscoelastic material in terms of the loss factor (η). The analytical evaluation of the experimental data was based on the natural frequencies and the half-power bandwidth of each mode. Then the Young's modulus and the loss factor of the damping material were calculated. Lastly, numerical analysis was performed on the dynamic behavior of the Glare 2A without and with PCL layers in order to show the damping behavior. The damped Glare 2A behavior depends on the vibration-damping properties of the PCL determined as the constants/factors of Rayleigh's model for damping (α and β). The numerical analysis employed a shell element computation developed at SLA Institute. The viscoelastic material was located in two different positions, on the outer layers and the inner layers between metal and the composite glass fiber/epoxy layers, as well as, varying the PCL thickness. The numerical results evaluated the correlation of some parameters (i. e., thickness and the location of the damping layers) to increase the damping capacity of the Glare 2A. The results showed a significant gain on the vibration damping behavior of the beams, the thicker thickness showed higher frequency and low amplitude and the viscoelastic material position affects the dynamic behavior of the Glare 2A.

Keywords: Fiber metal laminate, viscoelastic material, resonance frequency, loss factor, damping behavior

1 Introduction

Fiber metal laminates (FML) have been used in aircraft structures at lower wing skin and fuselage. FMLs are composed of laminae of reinforced polymeric composites and

Acknowledgement: The authors would like to acknowledge the PhD researcher, Zamaan Sadeghi (SLA Institute), and the technicians, Werner Just and Lars Moosdorf, from SLA Institute.

Mayara Bortolotti Rossini, Narasimha Rao Mekala, Mauricio Chaves-Vargas, Kai-Uwe Schröder, RWTH Aachen, Aachen, Germany, e-mails: mayara.rossini@sla.rwth-aachen.de, kai-uwe.schroeder@sla.rwth-aachen.de

https://doi.org/10.1515/9783110584455-016

aluminum alloys. Their main improvements in structure application are in fatigue resistance, resistance to impact and environmental conditions. For application in aircraft structures, the vibration response of the structure can be an issue when the frequency of excitation corresponds to the resonance frequency [1]. Additional damping material has been used on structures to control the vibration response of the system. Some studies were done on the efficacy of treating composite materials with damping layers [2]. For damping treatments, viscoelastic materials are used due to their creep behavior, which is represented by an increasing in deformation under constant load [3]. The mechanical properties of viscoelastic materials are frequency and temperature dependent and described by time dependence linearity, as well as, stress-strain linearity [4]. For viscoelastic materials, the linear viscoelastic stress-strain law is applied. The stress calculation depends on the storage modulus in shear or tension (G' or E') and the loss modulus in shear or tension (G'' or E'') [3]. The storage modulus represents the energy restored, and the loss modulus is the energy dissipated [4]. Thermoplastic materials are applied as viscoelastic material to increase damping properties and can be used to increase vibration properties of fiber metal laminates. From FMLs available, the Glare 2A – 3/2 – 0.3 shows superior dynamic properties under high frequency bending tests, as exemplified in Figure 1 [5]. For thermoplastic material, Polycaprolactone (PCL) has slow degradation rate and distinct viscoelastic properties, making it satisfactory for long term application [6].

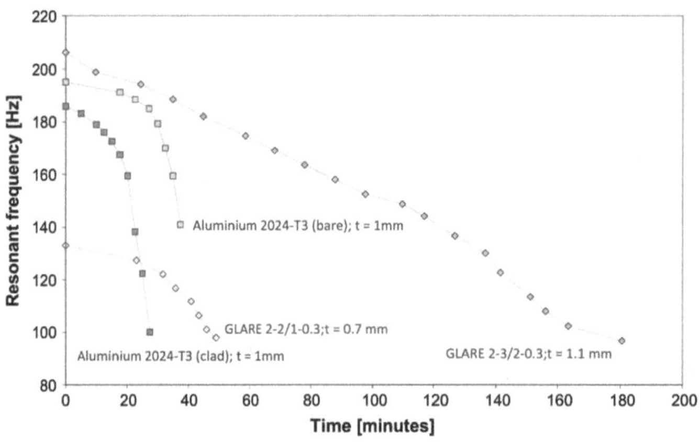

Figure 1: Plot frequency versus time for shaker test comparing glare and monolithic aluminum [5].

Therefore, the objectives of the present study relate to demonstrate the gain in damping properties adding PCL layers on Glare 2A – 3/2 – 0.3, as well as, to verify the contribution in damping behavior varying the damping layer thickness and position in Glare 2A. The methodology used to perform this study was based in experimental and numerical analyses. The experimental was conducted by following the standard ASTM

E756-05 to determine the loss factor of the viscoelastic layer for extensional module. As the methodology in the numerical part was based on the first-order transverse shear deformation (FOSD) equivalent single layer modeling, it was assumed same loss factor for shear modulus. Further, the effect of the position and the thickness of the PCL in the considered FML beams were evaluated.

2 Fiber metal laminate and viscoelastic material

The family of fiber metal laminate (FML) materials consists of thin metal sheets bonded alternately to composite layers of fiber/epoxy resin. The hybrid material derives a lightweight structure with superior mechanical properties compare to the monolithic aluminum or fiber-reinforced laminae isolated. The first FML introduced was Aramid reinforced aluminum laminate (Arall), accompanied by the development of carbon reinforced (Carall) and the glass reinforced (Glare) aluminum laminate in order to increase the fatigue performance [7]. These three types comprise the most common known FMLs [8].

FML has been applied in aeronautical structures due to its high fatigue and corrosion resistance as well as damage tolerance, high specific strength and impact resistance. The most important application in aeronautical structure is the assembly of the upper fuselage and the stabilizer's edges of A380 airplane using Glare. Glare is manufactured by thin sheets of aeronautical aluminum alloy bonded to S-glass fiber-reinforced epoxy layers [9]. Glare can be adjusted to outfit a vast number of applications changing some parameters as the number of layers, the thickness, stacking sequence, fiber orientation, surface treatment and others [10]. The Glare 2A – 3/2 – 0.3 consists of three Aluminum 2024-T3 layers with thickness of 0.3 [mm] and two S2-glass/FM94 epoxy layers each one with two prepreg layers orientated in θ° (parallel to the metal rolling direction) with a thickness of 0.133 [mm], as the representation in Figure 2 [11, 12].

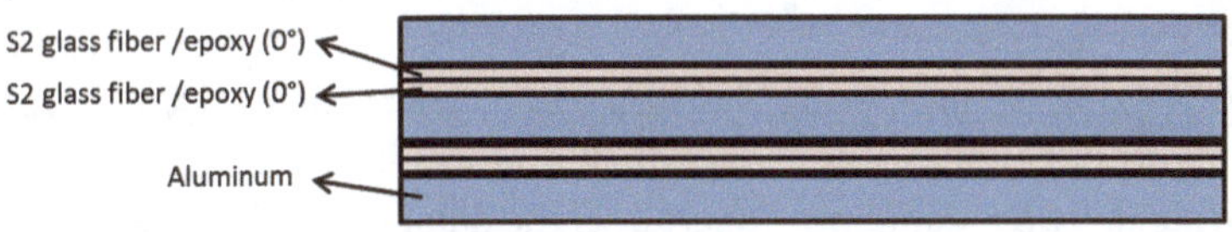

Figure 2: Glare 2A – 3/2 – 0.3 layup.

Laminate materials, as Glare, with adhesive bonded show improved vibrational damping properties compared to monolithic metal due to the viscoelastic layers. The internal damping efficiency reduces the stresses more than in the monolithic metal [5].

Viscoelastic materials yield energy dissipation as a delayed response between stress-strain under steady-state harmonic vibration. In order to increase artificially the structure damping, viscoelastic materials are added as a passive damping treatments and act to reduce the vibration propagation [13].

The incorporation of viscoelastic material in the structure can be done by free-layer damping, constrained-layer or sandwich-layer damping treatment (Figure 3) and tuned viscoelastic damper. In the sandwich-layer method, the viscoelastic material is placed between two elastic layers and deformed in shear because of the stiff of one layer when the other is bending by vibration. The energy consumed and dissipated into heat during shear mode in the viscoelastic layer for the sandwich-layer method is higher than in the free-layer method [14].

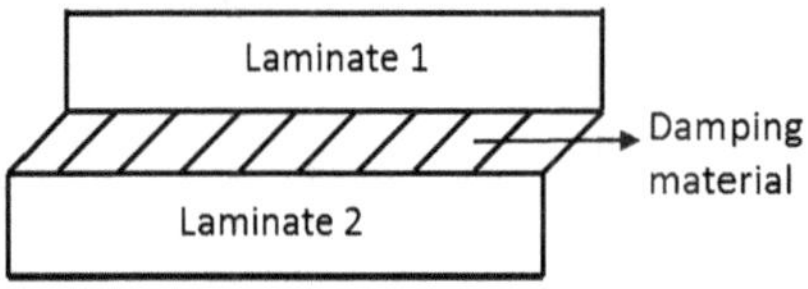

Figure 3: Sandwich-layer.

For a damping layer in a sandwich beam, the thickness of the viscoelastic material and its properties, such as shear modulus and loss factor, have significant influence on the dynamic stiffness and vibration-damping potential of the sandwich structure. In case of multiple damping layers in composite beams, the location of the damping layers has also effect on the dynamic stiffness and damping behavior of the composite structure [15].

3 Experimental analysis

A cantilever beam test is performed to define the viscoelastic material damping properties in terms of the loss factor (η), in addition to the Young's modulus (E). The ASTM E756-05 was used for the test procedure. The test has a good precision on the frequency range of 50 to 5000 [Hz]. A uniform beam was tested to measure the damping property of the self-supporting Aluminum base beam including its resonance frequencies, which was used to compute the damping properties of the damped configuration. The configuration of the dampened specimens is depicted in Figure 4 and two specimens were tested. The test determines the Young's modulus and loss factor of the damping material under extensional deformation [16].

The specimen preparation initiated with cleaning the surface of the materials using Acetone, any particles were removed for better adhesion of the adhesive (LOCTITE 454) between layers. The roots were bonded to the beam with the same adhesive. The

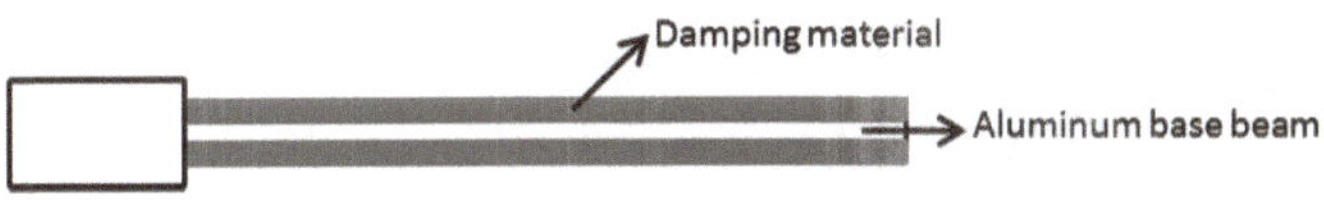

Figure 4: Test specimen configuration for beam damped in both sides [16].

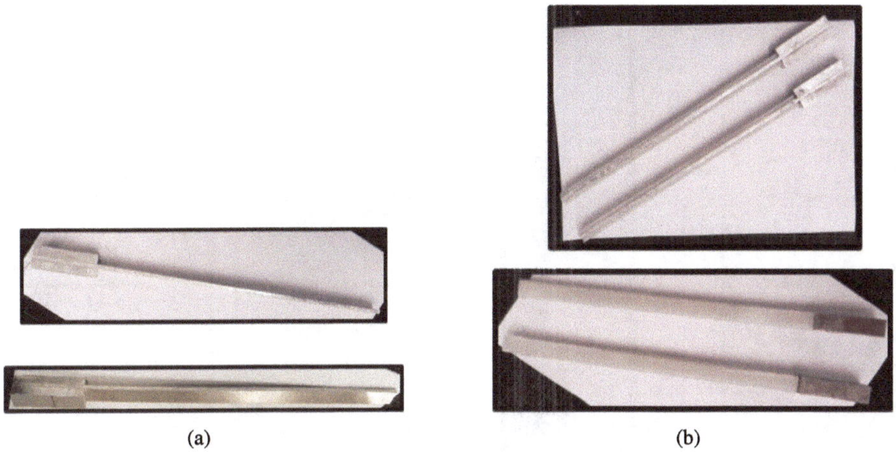

Figure 5: Prepared specimens (a) Uniform beam and (b) Beam damped in both sides.

material used for root and the beam base is Aluminum 5083, and PCL for the viscoelastic layer. The specimen configurations are represented in Figure 5.

The test was conducted in an environmental chamber under 20 °C of temperature and humidity control of 20 % RH. The specimens were fixed in one end and the other was free in a vertical orientation. Two contactless transducers were used, one to produce the sinusoidal excitation signal and other to measure the response signal. Magnetic coils were used as the contactless transducers to generate the excitation on the Aluminum base beam. Four small pieces of magnetic material were glued on the free end of the specimens and on the lateral of the beams, both in front of the magnetic coils as depicted in Figure 6. The setup describing the apparatus for the tests is represented in Figure 7 [16].

The test procedure started placing the clamped apparatus with the specimen and the fixed magnetics coils into the climate chamber with the defined environmental conditions: Figures 8(a), 8(b) and 8(c). Magnetic pieces were placed in front of the magnetic coils, and a thermocouple in contact with the specimen's surface was used to control the specimen's temperature; Figure 8(d). The two-channel oscilloscope was set with frequency range from 5 to 5000 [Hz], amplitude of 10 V_{pp} (peak-to-peak) and sweep of 6 [s]; Figures 8(e) and 8(f). The software to record the data was NI-Dasylab.

The plot generated in MATLAB, amplitude versus frequency [Hz] is calculated by Fast Fourier transform (FFT) of the input and output signals from experiments. On the plot, a half-power bandwidth method is used to determine the loss factor of a vibrat-

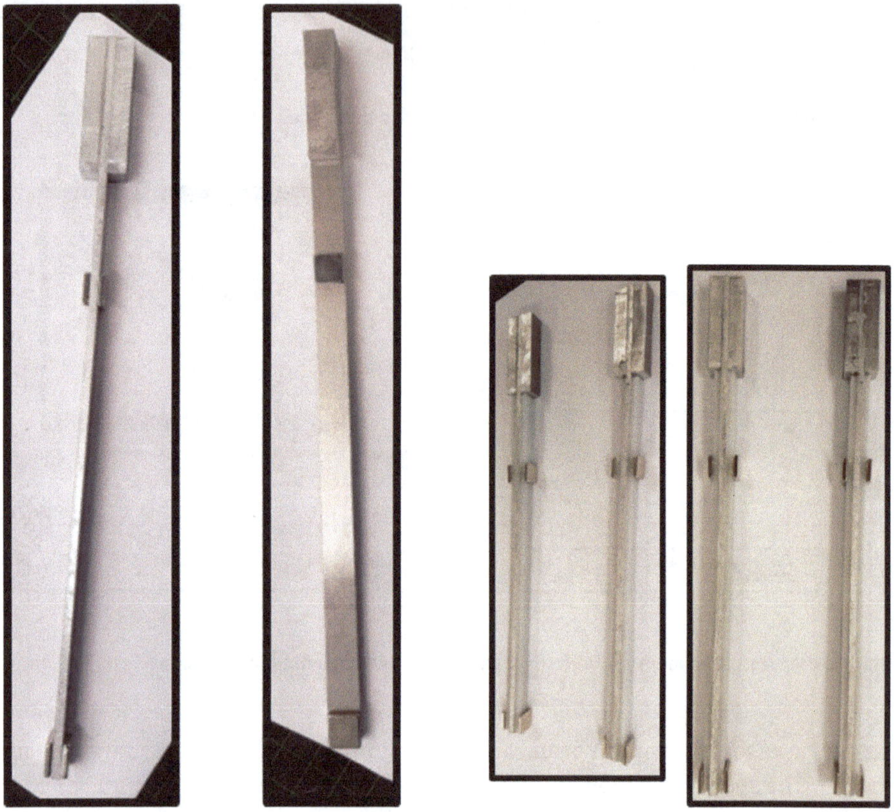

(a) (b)

Figure 6: Specimens with magnetic pieces.

ing beam. The method is executed for each vibration mode and extracted from the difference between the above and the below frequencies from the resonant frequency reducing the amplitude by ($\frac{1}{\sqrt{2}}$), as depicted in Figure 9. The loss factor calculation is defined by equation (1), that represents the ratio of the half-power bandwidth and the resonant frequency [5, 16].

Case a

$$\eta = \frac{\Delta f_n}{f_n} \tag{1}$$

For an uniform metal beam under vibration (i. e., Aluminum beam) the Young's modulus of the beam material is calculated by the expression below (equation (2)) [16].

$$E = \frac{12\rho l^2 f_n^2}{H^2 C_n^2} \tag{2}$$

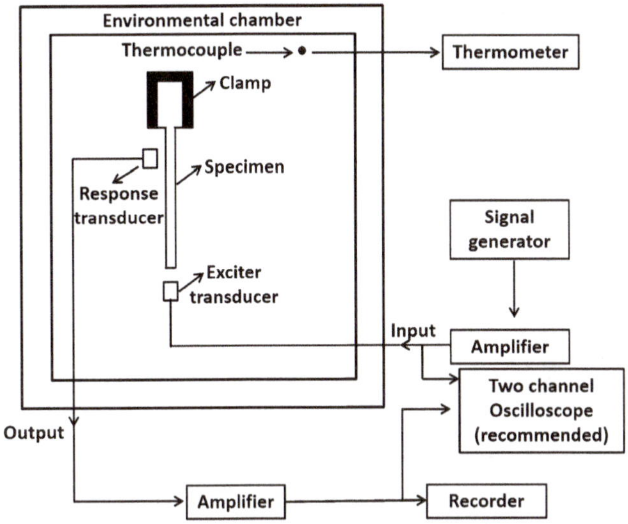

Figure 7: Set-up of the experiment under sinusoidal excitation signal [16].

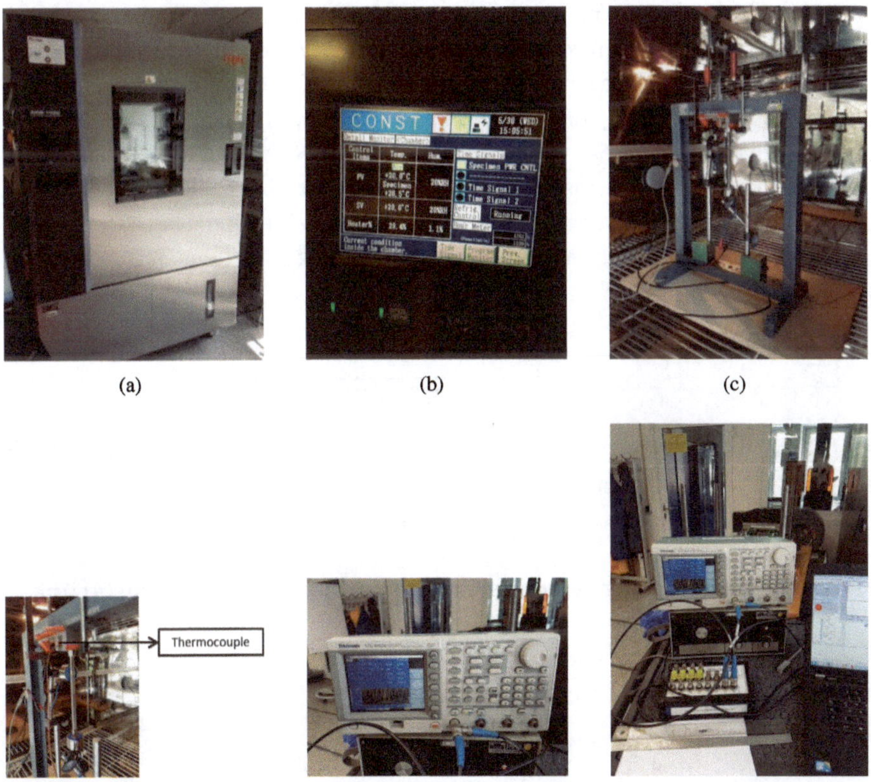

Figure 8: Apparatus used for tests.

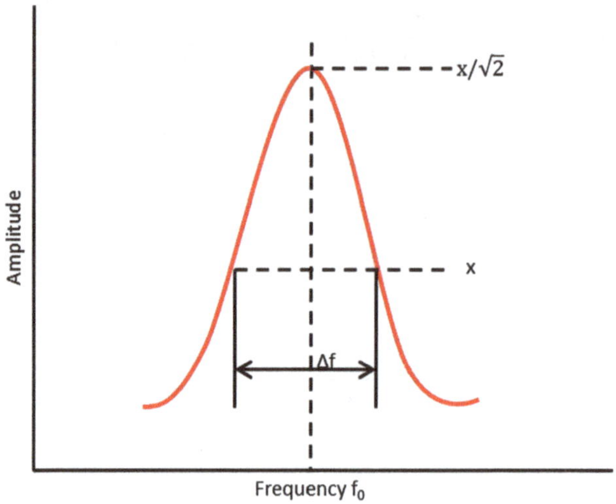

Figure 9: Half-power bandwidth method to calculate the loss factor [5].

where C_n is the coefficient for mode n for clamped-free beam; E is the Young's modulus of the beam [Pa]; H is the thickness of the beam in vibration direction [m]; l is the length of the beam [m]; n is the mode number $(1, 2, 3, \ldots)$; f_n is the natural frequency for mode n of the base beam [Hz]; Δf_n is the half-power bandwidth of mode n [Hz]; η is the loss factor of the beam material and ρ is the density of the base beam [Kg/m^3] [16]. The damping property of a viscoelastic material is determined by a beam with damping material in the configuration described in the experimental section, the beam damped in both sides (Case b). Equation (3) calculates the Young's modulus of the damping material [16]:

$$E_1 = E\frac{[(\frac{f_m}{f_n})^2(1 + 2DT) - 1]}{8T^3 + 12T^2 + 6T} \tag{3}$$

where D is the density ration $(\frac{\rho_1}{\rho})$; ρ_1 is the density of the viscoelastic material [Kg/m^3]; E_1 is the Young's modulus of the viscoelastic material [Pa]; H_1 is the total thickness of one side of the viscoelastic material (both sides are the same thickness) [m] and T is the thickness ratio $(\frac{H_1}{H})$ [16].

Therefore, the loss factor of the damping material is given by equation (4) [16].

Case b

$$\eta_1 = \eta_m + \left[\frac{E\eta_m}{E_1(8T^3 + 12T^2 + 6T)}\right] \tag{4}$$

where m is the index number $(1, 2, 3, \ldots, m = n)$, fm is the natural frequency for mode m of the composite beam [Hz], Δf_m is the half-power bandwidth of mode m of the composite beam [Hz], η_m is the loss factor of the composite beam ($\frac{\Delta f_m}{f_m}$) and η_1 is the loss factor of the viscoelastic material [16].

3.1 Resonant frequency

An initial evaluation of the resonant frequencies was performed by FE model for each configuration tested and compared with the results obtained experimentally. The model consists of shell elements to discretize the metal beam and viscoelastic layer. The materials physical and mechanical properties are defined in the Table 1. The structure is clamped in one extremity and free on another. The lumped mass position of the four magnet pieces were defined in 37 [mm] from the end of the clamped region and in the end of the beam. They were classified as nonstructural mass on the FE program.

Table 1: Physical and mechanical properties of the materials used on FE model.

Aluminum 5083	
Physical properties	
Root length [mm]	40
Root width [mm]	10
Root thickness [mm]	6 (configuration Fig. 4)
Beam length [mm]	180
Beam width [mm]	10
Beam thickness [mm]	3
Density [ton/mm^3]	2.66e-009
Mechanical properties	
Young's modulus [MPa]	70000
Poisson's ratio	0.33
PCL	
Physical properties	
Layer length [mm]	180
Layer width [mm]	10
Layer thickness [mm]	3 (configuration Fig. 4)
Density [ton/mm3]	1.1e-009
Mechanical properties	
Young's modulus [MPa]	120
Poisson's ratio	0.3
Magnets	
Physical properties	
Length [mm]	10
Width [mm]	10
Mass [g]	1.8

The resonant frequencies results from FE method and the experiments are given in Table 2 for each configuration. “Case a” is the uniform beam without damping material, “Case b” is the dampened beam in both sides.

Table 2: Resonant frequencies comparison between FE model and experiments.

Vibration mode	FE method [Hz]	Experiments average [Hz]	Error [%]
Case a			
Mode 1	55.145	54.8	0.6
Mode 2	384.08	364.4	5.4
Mode 3	1007.1	965.5	4.3
Case b			
Mode 1	47.163	48.525	3
Mode 2	309.94	302.25	2.5
Mode 3	831.96	793.35	5

The results from the experiments to determine the damping properties of the viscoelastic material using the equations described in this section are presented in the Table 3. The Young’s modulus and the loss factor are calculated for the first three vibration modes and defined as the average value from the test of two specimens in each configuration.

Table 3: Properties of the viscoelastic material from experiments.

Experiments	
Case a	
E_{Aluminum} [Pa]	3.7309e+10
η_1	0.0108
η_2	0.0012
η_3	0.0003
Case b	
E_{Damping} [Pa]	4.3795e+08
η_1	0.0403
η_2	0.0105
η_3	0.0013

4 Numerical modeling analysis

The 3-D FE method is a very good choice to model thin-walled composite structures, which can obtain relatively high accuracy results with large model size and high computation time. Due to small thickness of thin-walled structures, 2-D FE methods are more frequently used in theoretical and numerical analysis using various hypotheses. The present shell element presumes small strains and linear strain-displacement relations of [17, 18] in the framework of Reissner–Mindlin first-order transverse shear deformation (FOSD) hypothesis, which takes transverse shear strains into account. Analogously, the FOSD plate/shell theory can be understood as an extension of the Timoshenko beam theory with assuming that the straight lines normal to the mid-surface remain straight after deformation, but not necessarily normal. This theory assumes that a linear variation of displacement is distributed across the shell thickness. The numerical implementation of Gruttmann and Wagner [19] is considered in the present article. However, the element presented in Gruttmann and Wagner [19] work deals with geometrically nonlinear finite element analysis of unidirectional composite structures, whereas the present paper is focused only on the small displacement analysis in order to study the effect of the modeling FML/polymer composites under mechanical loads.

The local a and the orthonormal material t_i coordinate systems of a four node shell element are shown in Figure 10. x_i denotes the position vector to the reference surface in the initial configuration at node I, d is the inextensible director which is a unit vector perpendicular to the reference surface in the undeformed configuration, X is the position vector to an arbitrary point on the director and z denotes the thickness coordinate ranging from the bottom to the top of the considered shell [19, 20].

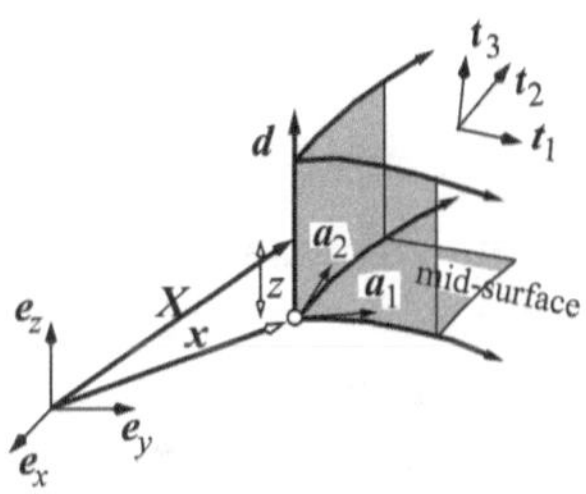

Figure 10: The geometry of the shell.

The strain tensor components in the framework of small strain-displacement relations and FOSD hypothesis are given by

$$\varepsilon_{\alpha\beta} = \overset{0}{\varepsilon}_{\alpha\beta} + z\overset{1}{\varepsilon}_{\alpha\beta} + z^2\overset{2}{\varepsilon}_{\alpha\beta} \tag{5}$$

$$\varepsilon_{\alpha 3} = \overset{0}{\varepsilon}_{\alpha 3} + z\overset{1}{\varepsilon}_{\alpha 3} \tag{6}$$

$$\varepsilon_{33} = \overset{0}{\varepsilon}_{33} \tag{7}$$

Because of the thinness of the considered shells and the inextensible director assumption, the components $\overset{2}{\varepsilon}_{\alpha\beta}$ and $\overset{0}{\varepsilon}_{33}$ are neglected. The components of the other strain-displacement terms in equation (5) and equation (6) are

$$\begin{aligned}2\overset{0}{\varepsilon}_{\alpha\beta} &= x_{,\alpha}(\bar{x}_{,\beta} - x_{,\beta}) + x_{,\beta}(\bar{x}_{,\alpha} - x_{,\alpha}),\\ 2\overset{1}{\varepsilon}_{\alpha\beta} &= x_{,\alpha}(\bar{d}_{,\beta} - d_{,\beta}) + d_{,\beta}(\bar{d}_{,\alpha} - d_{,\alpha}) + x_{,\beta}(\bar{x}_{,\alpha} - x_{,\alpha})\\ 2\overset{0}{\varepsilon}_{\alpha 3} &= x_{,\alpha}\bar{d} + \bar{d}_{,\beta}\bar{x}_{,\alpha}\end{aligned} \tag{8}$$

In equation (8), the deformed vectors are represented by an overbar. The strain field vector can be rewritten as

$$\varepsilon = Z_s \varepsilon_\varepsilon \tag{9}$$

where $\left\{\overset{0}{\varepsilon}_{11}, \overset{0}{\varepsilon}_{22}, 2\overset{0}{\varepsilon}_{12}, \overset{1}{\varepsilon}_{11}, \overset{1}{\varepsilon}_{22}, \overset{1}{\varepsilon}_{12}, 2\overset{0}{\varepsilon}_{13}, 2\overset{0}{\varepsilon}_{23}\right\}^T$ and,

$$Z_s = \begin{bmatrix} 1 & 0 & 0 & z & 0 & 0 & 0 & 0\\ 0 & 1 & 0 & 0 & z & 0 & 0 & 0\\ 0 & 0 & 1 & 0 & 0 & z & 0 & 0\\ 0 & 0 & 0 & 0 & 0 & 0 & 1 & 0\\ 0 & 0 & 0 & 0 & 0 & 0 & 0 & 1 \end{bmatrix}$$

4.1 Finite element implementation

The variation of the position vector $\delta\bar{x}_{,\alpha} = \delta v_{,\alpha}$ and the variation of the director vector can be calculated as

$$\delta\bar{d}_1 = \text{skew}\,\bar{d}A\delta\theta_1 \tag{10}$$

where $A = [\bar{a}_1\bar{a}_2]$ and $\bar{d}$ are tracked by using Rodrigues' formula for small displacements and rotations (see [18, 19]). The vector θ_I contains local nodal rotations.

To overcome the locking effects in the shell finite element, an assumed natural strain (ANS) method which was first proposed by Macneal [21] and Hughes and Tezduyar [22] is considered in the present formulation along with enhanced assumed strain (EAS) method (Rao [20, 23] and Simo and Rfiai [24] for additional details). An enhanced strain "$\hat{\epsilon}$" is the summation of the kinematically obtained value "ϵ" and the enhancing part "$\tilde{\epsilon}$." Thus,

$$\tilde{\epsilon} = \epsilon + \hat{\epsilon} \tag{11}$$

$$\delta\varepsilon_{\tilde{\epsilon}} = B_\alpha \delta\alpha \tag{12}$$

The internal virtual work of the FML/polymer composites can be expressed as

$$\delta W_i = \int_V (\sigma : \delta\hat{\epsilon})\, dV = \int_\Omega \delta\varepsilon^T(\mathbb{C})\varepsilon d\Omega \tag{13}$$

and $\mathbb{C} = \int_{-z/2}^{+z/2} Z_s^T c Z_s\, dZ$.

The external virtual work can be obtained using the following relation:

$$\delta W_e = \int_V \rho b \delta \upsilon \, dV + \int_A (T\delta \upsilon) \, dA \tag{14}$$

where T is the traction vector on the boundary, ρ is the mass density, b is the body force density, $\delta \upsilon$ and $\delta \phi$ are the virtual displacement vector, respectively.

In the FEM approximation after expressing the generalized displacements and electric potentials with linear interpolation of shape functions, the variation of strain and electric fields for the four-node element are defined as

$$\delta \varepsilon_\varepsilon = \sum_{I=1}^{4} B_I^{\upsilon} \delta \upsilon_I = B\delta u \tag{15}$$

where contains derivatives of shape functions and generalized displacement and electric potential vectors. Substituting equation (14) and equation (15) in equation (12) and equation (13), one obtains the following static system of equations on element level:

$$M\ddot{U} + K\Delta U = F \tag{16}$$

where $K = \int_\Omega B^T (\mathbb{C}) B d\Omega$, and $\delta U = \{\delta u \ \ \delta \alpha\}^T$ and $B = \{Bu \ \ B\alpha\}^T$, the force vector with mechanical forces including in-balance forces corresponding to the additional degrees freedom in ANS method is $F = \{Fu \ \ F\alpha\}^T$ and K is the stiffness matrix and M is the mass matrix.

With this linear approach, which was initially introduced by Rayleigh [25], it is supposed that the damping matrix is in a linear combination of the mass and stiffness matrices. Although this idea was suggested for mathematical convenience only, it allows the damping matrix to be diagonalized simultaneously with the mass and stiffness matrices, preserving the simplicity of uncoupled, real normal modes as in the undamped case [26]. The relation is

$$M\ddot{U} + C\dot{U} + K\Delta U = F \tag{17}$$

$$C = \alpha M + \beta K \tag{18}$$

where α and β are real scalars that must be determined.

Considering modal superposition method, it can have for the mode i the form

$$\xi_i = \frac{\alpha}{2\omega_i} + \beta \frac{\omega_i}{2} \tag{19}$$

Here, two values for the eigenfrequency ω_i with the corresponding values of ξ_i are needed to find out the scalars α and β. Relevant are the first few modes for which there is a significant mass participation, i. e., a mass participation which is not less than 95 %. To find out α and β, two methods are presented and then applied to the smart beam to find its damping characteristics. These are the method of Chowdhury and Dasgupta [27], and the method of damping from normalized spectra, also known as the half-power bandwidth method [28, 29].

One of the most convenient methods used to do that is the half-power bandwidth method [30]. Although this method is applicable only to lightly damped single DOF systems, yet it is frequently applied to well-separated modes of multi-DOFs [28]. Here, the half-power bandwidth method is chosen for the further evaluation of damping constants:

$$\xi = f_2 - \frac{f_1}{2f_{\text{res}}} \tag{20}$$

Calculating two values for ω_i in two different eigenmodes with the corresponding values of the frequency ranges and substituting them in equation (19) yields α and β.

5 Results and discussion

The numerical analyses comprise the dynamic behavior of the Glare 2A beam interleaved with PCL viscoelastic material located between the metal and the composite layers. The influence of the damping material position and its thickness are evaluated in terms of frequency response.

A cantilevered PCL interleaved Glare 2A-3/2 laminate beam of length 500 mm and width 25 mm is considered to show the effect of damping for different thickness of the PCL damping layers. The thickness of the aluminum layer is 0.3 [mm] each and the S2 glass/epoxy layers are 0.133 [mm] each. All the layers are perfectly bonded with negligible adhesive thickness. The material properties of the S2 glass/epoxy are $E_1 = 43$ [GPa], $E_2 = 6.2$ [GPa], $v_1 = 0.25$ [GPa] and $\rho_M = 1850$ [Kg/m^3]. The material properties of aluminum and PCL material are given in Table 1. Two different cases of location of damping layers along the thickness of the structure are considered as shown in Figure 11. The constants of Rayleigh's model were selected in such a way that the response matches the measured/experimental one of the PCL material. The beam is modeled with 50×2 elements with enhanced assumed natural strain scheme of 774 (see [20]). For time-response analysis, a Newmark scheme is used with a time step of 0.001 s.

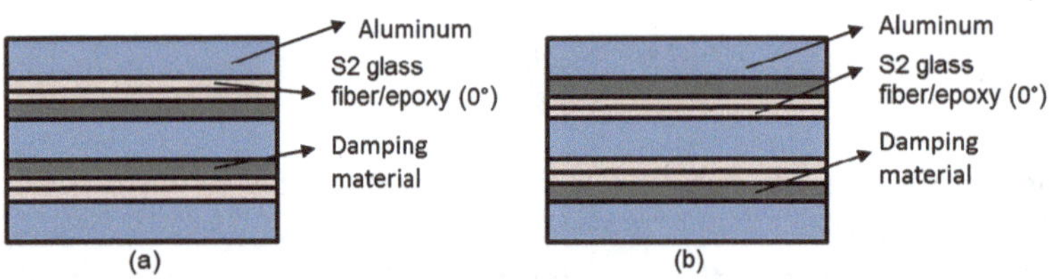

Figure 11: Different stacking sequence of PCL interleaved Glare 2A – 3/2 laminate.

At first, in order to study the effects of absolute transverse deflections on the tip of the beam at free end and resonant frequencies, a frequency response analysis has been performed. Numerical simulations were executed by varying different thickness of the PCL layers (h_{pcl}) such as 0.127 [mm], 0.3 [mm] and 0.75 [mm] on the case (a) showed in Figure 11, and compared with the case without damping material. Here, the numerical simulations are performed for the beam with initial bending conditions, the transverse deflection at the tip of the beam is 6.50 [mm] and later the beam is kept at free vibration.

The time-response curves for the different values of the damping layer thickness with and without damping properties are shown in Figure 12. From Figure 12, it has been observed that by varying the thickness of the damping layer can adjust the damping behavior. The frequency response of the beam structure for different thickness of the PCL is depicted in Figure 13. From Figure 13, it has been observed that there are clear variation of the responses in each plot, i. e., frequency of excitation and corresponding peak values for different values of PCL thickness. Figure 13 also describes that the absolute amplitude decreases as the thickness of PCL increases, whereas the corresponding resonant frequencies increase. This is expected since the numerical is developed with the assumptions of FOSD theory along the thickness. This means that the straight lines normal to the mid-surface remain straight after deformation, but not necessarily normal and equivalent single layer model is adopted. This makes the sandwich structure with viscoelastic layers behaves like an unconstrained layer damping as prescribed in [31]. To achieve more damping, high amount of thickness of the damping material is required in this case of equivalent single layer model. The above results also conclude the same that higher the thickness of the viscoelastic layer higher is the damping.

Next, the frequency response simulations are performed for two different cases of stacking sequence Figure 11 (a) and (b) for the PCL thickness of 0.127 [mm]. Figure 14 depicts the plots of absolute amplitude at the tip of the beam versus the excitation frequencies for two different positions of the PCL layers. The position of the damping layers shows some amount of influence in predicting the absolute amplitudes and resonant frequencies. From Figure 14, it has been observed that the resonant frequencies for the case (b) is slightly lower than the case (a), this is because of the fact that the stress distribution along the thickness different for the case (a) and case (b), hence the total stiffness is different in each case and the predicted resonant frequencies are different. The assumed numerical modeling and methodology described above attributes the deformation of PCL layer on the local deformation (curvature) of the surface to which it is applied. That means loss factor depends on the extension of the damping material. The damping predicted by manipulating the position of the viscoelastic layer away from the neutral axis of the beam increases the mechanical deformation of PCL layer. This results in higher damping for the same amount of material or conversely same amount of damping with less PCL material

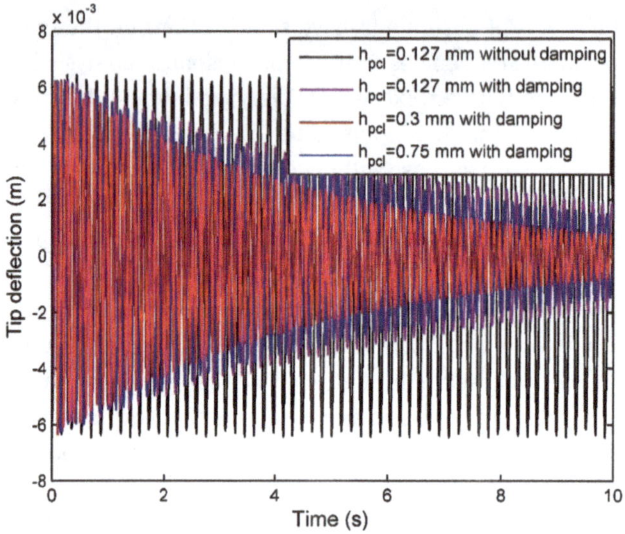

Figure 12: Tip deflections over time for different thickness of the PCL layers.

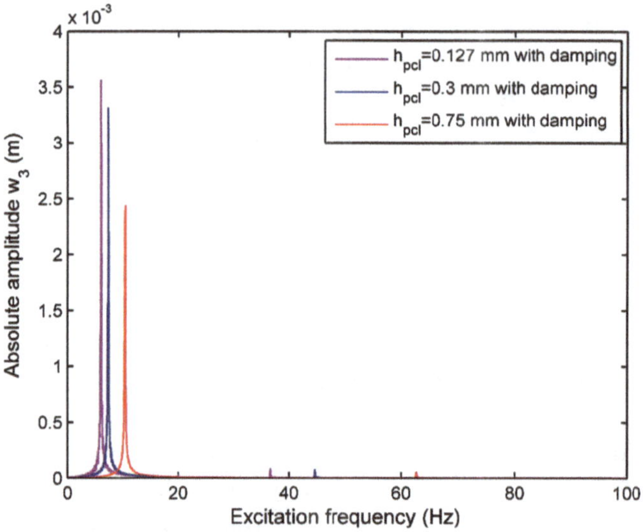

Figure 13: Absolute amplitudes of tip deflections of the beam over the resonant frequencies.

6 Conclusion

The article proposes the influence of PCL layers on the dynamic behavior of Glare 2A – 3/2. Then vibration tests are performed and the loss factor of PCL calculated to model the damping behavior of the PCL located between the metal and the composite layers

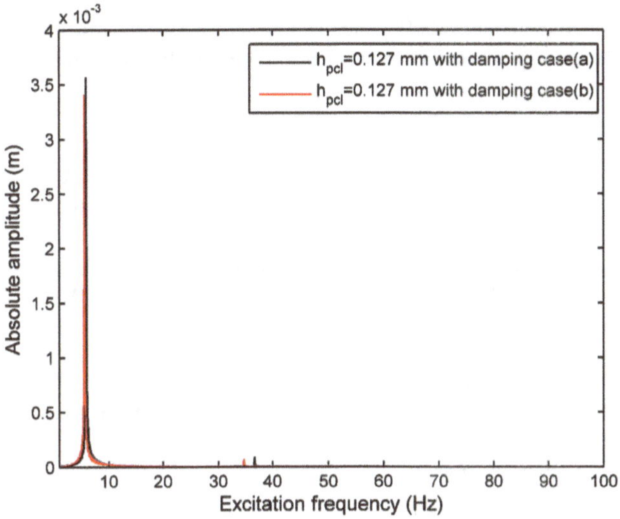

Figure 14: Absolute amplitudes of tip deflections of the beam over the resonant frequencies of two different locations of PCL layers for $h_{\mathrm{pcl}} = 0.127$ [mm].

in Glare 2A – 3/2. A four-node shell element is employed for finite element discretization of such Glare 2A – 3/2 structures. The transient analysis reveals the PCL material layers influence damping behavior of the PCL interleaved Glare 2A – 3/2 structure by selecting the suitable thickness of the PCL material layer. A frequency response analysis confirms that the system has gained damping due to the use of PCL layers in the Glare 2A – 3/2. Lower amplitudes and higher resonant frequencies are obtained by increasing the thickness of the PCL layers in Glare 2A – 3/2 structure. Further, the study of changing the location of damping layers in the stacking sequence of Glare 2A – 3/2 structure reveals that higher damping is achieved if the damping layers are located away from the neutral axis of the laminate GLARE 2A – 3/2 structure.

Bibliography

[1] Ghashochi-Bargh, H., Sadr, M. H.: Vibration Optimization of Fiber-Metal Laminated Composite Shallow Shell Panels Using an Adaptive PSO Algorithm. Mech. Adv. Compos. Struct. 4, 99–110 (2017).

[2] Ramesh, T. C., Ganesan, N.: Influence of constrained damping layer on the resonant response of orthotropic cylindrical shells. J. Sound Vib. 185(3), 483–500 (1995).

[3] Goodman, L. E.: Material damping and slip damping. In: Shock and Vibration Handbook. McGraw-Hill Book Company, Inc., New York (1961).

[4] Vasques, C. M. A., Moreira, R. A. S., Rodrigues, J. D.: Viscoelastic Damping Technologies-Part I: Modeling and Finite Element Implementation. J. Adv. Res. Mech. Eng. 1(2), 76–95 (2010).

[5] Alderliesten, R.: Fatigue and Fracture of Fibre Metal Laminates. Springer (2017).

[6] Baker, S. R., Banerjee, S., Bonin, K., Guthold, M.: Determining the mechanical properties of electrospun poly-ε-caprolactone (PCL) nanofibers using AFM and a novel fiber anchoring technique. Mater. Sci. Eng. C 59, 203–212 (2016).
[7] Liu, Q., Ma, J., Kang, L., Sun, G., Li, Q.: An experimental study on fatigue characteristics of CFRP-steel hybrid laminates. Mater. Des. 88, 643–650 (2015).
[8] Giasin, K., Ayvar-Soberanis, S.: Evaluation of workpiece temperature during drilling of GLARE fiber metal laminates using infrared techniques: Effect of cutting parameters, fiber orientation and spray mist application. Materials 9(8) (2016).
[9] Almeida, C. F. G., Bastian, F. L., Castrodeza, E. M.: Comparison of J-R curves and Jc values of C(T) and M(T) specimens of bidirectional GLARE 3 5/4 0.3 fiber-metal laminates. Eng. Fract. Mech. 159, 79–89 (2016).
[10] Qi, C., Zhidong, G., Zengshan, L., Zhaojie, J., Yue, Z.: Experimental investigation on impact performances of GLARE laminates. Chin. J. Aeronaut. 28(6), 1784–1792 (2015).
[11] Vlot, A., Gunnink, J. W. (eds.): Fibre Metal Laminates: An Introduction. Springer, Netherlands (2001).
[12] Giasin, K., Ayvar-Soberanis, S., French, T., Phadnis, V.: 3D Finite Element Modelling of Cutting Forces in Drilling Fibre Metal Laminates and Experimental Hole Quality Analysis. Appl. Compos. Mater. 24(1), 113–137 (2017).
[13] Vasques, C. M. A.: Vibration Control of Adaptive Structures: Modeling, Simulation and Implementation of Viscoelastic and Piezoelectric Damping Technologies. Universidade do Porto (2008).
[14] Rao, M. D.: Recent applications of viscoelastic damping for noise control in automobiles and commercial airplanes. J. Sound Vib. 262, 457–474 (2003).
[15] Rao, M. D., He, S.: Dynamic Analysis and Design of Laminated Composite Beams with Multiple Damping Layers. AIAA J. 31(4), 736–745 (1993).
[16] ASTM International, ASTM E756-05: Standard Test Method for Measuring Vibration-Damping Properties of Materials, West Conshohocken, PA (2017).
[17] Habip, L. M., Ebcioglu, I. K.: On the equations of motion of shells in the reference state. Ing.-Arch. 34(1), 28–32 (1965).
[18] Dvorkin, E. N., Onte, E., Oliver, J.: On a non-linear formulation for curved Timoshenko beam elements considering large displacement/rotation increments. Int. J. Numer. Methods Eng. 26(7), 1597–1613 (1988).
[19] Gruttmann, F., Wagner, W.: A linear quadrilateral shell element with fast stiffness computation. Comput. Methods Appl. Mech. Eng. 194, no. 39-41, 4279–4300 (2005).
[20] Rao, M. N., Schmidt, R., Schröder, K. U.: Finite rotation FE-simulation and active vibration control of smart composite laminated structures. Comput. Mech. 55(4), 719–735 (2015).
[21] Macneal, R. H.: A simple quadrilateral shell element. Comput. Struct. 8 (1978).
[22] Hughes, T. J. R., Tezduyar, T. E.: Finite elements based upon Mindlin plate theory with particular reference to the four-node bilinear isoparametric element. J. Appl. Mech. 48, 587–596 (1981).
[23] Rao, M. N., Schmidt, R., Schröder, K.-U.: Geometrically nonlinear static FE-simulation of multilayered magneto-electro-elastic composite structures. Compos. Struct. 127, 120–131 (2015).
[24] Simo, J. C., Rfiai, M. S.: A class of mixed assumed strain methods and the method of incompatible modes. Int. J. Numer. Methods Eng. 29, 1595–1638 (1990).
[25] Rayleigh, L.: Theory of Sound. (2 Volumes). Dover Publications (1877).
[26] Adhikari, S., Woodhouse, J.: Identification of damping: Part 1, viscous damping. J. Sound Vib. 243(1), 43–61 (2001).
[27] Chowdhury, I., Dasgupta, S.: Computation of Rayleigh damping coefficients for large systems. Electron. J. Geotech. Eng. 8 (2003).

[28] Butterworth, J., Lee, J. H., Davidson, B.: Experimental determination of modal damping from full scale testing. In: Proc. 13th world Conf. Earthq. Eng. (13 WCEE), vol. 310, p. 15 (2004).
[29] Ewins, D. J.: Modal Testing. John Wiley & Sons Inc (1984).
[30] Clough, R., Penzien, J.: Dynamics of Structures. Computers & Structures (1995).
[31] Kerwin, E. M.: Micromechanics of damping in composite structures. ASTM Special Technical Publication, vol. 378. ASTM (1965).

Abderrahmane Benchirouf and Olfa Kanoun

Piezoresistivity of low carbon nanotubes content in elastomeric polymer matrix

Abstract: This work explores the realization of a nanocomposite based on elastomer silicone-organic polymer, with a low filler concentration and investigates their piezoresistive behavior. For this purpose, Poly(dimethylsiloxane) (PDMS) and multiwalled carbon nanotubes (MWCNT) are mixed using a solution-mixing method to prepare nanocomposite pressure-sensitive films. To accomplish a good homogeneous aqueous solution with uniform carbon nanotubes within the polymer matrix, the fabrication parameters such as filler content, the mixing ratio between the PDMS and MWCNTs, and the deposition volume are optimized. An organic solvent-isopropanol is used to dissolve the MWCNTs, which is then easily mixed with PDMS base polymer. The percolation threshold is reached at low CNT content at about 0.1 wt.%. The films at concentrations of 1 wt.% MWCNT exhibit good piezoresistive properties under applied pressure force with good repeatability and longtime stability. The fabrication approach used in this work demonstrates feasibility to use a simple method to deposit films that act as pressure sensors. Furthermore, as the prepared nanocomposite has a high viscosity, printing techniques such as screen printing can be easily adapted to deposit and manufacture the films.

Keywords: MWCNTs, PDMS, pressure sensor, nanocomposite, percolation theory, elastomer

1 Introduction

The integration of nanomaterials in polymers leads to the development of nanocomposites – a class of material having a wide range of application due to their outstanding electrical, thermal and mechanical characteristics. Among manifold applications, the nanocomposites are an excellent candidate for the development of sensors and actuators, such as strain or pressure sensors for healthcare and structural health monitoring, and flexible electronics due to the improved electrical and mechanical properties [1–8]. The use of carbon fiber composites has been investigated since 1961, and it acquired several more years to be used in a number of industries, including aerospace, automotive and infrastructure applications [9]. Until now, their usage was largely restricted by higher filler concentrations. Dissimilar to the traditional polymer composites, the desired properties can be achieved at a very low filler concentra-

Abderrahmane Benchirouf, Olfa Kanoun, Technische Universität Chemnitz, Chemnitz, Germany, e-mails: benchirouf@ieee.org, olfa.kanoun@etit.tu-chemnitz.de

https://doi.org/10.1515/9783110584455-017

tion in nanocomposites. For instance, nanofillers in the range of 3–5 wt.% reach the same mechanical reinforcement properties as 20–30 wt.% of traditional fillers known also as microsized fillers [10]. Consequently, nanocomposites are lightweight compared to the conventional composites, referred to its specific interfacial area, which increases the higher interfacial interactions between the nanofiller and the polymer, and thus improves its modulus. Some of the commonly used nanomaterials for the development of nanocomposites includes nanoparticles [11], nanowires [12], carbon nanotubes (CNTs) [13, 14] and graphene [15, 16]. Among them, CNTs have attracted numerous constituents in such nanocomposites owing to its high aspect ratio, excellent electrical conductivity, longitudinal polarizability, mechanical and thermal properties, which results in a significant enhancement of the composite properties at a very low electrical percolation threshold [15–17]. Based on the number of the graphene layers in its structure, CNTs can be categorized as single-walled (SWCNTs) or multiwalled (MWCNTs). While SWCNTs can be either semiconducting or metallic depending on its tube diameter and chirality, MWCNTs are only metallic [18]. Some experimental results show that the maximum current density of an individual SWCNT at room temperature is in the range of 10^9–10^{10} $A.cm^{-2}$, which is 1000 times higher than a copper and even higher than of some superconductors – with a maximum mean free path at the room temperature about 1 µm, compared to only 40 nm of copper [19]. Apart from the unique electrical properties, their mechanical properties show outstanding characteristics, such as tensile moduli of 0.3–1.47 TPa, tensile strengths between 10–100 GPa with a toughness of about 770 $J.g^{-1}$, and tensile strains varying between 6–12 % [19, 20]. However, the values of modulus and strength under uniform compression of a nanocomposite-based Epoxy and CNTs are even higher [21]. Moreover, the thermal conductivity at room temperature of an isolated SWCNT and MWCNT in the axial direction is 3500 and 3000 $W\,m^{-1}\,K^{-1}$, respectively [20]. Anyhow, MWCNT is mostly utilized in the nanocomposite blends due to its high purity and low costs; in addition, MWCNT achieves better dispersion than SWCNT, and it offers better conductivity [22]. Another advantage of CNTs in comparison to carbon black and metal nanoparticles is the low percolation threshold that can be reached at a very low weight percentage about 0.0025 % [23]. The percolation threshold is described as the minimum nanofiller volume fraction known also as critical filler content required to form a conductive path within the polymer matrix that increases the electrical conductivity of the polymer matrix dramatically by several orders of magnitude, it changes from an insulating state to a conductive state. Different parameters are attributed to the change in the electrical percolation threshold (Φc) of CNT within the polymer – some are related to CNTs such as quality and purity of the used CNTs, aspect ratio, dispersion quality and functionalization type of CNTs, that is, covalent or noncovalent functionalization and others depend strongly on the chemistry of polymer matrices that is, the type of the used polymer and the processing method. Generally, the increase in conductivity as a function of filler loading once electrical percolation has been achieved can be modeled by a simple power-law expression (scaling law) as shown in equation (1) [24], and the general behavior of the percolation threshold as a function of the nanofiller content is described in Figure 1:

$$\delta_c = \delta_0(\Phi - \Phi_c)^t, \quad \text{for } \Phi > \Phi_c \tag{1}$$

where δ_0 is the initial electrical conductivity of the polymer, Φ is the conductive filler volume fraction, and t is a scaling exponent.

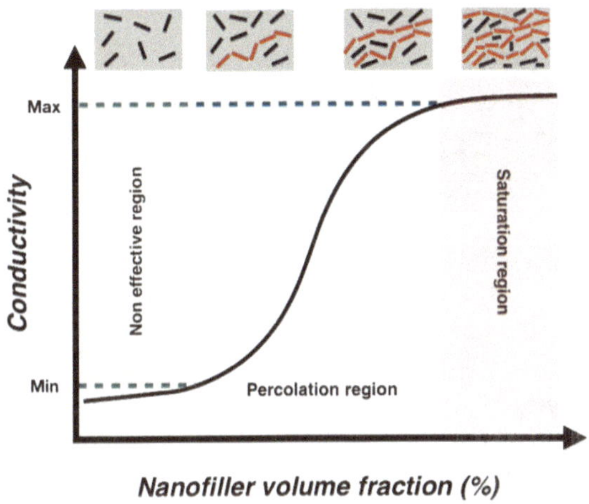

Figure 1: Description of the percolatigon threshold.

Although CNTs have excellent properties, the use of CNTs in real-time applications is limited due to their low processability and insolubility caused by the agglomeration of the CNTs, as shown in Figure 2. This agglomeration results from the strong van der Waals (vdW) attraction forces between the CNT tubes, which are considered to be around 0.5 eV/nm [25]. Accordingly, the interaction between the polymer matrix and the CNTs is the main key in achieving a homogeneous nanocomposite. Hence, the excellent properties of CNTs disappears in the transition of polymer improvement [13]. CNTs-based insulating polymers such as Epoxy, PAN, PBO, PET and PU have been profoundly investigated for force, strain, pressure, optical, chemical and biological sensors [26, 27]. Additionally, nanocomposites fabricated based on CNTs and intrinsically conductive polymers (ICPs), such as PA, P3AT, PANI, PEDOT, have been also deeply explored for different application fields such as transparent electrodes, layer injection in solar cells, photovoltaic cells, field emitters, supercapacitors and electrocatalysis [28, 29]. Among others, Polydimethylsiloxane/multi-walled carbon nanotube nanocomposite (PDMS:MWCNT) has attracted enormous attention and has been intensively studied because of their enticing electronic, chemical and mechanical properties due to the biocompatibility of this elastomer polymer [30]. PDMS silicon-based organic polymer (elastomers) consisted of a repeating $[SiO(CH_3)_2]_n$ unit, where n is

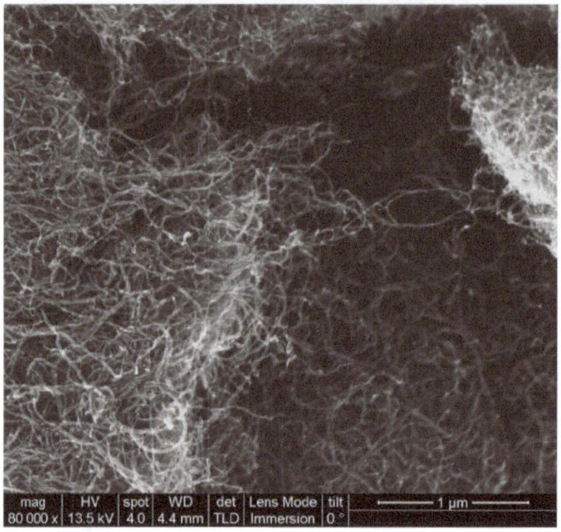

Figure 2: SEM image of the used MWCNTs in this work in its aggromolates form.

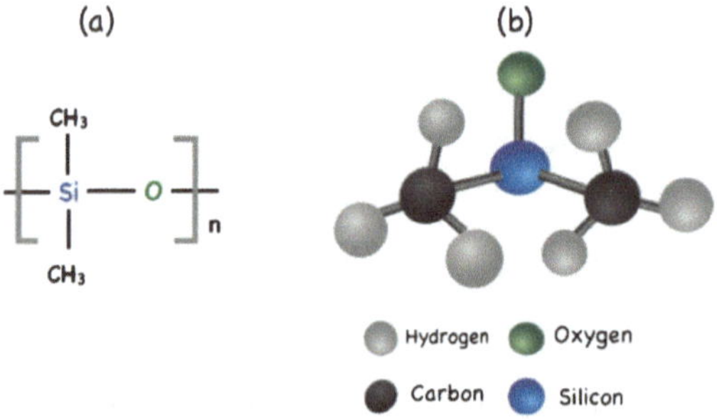

Figure 3: (a) Chemical formula and (b) 3D representation of the PDMS polymer.

the number of recurring monomer with an average value between 90 and 410 [31]. Figure 3 shows a chemical structure and three-dimensional (3D) molecule structure of the PDMS monomer. The glass temperature of the PDMS is less than −120 °C; hence, it is a rubbery liquid at room temperature.

2 Nanomaterials and nanocomposite fabrication

The road map for the PDMS:MWCNT nanocomposites fabrication has mainly four components – CNT as nanofiller, a solvent to functionalize the CNT, PDMS as base polymer and the curing agent that activates the polymer cross-linking. Primarily, the effective implementation of the CNT nanocomposites in different areas of application depends strongly on the ability of the CNTs to be well dispersed within a matrix. As mentioned previously, CNTs tend to form agglomerations due to the attraction forces between the tubes, thus makes fabricating CNT nanocomposites in viscoelastic polymer matrix very challenging. A good homogeneous dispersion quality is a crucial prerequisite, which is described by the absence of the macroscopic clusters of the nanotubes within the polymer improving the interfacial interaction between the CNT nanofiller and the polymer matrix. Numerous techniques have been established to promote the dispersion of the nanotubes; this involves surface modification, energetic agitation either mechanical or ultrasonic and shear mixing and ball – or microbead milling [32] [33]. Among them, the ultrasonic agitation is the most popular method, as it includes mixing the different CNTs concentrations in a solvent and then treating the concentration with ultrasonication energy. Typically, two different sorts of ultrasonic waves are used, depending on the frequency of the delivered waves: (1) low-frequency ultrasonic (between 20 and 24 kHz); and (2) high-frequency ultrasonic (between 42 and 50 kHz) [34]. Although treating the aqueous dispersion by the ultrasonication energy decreases the agglomeration size of the CNTs and improves its homogeneity, excessive amounts of energy might damage the CNTs and introduce defects on its sidewall [35]. The degree of homogeneity can be characterized by different techniques – namely, UV-Vis spectroscopy, scanning and tunneling electron microscopy, Raman spectroscopy, and particle size analyzer [36]. After a stable and homogenous aqueous dispersion is reached, the base PDMS polymer is added to the nanotube-organic solvent dispersion, which is followed by either a mixing process or by using an energy agitation. However, the mixing ratio between the CNT to PDMS is designated dependent on the desired application. For example, in pipelines fiber-reinforcement and strengthening automobile portions, a high CNT content is required to reach a high strength of the nanocomposite; and in order to apply pressure and strain sensors and to enhance the sensitivity, a lower CNT content is needed [37]. The activation of the cross-linking between the polymer chains is usually completed after the pure polymer and the nanotube are mixed together. Typically, a mixing ratio of 10:1 between the base PDMS polymer and the curing/cross-linking agent is used. This step is followed by a mixing process under temperature to improve the homogeneity of the final nanocomposite and to evaporate all the solvent used to disperse the CNTs. Once the organic solvent is totally evaporated from the nanocomposite, a degassing process under vacuum condition should be performed to remove all the air bubbles from the nanocomposites. It is important to mention that adding the curing agent will decrease the flexibility of the nanocomposites and increase its hardness [30, 32]. Consequently, the curing agent affects strongly the mechanical properties of the resulting nanocomposites. Figure 4 shows the standard steps used for the fabrication of the elastomeric nanocomposites. Microscopy techniques are also used to study other features such as morphology and topography, which determine the qualitative grade of affinity between CNTs and the polymer matrix. As depicted from Figure 1, the electrical conductivity of films-based nanocomposite increases with the increasing CNT concentration up to a critical filler concentration with the base polymer, which is proportional to the number of the conduction paths available within the nanocomposite matrix but also on its correspond-

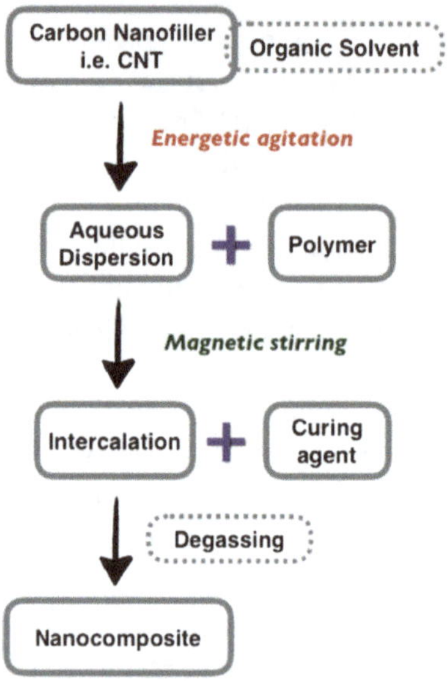

Figure 4: Schematic of the preparation method of PDMS:MWCNT nanocomposite.

ing morphology as the scattering effect occurring in the nanocomposite affect negatively on its conductivity, thus the electron conduction is enabled through a "hopping" or "tunneling" mechanism as it is shown in Figure 5. At this concentration, the high conductive CNT clusters embedded in the high-resistance polymer PDMS matrix forms a randomly 3D network, hence an electron can hop and/or tunnel from one CNT to another one to overcome the high resistance offered by an insulating polymer matrix (see Figure 5), this type of networks is known also as disordered systems. In this disordered system, a metal-insulator transition occurs due to the hopping conductivity mechanism [15, 38, 39]. In this mechanism, the conduction is regulated by the electron hopping between nanocarbon clusters. The hopping conductivity (σ_{hopp}) of the electron jumping between carbon nanotube clusters is evaluated using [40]:

$$\sigma_{\text{hopp}} = \sigma_0 \exp\left(-\frac{4}{3}\left(\frac{4\alpha r_{\text{tun}}}{a}\right)^{\frac{3}{4}}\left(\frac{W_0}{\kappa T}\right)^{\frac{1}{4}}\right) \tag{2}$$

where σ_0 is the pre-exponential factor is calculated taking into account the character of localization centers distribution, α is the characteristic "borous radius" of the considered "doping" center, $a = 0.70$ is an empirical constant resulting from Monte-Carlo numerical simulations [40], W_0 is the characteristic potential barrier for electron tunneling, κ is the Boltzmann constant, T is the sample temperature, r_{tun} is the tunneling path length of the electron, which is the distance between two neighboring nanotubes, σ_0 is the pre-exponential normalization factor referred to the conductivity of monolithic dielectric medium. This pre-exponential normalization factor can be

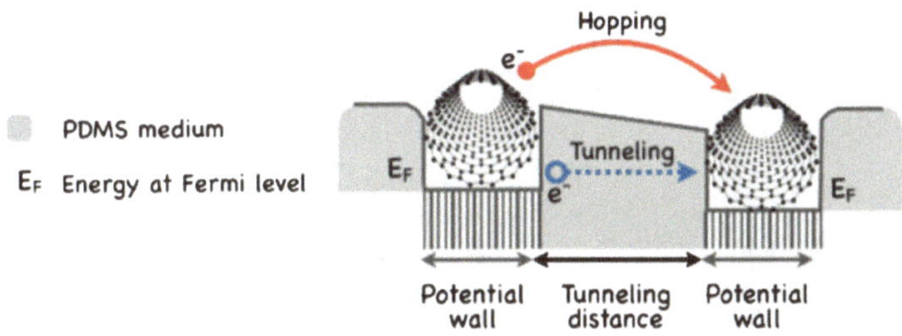

Figure 5: Potential model for hopping/tunneling mechanism in PDMS:MWCNT nanocomposite.

obtained using the following equation [40]:

$$\sigma_o = \frac{16e^3 r_{\text{tun}}^3 N}{\pi \hbar a_o} \tag{3}$$

where N is the concentration of the carbon nanotube localization centers. For arbitrary distribution of localization centers $r_{\text{tun}}^3 \times N = 0.24 \times 3$ [40].

By increasing the nanofiller concentration, the embedded nanofiller get close enough to each other. When they reach a range of 1–2 nm, the electrons can tunnel from one filler to another through the polymer medium, in which they have different energies at the Fermi level – this is a typical quantum phenomenon called tunneling [41]. The value of the tunneling resistance can be estimated using Simon's formula [41]:

$$R_{\text{tunneling}} = \frac{V}{AJ} = \frac{h^2 \times d}{A \times e^2 \times \sqrt{2m\lambda}} \times \exp\left(\left(\frac{4\pi d}{\hbar}\right)\sqrt{2m\lambda}\right) \tag{4}$$

where V is the potential difference, A is the cross-sectional area of the filler, J is the tunneling density, h is the height of the electrical barrier of the polymer, d is the distance between nanofillers, e is the electron charge and m is the electron mass, λ is the wavelength and $\hbar$ is Planck's constant.

As it is very difficult to measure the tunneling resistance, many calculations were conducted to estimate this resistance. For a randomly distributed network with polymer matrix and carbon nanocluster, different parameters affect the change in the tunneling resistance such as contact area, contact gap, junction type (metallic/metallic or metallic/semiconducting) (Figure 6 (a)), polymer molecule size and degree of dispersibility. However, the intrinsic tube resistance is significantly smaller compared to the tunneling resistance hence negligible. Furthermore, other issues, such as orientation and alignment of the CNTs, curvature, twisting and curling of tubes within the polymer, further degrades the total conductivity of nanocomposites. The spatial distribution of nanocarbon clusters is higher in the longitudinal electric field orientations and lower in the transverse ones. When the films made of PDMS:CNT undergo an applied stimulus such as pressure, the resistance of the film changes dramatically;

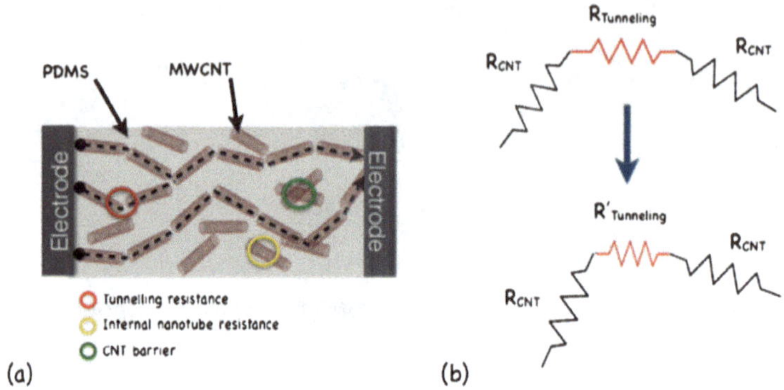

Figure 6: Model of nanocomposite polymer material with carbon nanocluster inclusions (a) Schematic of the CNT random network and (b) Resistance representation of the CNT network before (upper) and after (lower) applied force.

this characteristic is known as piezoresistivity (Figure 6 (b)). Generally, the piezoresistive behavior of the nanocomposite is affected by three main working mechanisms: (1) change of the internal conductive network formed by the CNTs due to the change, deterioration and reformation of the CNT network, (2) tunneling effect among neighboring CNTs, (3) internal CNTs' piezoresistivity.

It is found that the first and the second working mechanisms (i. e., the change of the internal conductive network and the tunneling effect) play a major role on the piezoresistivity of the nanocomposite strain and pressure sensors (Figure 6 (b)), where the influence from the CNTs' piezoresistivity is quite small [41–43]. The piezoresistive sensors based on PDMS:CNT nanocomposites can be regarded as a three-dimensional (3D) resistor network, as the whole resistance is the sum of three components: the resistance of the contact electrodes, the change of internal CNT conductive network, and the change in the tunneling distance between the nanotube clusters, which can be represented as follows:

$$R_{\text{piezoresistive}} = R_{\text{contact}} + R_{\text{CNT}} + R_{\text{Tunneling}} \tag{5}$$

As the contact electrodes are metal, their change resistance created by applied pressure is very small, thus ignorable. Generally speaking, piezoresistive sensors are suitable for use over broad pressure ranges, making it possible to be used over a large application area, as shown in Figure 7. Subsequently, fabricating PDMS:CNT pressure sensors with an optimized sensitivity is critical. Therefore, different efforts have been dedicated to improve the pressure sensor sensitivity. However, and due to the viscoelasticity of the PDMS, slow response and long relaxation time are two serious exiting problems for such kinds of sensors. The aim of this work is to provide flexible piezoresistive pressure sensor with a pressure range of 1 MPa, reduced MWCNT concentration that is cost effective, and improved fabrication process of heterogeneous nanocomposites that offer better sensitivity for change in pressure. This type of sen-

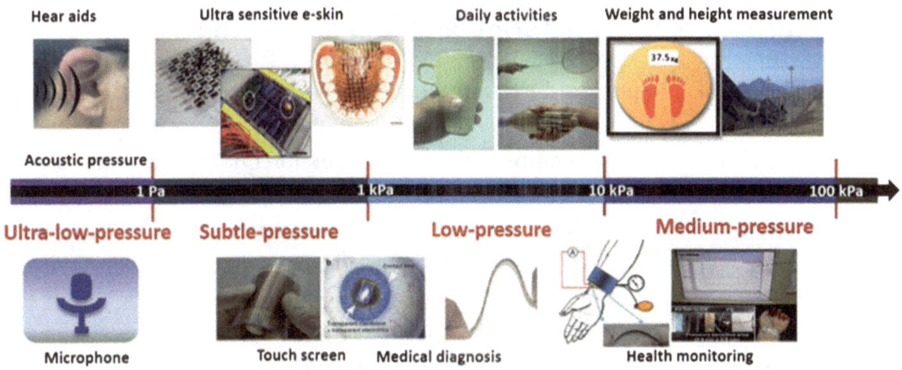

Figure 7: Diagram of pressure regimes and their relevant applications [40].[1]

sors could be used in biomedical and health monitoring applications where tactility and flexibility play an important role.

3 PDMS:CNT nanocomposite preparation and its characterization

In this section, the materials used to prepare the PDMS:MWCNT nanocomposites and the characterization technique setup is explained. Here, the solution processing method as described in Figure 4 is implemented to prepare the nanocomposite, and the physical liquid deposition is adopted to fabricate the films based on PDMS:MWCNT nanocomposite.

3.1 Material characteristics

In this experiment, the used MWCNTs were purchased from Southwest Nano Technology with >95 % degree of purity, outer diameter of 6–9 nm, and length less than 1 µm. The PDMS base polymer and the curing agent were Silicon Elastomer Sylgard 184 from Dow Corning, Inc. (Midland, MI, USA) – both are colorless with low viscosity and having specific gravity of 1.11 and 1.03 g.cm^{-3}, respectively. They are chemically stable and they do not form any hazardous polymerization. It is highly recommended to mix the base polymer and the base with a ratio of 10:1 then exposed to thermal curing in order to cross-link the PDMS. However, the mixing ratio could be modified in order to adjust the mechanical properties of final PDMS, that is, Young's modulus. So, after the cross-linking, PDMS samples present an external hydrophobic surface.

1 Picture licensed by Royal Society of Chemistry.

3.2 Nanocomposite preparation and film deposition

The nanocomposites are prepared by dissolving different proportions of MWCNT 0.1, 0.5, 0.75 and 1 wt.% in an organic solvent. Determining the organic solvent to dissolve the CNTs is crucial for determining the final dispersion quality. Solvents including toluene, tetrahydrofuran (THF), chloroform and dimethylformamide (DMF) have been reported to reach a great dispersion quality [44]. Among them, using Isoprpanol (Propan-2-ol) at low temperature is reported as a good solvent to prepare the CNT colloids to avoid substantial damage to CNTs during the sonication process, and it led consequently to more homogeneous and stable dispersions as the CNTs are largely individualized in the solution [45]. To provide a high shear forces to break down the CNT bundles, the aqueous dispersion is sonicated using a high-frequency horn-tip probe 42 kHz in pulse mode (0.5 s on and 0.5 s off cycle) for a total time of 30 minutes at 15 % amplitude. The ultrasonic probe is immersed until the middle of the tube containing the CNT and the solvent. This burst mode is selected to minimize damages on the CNT sidewalls [45]. The maximum energy delivered to the system is 15.3 kJ during the entire sonication process. All of the processes are conducted in an ice bath to maintain the temperature low and avoid overheating which could also cause damaging of CNTs. Afterwards, the base PDMS polymer is added to the MWCNT dispersion with mixing ratio of 4:4, 5:3 and 6:2 vol. and then the dispersion is mixed in a magnetic stirrer at 800 rpm for 1 h at 65 °C to improve the attachment of the PDMS base on the CNTs. Afterwards, the curing agent is added in the ratio 10:1 that is, ten parts of MWCNT and PDMS suspension and one part of curing agent and then it was magnetically stirred for 20 min at 800 rpm and at 65 °C, so that the solvent is completely evaporated during the mixing process, which could be visually determined by weight, or by calculating the volume of the nanocomposite. Once the process is completed, the nanocomposite is placed for 2 h in a degassing chamber under vacuum to remove any air bubbles formed during the cross-linking stage. A detailed illustration of the preparation process and how the solvent and the polymer is wrapped by the CNTs is illustrated in Figure 8. The flexible polyimide Kapton HN 500 having a thickness of 125 μm is used as substrate to deposit the nanocomposite. Prior deposition, the substrate is pretreated to remove all the contaminations from its surface and to improve the wettability of the nanocomposite on its surface. First, the substrate is immersed in an isopropanol bath and sonicated for 10 min. This is followed by distilled water washing step and dried by a nitrogen flow. The substrate is covered with a polyimide mask having a dimension of 2.5 × 1 cm with the help of a physical liquid deposition known also as drop casting, the aqueous dispersion is deposited to form the film. To improve the homogeneity and reproducibility, volumes of 50, 62.5, 75 and 87.5 μL are used to form the films to define the optimized volume. After deposition, the films are annealed at 100 °C for 45 min. To make the connections, the mask is carefully removed and the metallic contacts are deposited on the sample surface using a silver conductive paste. This method is easy, fast, and reproducible and does not require any special and expensive equipment such as thermal

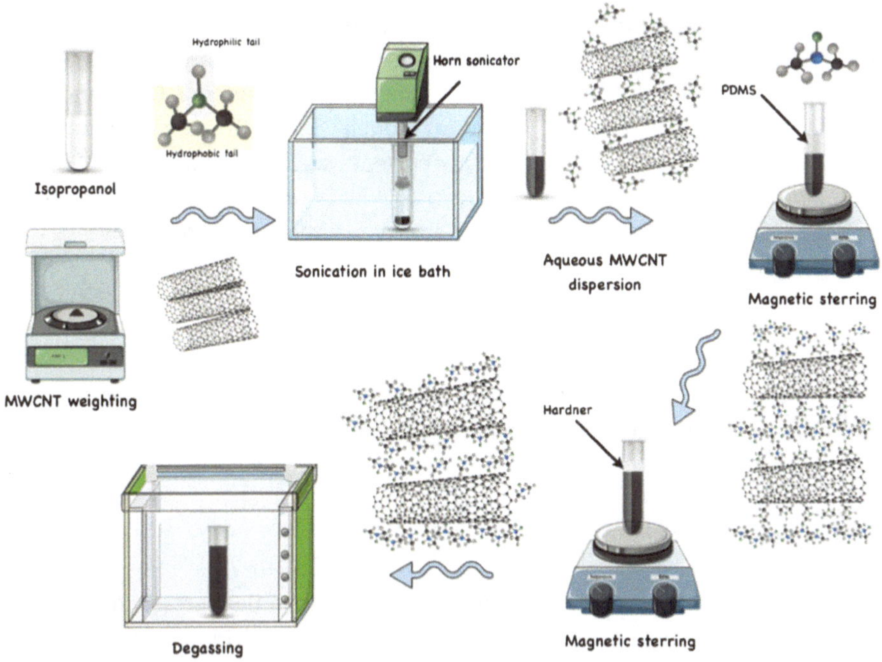

Figure 8: Schematics of PDMS:MWCNT hybrid nanocomposite fabrication steps with an illustration of how the solvent is detached and the wrapping of PDMS polymer on the CNTs.

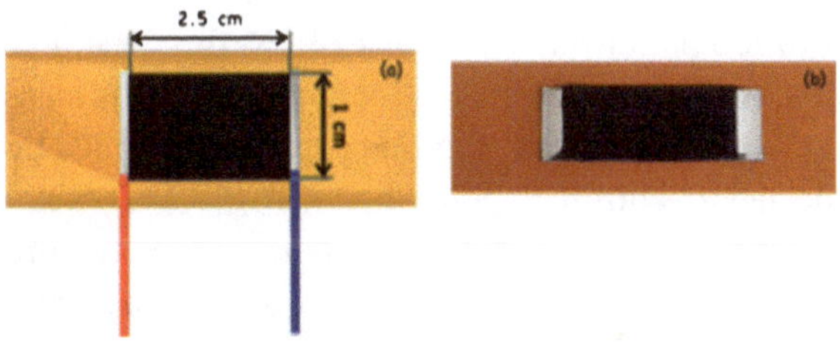

Figure 9: (a) Schematic and (b) snapshot of PDMS:MWCNT hybrid nanocomposite film.

evaporation or sputtering techniques. The conductive paste is kept for several minutes to dry as the solvent is evaporated. Figure 9 shows a schematic and snapshot of the PDMS:MWCNT film.

3.3 Films based on nanocomposite characterization techniques

3.3.1 Electrical characterization

The electrical measurements are performed for each specimen using two-wire resistance measurement technique instead of the four-wire technique used typically for highly conductive materials [46]. In order to perform I-V measurements, the specimen is places on flat surface to avoid the effect of any bending and connected to a Keithlay 2602 dual-channel source meter (Keithley Instruments Inc., Cleveland, OH, USA) which is connected to a host computer through GPIB/USB cable, and by applying voltage from −0.5 to +0.5 V, the resulting current was recorded and the DC-ohmic resistance is estimated at room temperature using the Ohm's law (equation (6)) and at least three independent specimens of each fabricated nanocomposite are investigated to ensure high reproducibility:

$$R = \frac{\Delta V}{\Delta I} \tag{6}$$

with $\Delta V = V_1 - V_2$ and $\Delta I = I_1 - I_2$ and R is the film ohmic resistance in Ohm; V is the applied voltage in Volt and I is the measured current in amps.

3.3.2 Piezoresistance characterization under pressure

The pressure test is conducted by placing the specimen on flat surface to avoid the effect of any bending that can cause changes to the initial resistance of the film. The piezoresistive evaluation is carried out at room temperature using universal test machine Inspekt 10 table (Hegawald and Peschke, Me*β* – und Prüftechnik GmbH). Using a compressive load cell which has a surface area of 1 cm^2, a force is applied in the vertical direction (out-plane) on the surface of the nanocomposite film at a constant load speed of 1 mm/min from 0 N to 100 N and with a step size of 5 N, this generates a maximum pressure force of 1 MPa. At each step force, the I-V characteristic is measured by applying voltage from −0.5 to +0.5 V and the resulting current is recorded by a Keithley 2602 sourcemeter connected to a host computer through GPIB/USB cable. Later, the resistance profile of the specimen is calculated. Schematic of the pressure measurement setup is depicted in Figure 10. The measurement process of all specimen is accomplished in the same way.

4 Results and discussion

4.1 DC electrical characterization

The PDMS:MWCNT specimen fabricated using 1 wt.% MWCNT and mixing ratio of 5:3 between PDMS to MWCNT by different drop volume show a linear response, in the applied voltage range ±0.5 V (mA range under 0.5 V), as shown in Figure 11(a). It is clearly seen that increasing the amount of the deposition volume decreases the resistance of the film as shown in Figure 11(b). This is referred to the increase in the number of conduction paths formed by the CNTs within the PDMS polymer matrix which

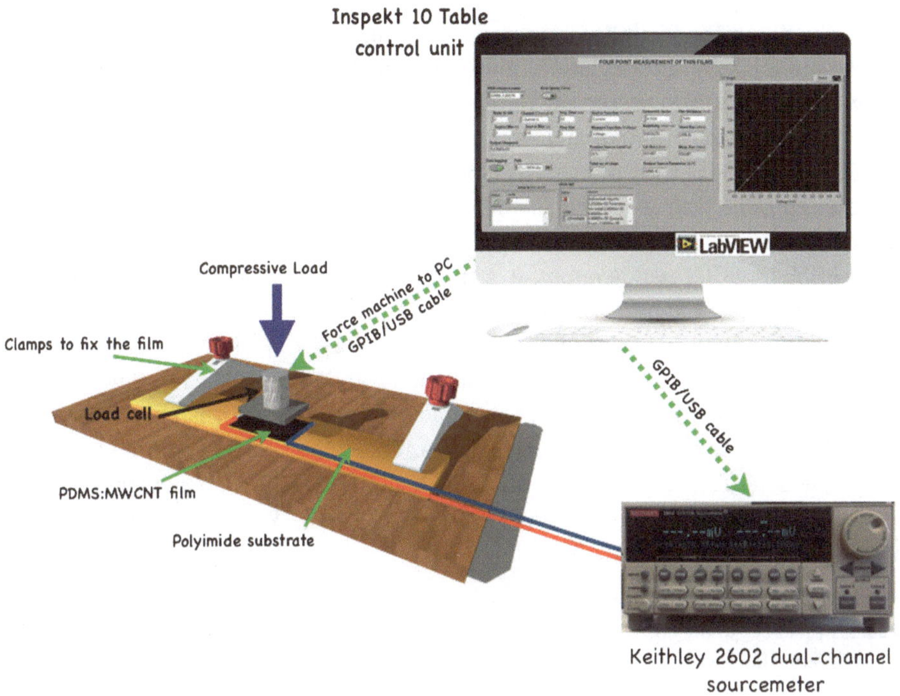

Figure 10: Measurement setup of the pressure used to evaluate the PDMS:MWCNT nanocomposite film.

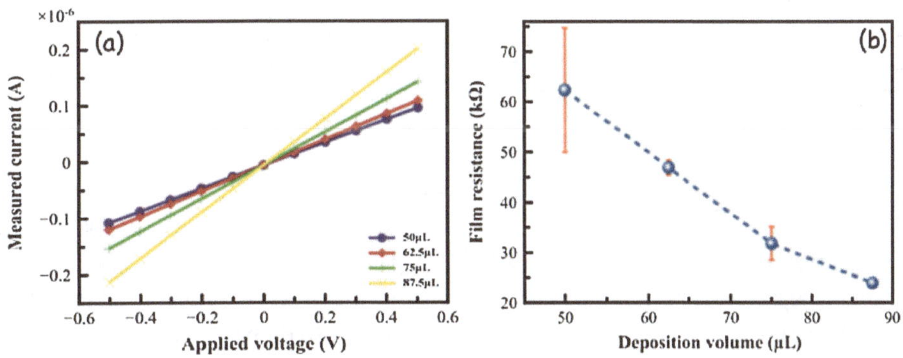

Figure 11: Linear response of PDMS + 1 wt.% MWCNT specimen at different deposition volume (a) I-V curves and (b) resistance of the films.

leads to decrease in the tunneling distances between the CNT clusters and thus more charge carrier, that is, electrons are delocalized between the clusters, consequently improving the film conductivity. The PDMS:MWCNT film resistance at 1wt.% MWCNT are measured to be 62.29, 46.87, 31.76 and 23.92 kΩ for 50, 62.5, 75 and 87.5 μL, respectively. The film reproducibility improved significantly which is well remarked by the decrease in the standard deviation from 24.7 % to 1.5 % for 50 µL and 87.5 µL, respectively. Therefore, and to ensure a high reproducibility of the fabricated films, a volume of 87.5 µL is used after that for all investigations (Figure 11).

The resistance of the PDMS:MWCNT specimen can be adjusted by using different CNT concentration range as well as different mixing ratio between the PDMS and the MWNCTs, as shown in Table 1. It is notable from Table 1 that the percolation threshold of PDMS:MWCNT nanocomposite is approximately at 0.1 wt.%, knowing that the initial resistivity of the pure PDMS matrix is the range of 10^{13} Ω. It is can be also concluded that the PDMS:MWCNT with mixing ratio of 5:3 at 1 wt.% MWCNT, gave good electrical characteristics with high reproducibility, and because the reproducibility of the electrical properties is decisive for stable mechanical performance, this mixing ratio is further chosen for the piezoresistivity investigation. Another reason to choose the 1 wt.% is, if the electrical resistance is too high, it becomes difficult to differentiate between the measured signal and the electrical noises. The 5:3 mixing ratio is referred in this case as the optimal mixing ratio.

Table 1: Resistance of the PDMS:MWCNT film at different mixing ratio between the PDMS and MWCNT and at different MWCNT concentrations.

CNT content (wt.%)	Resistance of films based on different mixing ratio of PDMS:MWCNT (kΩ)		
	4:4	5:3	6:2
0.1	400.1 ± 4.8 %	628.19 ± 21 %	994.16 ± 50.1 %
0.5	125.77 ± 34.92 %	181.02 ± 27.05 %	211 ± 10.9 %
0.75	34.33 ± 18 %	44.32 ± 5.1 %	45.97 ± 14.6 %
1	21.76 ± 11.9 %	**23.92** ± **1.5** %	29.13 ± 6.2 %

4.2 Piezoresistive evaluation

The piezoresistive performance of the PDMS:MWCNT film is shown in Figure 12 as a normalized change in resistance-pressure relationship under a vertical axial force for specimen having 1 wt.% of MWCNT at 5:3 mixing ratio with PDMS. A nonlinear piezoresistivity is clearly defined, which can be regressed by a power function:

$$Y = a \times \exp^{(bX)} \tag{7}$$

where a is the coefficient (7.66×10^{-6}), and b is the power exponent (1.95).

The linear regression model can be adapted to linearize the model expressed in equation (7), square of the correlation coefficient $R_2 = 0.998$ indicates a good linearity. The linearization equation used for PDMS:MWCNT films is as follows:

$$Y = 1.95X - 13.81 \tag{8}$$

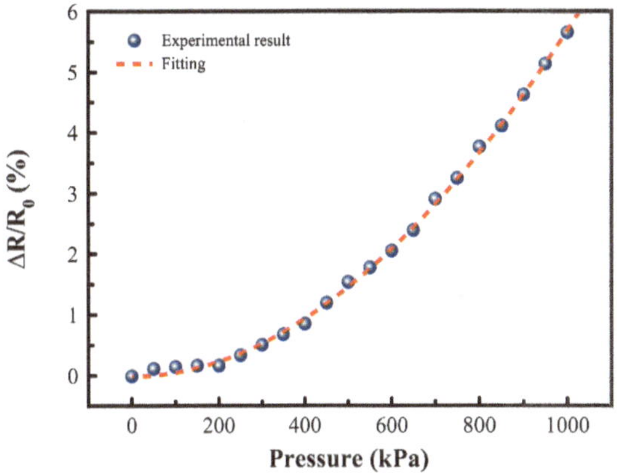

Figure 12: Piezoresistivity response of PDMS + 1 wt.% MWCNT specimen under applied pressure.

where *a* is the gain (1.95) and *b* is the coefficient (13.81). Equation (8) is used to obtain the pressure sensitivity of the PDMS:MWCNT nanocomposite.

The increase in the PDMS:MWCNT film resistance with increasing pressure is due to the breakup of the conductive paths, which results in the high sensor sensitivity. This breakup of the conductive paths within the PDMS matrix alters the tunneling distance between MWCNTs clusters during the compressive deformation of the nanocomposites, as it has been explained in Section 2, that consequently varies the interdistance tunneling distances between the CNTs exponentially; therefore, the resistance-pressure relationship is nonlinear [47–49]. The effect of pressure cycling is studied at room temperature, as it is depicted in Figure 13. Four cycles are applied to the PDMS:MWCNT film. A clear observation can be made of a drop in the piezoresistivity from 5.6 % at the first cycle to 4.8 % at the fourth cycle and then reaching stability. The hysteresis, on the other side, tend to decrease from 36.09 % at the first cycle to 17.08 % in the fourth cycle, which is a sign of a good recovery.

To investigate the creep, a constant load throughout the test on the prepared PDMS:MWCNT film are maintained. A constant force at 1 MPa for 30 min is applied and the change in the electrical resistance is monitored, as shown in Figure 14. It is noticeable that the creep of the PDMS:MWCNT specimen has a typical viscoelastic–viscoplastic behavior, whereas the creep gradually decreases with time and the applied pressure load [50]. A delay is observed in the resistance from 4.73 % in the first minute to about 2.25 % in 25 min, then it tends to stabilize.

5 Conclusion

In this study, an investigation is performed on the piezoresistive response of the PDMS:MWCNT film to demonstrate its feasibility as a pressure sensor under a large

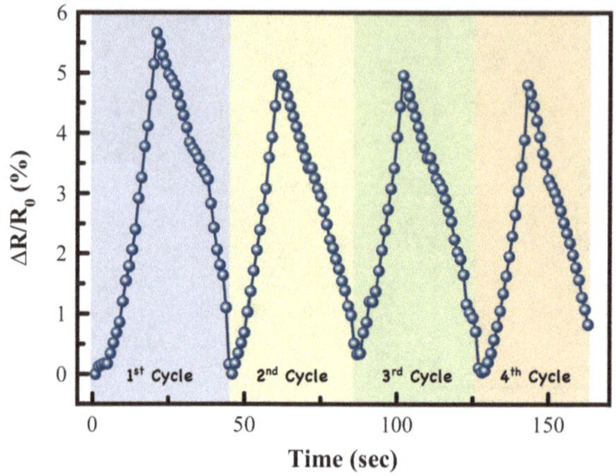

Figure 13: Cycling response of PDMS:MWCNT film under pressure loading.

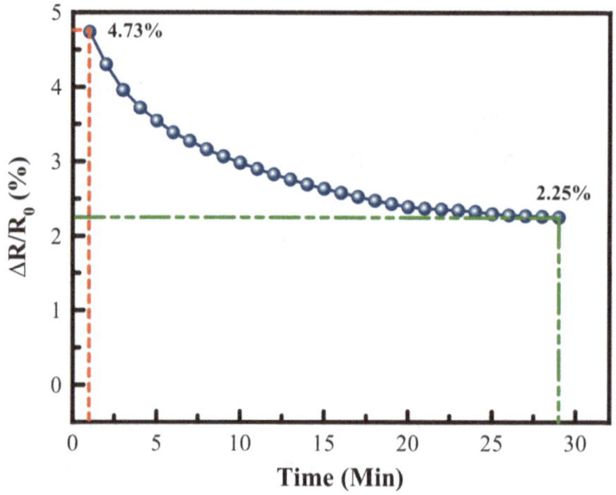

Figure 14: Creep of the PDMS:MWCNT film at a constant applied pressure.

compressive force until approximately 1 MPa. The PDMS:MWCNT nanocomposite is prepared using a simple cost-effective solution mixing method for human healthcare, motion detection sensors, and smart textiles. The experimental results reveal that the percolation threshold is reached at low CNT concentration at about 0.1 wt.% MWCNT. However, the 1 wt.% with a PDMS:MWCNT mixing ratio of 5:3 is chosen for the piezoresistive characterization, as it is has an adequate electrical resistance value that permits to differentiate between the measured signal and the electrical noises. Although the PDMS:MWCNT nanocomposite film exhibits a nonlinear piezoresistive behavior under pressure load, a good pressure sensitivity is obtained. The change in the resistance is proportional to the applied pressure, which is due to change in the tunneling distances between the CNT clusters. Besides, the hysteresis test showed that the loss in the energy falls to 50 % after the fourth cycle, while maintaining the same loading–unloading profile, which is an indication of a good repeatability. Whereas, the PDMS:MWCNT nanocomposite showed a typical viscoelastic–viscoplastic behavior under creep and longtime stability after 25 min.

Bibliography

[1] Kumar, A., Rajput, J. K.: Polymer Nanocomposites-Materials for Sensor Technology. In: Materials Science Forum. Trans Tech Publications, vol. 757, pp. 197–216 (2013).

[2] Kádár, R., Abbasi, M., Figuli, R., Rigdahl, M., Wilhelm, M.: Linear and nonlinear rheology combined with dielectric spectroscopy of hybrid polymer nanocomposites for semiconductive applications. Nanomaterials 7(2), 23 (2017).

[3] Silvestre, J., Silvestre, N., De Brito, J.: Polymer nanocomposites for structural applications: Recent trends and new perspectives. Mech. Adv. Mat. Struct. 23(11), 1263–1277 (2016).

[4] Tripathy, D. K., Sahoo, B. P. (eds.): Properties and Applications of Polymer Nanocomposites: Clay and Carbon Based Polymer Nanocomposites. Springer (2017).

[5] Benchirouf, A., Palaniyappan, S., Ramalingame, R., Raghunandan, P., Jagemann, T., Müller, C., Hietschold, M., Kanoun, O.: Electrical properties of multi-walled carbon nanotubes/PEDOT: PSS nanocomposites thin films under temperature and humidity effects. Sens. Actuators B, Chem. 224, 344–350 (2016).

[6] Benchirouf, A., Müller, C., Kanoun, O.: Electromechanical behavior of chemically reduced graphene oxide and multi-walled carbon nanotube hybrid material. Nanoscale Res. Lett. 11(1), 4 (2016).

[7] Sanli, A., Benchirouf, A., Müller, C., Kanoun, O.: Piezoresistive performance characterization of strain sensitive multi-walled carbon nanotube-epoxy nanocomposites. Sens. Actuators A, Phys. 254, 61–68 (2017).

[8] Bouhamed, A., Müller, C., Choura, S., Kanoun, O.: Processing and characterization of MWCNTs/epoxy nanocomposites thin films for strain sensing applications. Sens. Actuators A, Phys. 257, 65–72 (2017).

[9] Compositeslab. [Internet]. Available from: http://compositeslab.com/composites-101/history-of-composites/ [Accessed: 2020-07-26].

[10] Bhattacharya, M.: Polymer nanocomposites—a comparison between carbon nanotubes, graphene, and clay as nanofillers. Materials 9(4), 262 (2016).

[11] Mohanty, S., Nayak, S., Kaith, B., Kalia, S.: Chapter 9. In: Polymer Nanocomposites Based on Inorganic and Organic Nanomaterials, pp. 257–346. Scrivener Publishing (2015).

[12] Abramson, A. R., Kim, W. C., Huxtable, S. T., Yan, H., Wu, Y., Majumdar, A., Tien, C., Yang, P.: Fabrication and characterization of a nanowire/polymer-based nanocomposite for a prototype thermoelectric device. J. Microelectromech. Syst. 13(3), 505–513 (2004).

[13] Kanoun, O., Müller, C., Benchirouf, A., Sanli, A., Dinh, T. N., Al-Hamry, A., Bu, L., Garlach, C., Bouhamed, A.: Flexible carbon nanotube films for high performance strain sensors. Sensors 14(6), 10042–10071 (2014).

[14] Madni, A., Noreen, S., Maqbool, I., Rehman, F., Batool, A., Kashif, P. M., Khan, M. I.: Graphene-based nanocomposites: synthesis and their theranostic applications. J. Drug Target. 26, 1–26 (2018).

[15] Min, C., Shen, X., Shi, Z., Chen, L., Xu, Z.: The electrical properties and conducting mechanisms of carbon nanotube/polymer nanocomposites: A review. Polym.-Plast. Technol. Eng. 49(12), 1172–1181 (2010).

[16] Sathyanarayana, S., Hübner, C.: Thermoplastic nanocomposites with carbon nanotubes. In: Structural Nanocomposites, pp. 19–60. Springer, Berlin, Heidelberg (2013).
[17] Nan, C. W., Shen, Y., Ma, J.: Physical properties of composites near percolation. Annu. Rev. Mater. Res. 40, 131–151 (2010).
[18] Wong, H. S. P., Akinwande, D.: Carbon nanotube and graphene device physics. Cambridge University Press (2011).
[19] Lekawa-Raus, A., Patmore, J., Kurzepa, L., Bulmer, J., Koziol, K.: Electrical properties of carbon nanotube based on fibers and their future use in electrical wiring. Adv. Funct. Mater. 24(24), 3661–3682 (2014).
[20] Chiolerio, A., Castellino, M., Jagdale, P., Giorcelli, M., Bianco, S., Tagliaferro, A.: Electrical properties of CNT-based polymeric matrix nanocomposites. In: Carbon nanotubes-polymer nanocomposites. InTech (2011).
[21] Lourie, O., Wagner, H. D.: Evaluation of Young's modulus of carbon nanotubes by micro-Raman spectroscopy. J. Mater. Res. 13(9), 2418–2422 (1998).
[22] Grady, B. P.: Carbon nanotube-polymer composites: manufacture, properties, and applications. John Wiley & Sons (2011). Chapter 5.
[23] Sandler, J., Kirk, J. E., Kinloch, I. A., Shaffer, M. S. P., Windle, A. H.: Ultra-low electrical percolation threshold in carbon-nanotube-epoxy composites. Polymer 44(19), 5893–5899 (2003).
[24] Essam, J. W.: Percolation theory. Rep. Prog. Phys. 43(7), 833 (1980).
[25] Walters, D. A., Casavant, M. J., Qin, X. C., Huffman, C. B., Boul, P. J., Ericson, L. M., Smalley, R. E.: In-plane-aligned membranes of carbon nanotubes. Chem. Phys. Lett. 338(1), 14–20 (2001).
[26] Hwang, S. H., Park, Y. B., Yoon, K. H., Bang, D. S.: Smart materials and structures based on carbon nanotube composites. Carbon Nanotubes-Synthesis, Characterization, Applications. InTech (2011).
[27] Song, K., Zhang, Y., Meng, J., Green, E. C., Tajaddod, N., Li, H., Minus, M. L.: Structural polymer-based carbon nanotube composite fibers: understanding the processing–structure–performance relationship. Materials 6(6), 2543–2577 (2013).
[28] Kumar, V., Kalia, S., Swart, H. C. (eds.): Conducting Polymer Hybrids. Springer, (2017).
[29] Benchirouf, A., Palaniyappan, S., Ramalingame, R., Raghunandan, P., Jagemann, T., Müller, C., Hietschold, M., Kanoun, O.: Electrical properties of multi-walled carbon nanotubes/PEDOT: PSS nanocomposites thin films under temperature and humidity effects. Sens. Actuators B, Chem. 224, 344–350 (2016).
[30] Stöckelhuber, K. W., Das, A., Klüppel, M. (eds.): Designing of Elastomer Nanocomposites: From Theory to Applications. Springer (2017).
[31] Thakur, V. K., Thakur, M. K. (eds.): Handbook of Polymers for Pharmaceutical Technologies. Biodegradable Polymers, vol. 3. John Wiley & Sons (2015).
[32] Kim, J. H., Hwang, J. Y., Hwang, H. R., Kim, H. S., Lee, J. H., Seo, J. W., Lee, S. H.: Simple and cost-effective method of highly conductive and elastic carbon nanotube/polydimethylsiloxane composite for wearable electronics. Sci. Rep. 8(1), 1375 (2018).
[33] Tas, M. O., Baker, M. A., Masteghin, M. G., Bentz, J., Boxshall, K., Stolojan, V.: Highly Stretchable, Directionally Oriented Carbon Nanotube/PDMS Conductive Films with Enhanced Sensitivity as Wearable Strain Sensors. ACS Appl. Mater. Interfaces 11(43), 39560–39573 (2019).
[34] Khosla, A.: Nanoparticle-doped electrically-conducting polymers for flexible nano-micro Systems. Electrochem. Soc. Interface 21(3–4), 67–70 (2012).
[35] Yang, K., Yi, Z., Jing, Q., Yue, R., Jiang, W., Lin, D.: Sonication-assisted dispersion of carbon nanotubes in aqueous solutions of the anionic surfactant SDBS: The role of sonication energy. Chin. Sci. Bull. 58(17), 2082–2090 (2013).

[36] Njuguna, J., Vanli, O. A., Liang, R.: A review of spectral methods for dispersion characterization of carbon nanotubes in aqueous suspensions. J. Spectrosc. 2015, 1–11 (2015).
[37] Benchirouf, A., Müller, C., Kanoun, O.: Electromechanical behavior of chemically reduced graphene oxide and multi-walled carbon nanotube hybrid material. Nanoscale Res. Lett. 11(1), 4 (2016).
[38] Nakatsuji, S., Dobrosavljević, V., Tanasković, D., Minakata, M., Fukazawa, H., Maeno, Y.: Mechanism of hopping transport in disordered Mott insulators. Phys. Rev. Lett. 93(14), 146401 (2004).
[39] Matsutani, S., Sato, I.: A novel conductivity mechanism of highly disordered carbon systems based on an investigation of graph zeta function. Phys. Lett. A 381(36), 3107–3111 (2017).
[40] Bellucci, S., Shunin, Y., Gopeyenko, V., Lobanova-Shunina, T., Burlutskaya, N., Zhukovskii, Y.: Real time polymer nanocomposites-based physical nanosensors: theory and modeling. Nanotechnology 28(35), 355502 (2017).
[41] Hu, N., Fukunaga, H., Atobe, S., Liu, Y., Li, J.: Piezoresistive strain sensors made from carbon nanotubes based polymer nanocomposites. Sensors 11(11), 10691–10723 (2011).
[42] Chen, J., Zhu, Y., Jiang, W.: A stretchable and transparent strain sensor based on sandwich-like PDMS/CNTs/PDMS composite containing an ultrathin conductive CNT layer. Compos. Sci. Technol. 186, 107938 (2020).
[43] Hu, B., Hu, N., Li, Y., Akagi, K., Yuan, W., Watanabe, T., Cai, Y.: Multi-scale numerical simulations on piezoresistivity of CNT/polymer nanocomposites. Nanoscale Res. Lett. 7(1), 402 (2012).
[44] Liu, C. X., Choi, J. W.: Improved dispersion of carbon nanotubes in polymers at high concentrations. Nanomaterials 2(4), 329–347 (2012).
[45] Dumée, L., Sears, K., Schütz, J., Finn, N., Duke, M., Gray, S.: Influence of the sonication temperature on the debundling kinetics of carbon nanotubes in propan-2-ol. Nanomaterials 3(1), 70–85 (2013).
[46] Janesch, J.: Two-wire vs. four-wire resistance measurements: Which configuration makes sense for your application. no. May, 2–4.
[47] Yin, G., Hu, N., Karube, Y., Liu, Y., Li, Y., Fukunaga, H.: A carbon nanotube/polymer strain sensor with linear and anti-symmetric piezoresistivity. J. Compos. Mater. 45(12), 1315–1323 (2011).
[48] Kim, J. S., Kim, G. W.: Hysteresis compensation of piezoresistive carbon nanotube/polydimethylsiloxane composite-based force sensors. Sensors 17(2), 229 (2017).
[49] Ji, Y., Xing, Y., Li, X., Shao, L. H.: Dual-Stimuli Responsive Carbon Nanotube Sponge-PDMS Amphibious Actuator. Nanomaterials 9(12), 1704 (2019).
[50] Starkova, O., Buschhorn, S. T., Mannov, E., Schulte, K., Aniskevich, A.: Creep and recovery of epoxy/MWCNT nanocomposites. Composites, Part A, Appl. Sci. Manuf. 43(8), 1212–1218 (2012).

Joachim Hausmann, Janna Krummenacker, Andreas Klingler, and Bernd Wetzel

Improvement of fatigue strength of carbon fiber reinforced polymers by matrix modifications for ultrafast rotating flywheels

Abstract: The key to success for electrical power supply by renewable energy resources is a flexible, fast and reliable energy storage system. Flywheels are a well-fit technology for short-term energy storage and recovery of surplus energy in order to stabilize energy grids. The use of carbon fiber reinforced polymers instead of steel allows for higher rotational speed leading to a disproportionally increasing storage capacity. The combination of very high rotational speeds and fast loading–unloading cycles leads to design requirements such as high mechanical strength and fracture toughness with an emphasis on cyclic strength. A finite element analysis identified that the matrix-governed lamina properties are decisive for the cyclic component behavior. Based on the hypothesis that the toughness of the matrix is responsible for the fatigue strength of the composite, various modifications of a polymer matrix are analyzed in order to prolong the lifetime. The study comprises different types of modifiers such as core-shell particles, carbon nanotubes of different functionalization, rubber nanoparticles and self-assembling copolymers. The testing of the different matrix systems includes the determination of the tensile properties, characteristics influencing the processability and a fracture toughness analysis. The matrix-modification aims for a maximization of the ultimate tensile strain while viscosity and glass transition temperature still meet minimum requirements of manufacturing and service conditions. Based on these results, three different modified matrix systems are identified that show an elevated tensile strength as well as an optimized ultimate strain and thus offer the most promising results. In order to reproduce the loading condition due to rotation for the thick-walled flywheel cyclic split-disk tests are conducted. In preparation for the tests a detailed numerical analysis has been conducted to evaluate the stress state in the laminate. As a result to the split, a bending moment occurs and the specimen experiences a stress gradient in thickness direction similar to the stress gradient due to the rotational load case of the flywheel application. To the au-

Acknowledgement: The financial support of the Federal Ministry of Economic Affairs and Energy within the project "Cyclic loading resistant resins for energy storage applications" (funding reference 16KN037225) is gratefully acknowledged. Fibers and resin were kindly supplied by Enrichment Technology Company Limited (Jülich, Germany). This contribution is based on a paper of the proceedings of 21st International Conference on Composite Materials, Xi'an, 20-25th August 2017 [7].

Joachim Hausmann, Janna Krummenacker, Andreas Klingler, Bernd Wetzel, Leibniz-Institut für Verbundwerkstoffe, Kaiserslautern, Germany, e-mail: joachim.hausmann@ivw.uni-kl.de

https://doi.org/10.1515/9783110584455-018

thor's knowledge, this testing method has never been used under cyclic loading conditions before. Thus many issues such as minimization of friction and prevention of rotation of the specimen have to be addressed. The result of these tests will then lead to the selection of one modified matrix system that is used for component testing. The most obvious outcome of the study is the development of a matrix system that leads to an improved lifetime of a composite flywheel for energy storage application. Additionally, the presented approach will save a large amount of component testing time and allow for a broader material testing program and thereby a more productive, time saving and fruitful material development with a special focus on the material's application. In the future, the results along with detailed understanding of damage mechanisms can lead to optimized materials and a more profound estimation of component's lifetime.

Keywords: Carbon fiber reinforced plastic, matrix modification, fatigue testing, lifetime, flywheel

1 Introduction

1.1 Flywheel application

Flywheels made of CFRP are suitable for the application in short-term energy storage systems. They can be used for the stabilization of energy grids or for the recovery of surplus energy charged before. Very high rotational speeds are mandatory for a maximum storage capacity given that the kinetic energy (Ekin) increases by a power of two of the rotational speed or angular velocity Ekin = $I\omega 2$, where I is the moment of inertia, and thus dependent on the mass. It is noteworthy that the influence of rotational speed is more powerful than the mass of the rotor. Otherwise, the mass of the rotating material loads the structure by centrifugal forces and limits the maximum speed by the strength of the material. Therefore, the maximum rotational speed is given by the specific strength (tensile strength divided by density) of the material. Thus, the specific strength is the indicator for the performance of a material in flywheel applications. This property is superior for CFRP. For each loading and unloading of the flywheel, the material of the component is subjected to high circumferential tensile stresses. Therefore, a high cyclic strength under tensile mode is mandatory for the lifetime performance of the flywheel. The set-up of a commercial flywheel system for energy storage applications is presented in Figure 1. The rotating mass (rotor) is a thick-walled CFRP tube supported by magnetic bearings at both ends. To store and restore electric energy by acceleration or deceleration of the CFRP tube, a motor-generator unit is provided. To minimize frictional losses, all moveable parts are mounted in a vacuum vessel. A number of units can be assembled to a turn-key energy storage system.

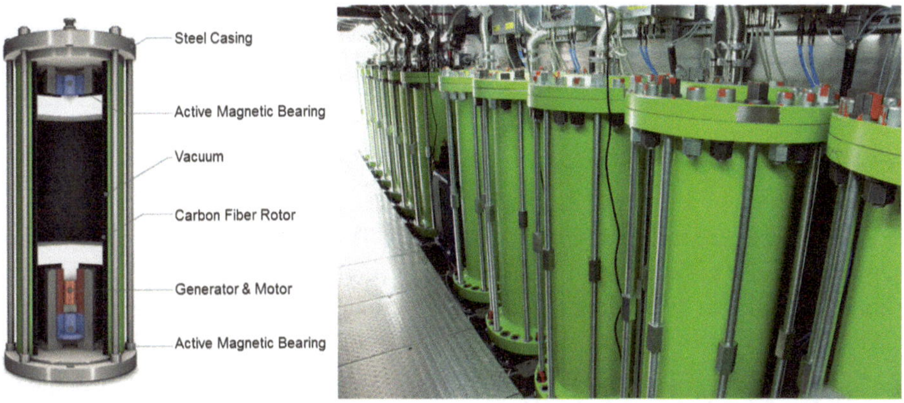

Figure 1: Cross-section of a flywheel unit (left) and assembled flywheels in container (right). (Stornetic GmbH).

1.2 Material behavior and testing

Recent studies have suggested the positive impact of matrix modifications on the fatigue performance of epoxy resins. By incorporating CSR nanoparticles into a carbon fiber reinforced epoxy resin (CF/EP) both the fracture toughness and the fatigue strength can be significantly increased [1]. Ozdemir et al. [2] found a remarkable toughening effect by introducing rubber particles into a CFRP composite. In [3], the addition of triblock copolymers into an epoxy resin led to an important gain of resistance to fatigue crack propagation. Therefore, in the presented study various matrix modifications were tested and characterized in order to augment the lifetime compared to a reference epoxy resin. The testing of the different matrix systems included the determination of the important mechanical properties as well as the characteristics influencing the workability. A common test method for the assessment of the fatigue strength of rotating components is the spin test. Major disadvantages of the spin test for the present component are the extensive testing time and costs. Also, a monitoring of stiffness degradation and the development of damage in the CFRP structure during the test is not possible. To the authors' knowledge, alternative test methods for the lifetime assessment of hoop wound cylinders have not yet been studied. For the determination of the quasistatic mechanical properties of hoop wound cylinders two test methods are state of the art: the split-disk test and the method for internal pressure testing. A recent publication [4] dealt with the comparison of these methods and came to the conclusion that the via finite element simulation adapted split-disk method provides reliable results for the strength in fiber direction whereas the internal pressure testing underestimated the strength values. The objective of this study is the development of a test method that can adequately replace a spin test for the assessment of the fatigue performance of a flywheel. The present study focuses on the lifetime assessment of the tube made of CF/EP and on the improvement of the fatigue resistance of the laminate by matrix modification. The other components of the flywheel are not further investigated here.

2 Matrix modification

CFRP flywheels are in serial production already. These are manufactured with carbon fibers and a commercial anhydride cured cycloaliphatic epoxy resin as matrix. This system will be designated as "reference system" in the following and provides the basis for the assessment of modifications. To augment the lifetime of the flywheel, the reference resin system was modified in order to improve its ductility. The flywheel component manufacturing is done by filament winding, and thus requires a low viscosity of the matrix system for a proper fiber impregnation. Therefore, a crucial parameter for the workability is to only alter the viscosity of the resin systems by varying its composition within narrow boundaries. In the presented study, in addition to the verification of the workability of the matrix modification the mechanical, fracture mechanical, viscoelastic and thermal properties of the modified resin systems were systematically examined. Since flywheels pre-damaged by spin tests showed serious interfiber cracking it was intended to suppress the crack initiation within the matrix. Thus, a main focus was placed on the ultimate strain for the assessment of the efficiency of the modification. To increase the ultimate strain of the matrix system various approaches were pursued, e. g., making use of additional micromechanical mechanisms introduced into the brittle epoxy system by a particulate or chemical modification. During this extensive modifier screening, three systems were identified as the most promising ones, namely an elevated ultimate strain and still meeting the other requirements. The first system, a core-shell rubber (CSR) toughened aromatic epoxy system, showed a tremendously increased ultimate strain due to crack pinning effects and plastic void growth of the matrix material. However, the glass transition temperature was reduced, but still meeting the requirements. A second approach was the synergistic reduction of the cross-link density and the introduction of CSR particles into the reference system. This doubled the ultimate strain with respect to the reference system. Finally, functionalized carbon nanotubes (CNT) were mechanically dispersed via a three-roll calendar into the reference system, which were found to be beneficial for improving the mechanical properties as well as the fracture toughness. However, the viscosity increased significantly, especially by the CNT modification. The analyzed properties of the selected resin systems are illustrated in Figure 2.

3 Development of a cyclic test method

The main objective of this study is the development of a test method that can adequately replace a spin test for a lifetime assessment of a flywheel. The herein studied

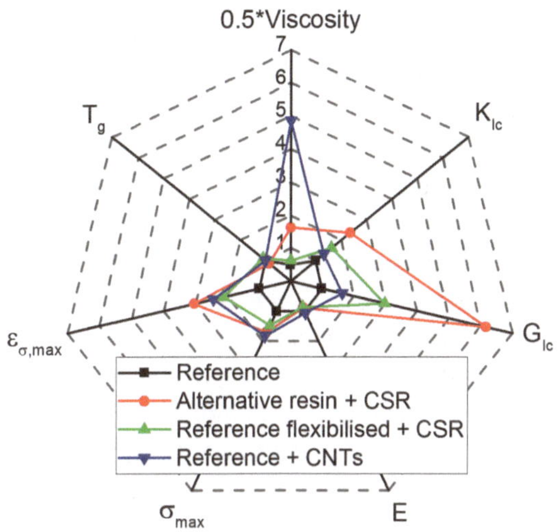

Figure 2: Relevant properties of modified matrices compared to the reference system.

flywheel is made of a commercial CF/EP and consists of hoop-wound and helical-wound layers. In order to obtain a high strength, the layers possess a very high fiber volume ratio (FVR). A manufacturing process that allows high FVR's is the winding technology. This is particularly true for circumferential bodies. By winding plane sheets, however, only a significantly lower FVR can be reached. Therefore, and in order to apply the same manufacturing process for the specimens as for the prototype, it was searched for a testing method that uses a circumferential specimen produced by winding technology. By means of a finite element analysis the stress state of the flywheel due to residual stresses, acceleration and deceleration as well as high rotational speed was studied. Whereas the helical-wound layers provide the longitudinal stiffness to prevent resonance in the flexural mode, the hoop-wound layers carry the circumferential load resulting of the centrifugal force. Compared to the stresses caused by the centrifugal force the residual stresses due to manufacturing and the stresses caused by acceleration and deceleration can be neglected. The centrifugal force leads to tensile stresses in fiber direction in the hoop-wound layers. These circumferential stresses result in tensile stresses perpendicular to the fiber direction as well as shear stresses in the helical layers. Furthermore, the differing stiffness in helical and hoop layers causes tensile stresses perpendicular to the fiber direction in the hoop layers. To summarize, in the load carrying hoop layers the effort due to fiber and interfiber-failure is 0.5 and 0.3, respectively. Given the fact that the fatigue resistance against interfiber failure decreases more rapidly than the resistance against fiber failure, it can be stated that the mode of failure depends on the applied stress-level. Therefore, an appropriate test specimen has to consist of helical as well as hoop-wound layers to reproduce the decisive stress condition. Under static loading, the split-disk method is a valid test method for the comparison of hoop-wound laminates. One characteristic of this method is the imposed bending moment at the split leading to a superposition with the tensile circumferential stress. This is the main reason for the rare application of this test method in the past. In the present study, the bending of the specimen in the split is even a desirable effect, given the fact that the circumferential tensile stresses due to a centrifugal force increase from the inside to the outside of a tube. This effect is even more pronounced in thick laminates as it is the case in the present study. Furthermore, a more recent study on glass-fiber and carbon-fiber reinforced hoop wound cylinders showed that by adapting the experimental results via a finite element analysis, the strength and even stiffness of a unidirectional laminate can be obtained [4]. As the main objective of the present study is the comparison of different matrix systems on the basis of their fatigue performance in a near-application loaded laminate, an adapted cyclic split-disk-test seems the adequate instrument.

4 Test set-up

The ASTM-standard D 2290 covers the regulation for the split-disk-method under quasistatic loading, and thus can be applied for the determination of the apparent hoop tensile strength of fiber-reinforced plastic tubes [5]. The test fixture is to be constructed in a manner that the influence of the bending moment is minimized meaning an initial split as small as possible.

4.1 Specimens

The standard offers two different types of specimens: (i) a ring with two notched sections located **180** apart, which is used for testing circumferentially fiber-reinforced thermosetting resins and (ii) a specimen with a full cross-section. A finite element study with the chosen laminate lay-up proves that in the notched section inter-fiber failure occurs before fiber-failure. Consequently, the circumferential strength cannot be determined by using a notched specimen, which has also been the finding for unidirectional reinforced tubes [4]. Therefore, a test set-up with constant cross-section specimens is chosen for the present study. For the specimen preparation, a tube is wound on a mandrel with a diameter of 146.0 mm. The laminate is composed of helical and hoop layers with scaled lay-up of the flywheel resulting in a laminate thickness of 3.5 mm. Compared to other split-disk-tests, this high thickness, however, results in an elevated aspect ratio thus causing a higher impact of the bending moment [6]. The specimens are then cut with a rotating diamond saw blade in rings with a width of 7.00 mm. To remove cutting marks, which can have a significant impact on the fatigue strength, the rings are then wet-sanded and polished. Given the great influence of the thickness of the hoop layer on the resulting stress distribution, the thickness and FVR is determined by using microscopy. Figure 3 presents two microscopic scans of the same specimen illustrating the considerable variation of thickness of the hoop layer complicating an exact assessment of the resulting stresses. Thus it can be stated that the specimen thickness cannot be measured manually, but needs to be determined with microscopic images.

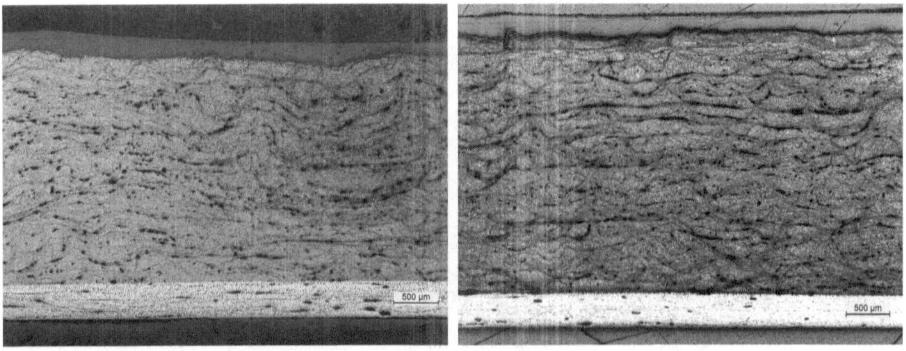

Figure 3: Cross-sections of filament wound samples to determine real wall thickness. Helical layers are brighter at the bottom of the images.

4.2 Set-up

The specimen is mounted on two disks and subsequently secured on the top and the bottom to avoid movement during the cyclic testing. A special focus has to be placed on an exact parallel alignment of the specimen with regard to the disks to avoid an unintentional bending. The disks are then fixed with bolts in two brackets that can be pulled apart. Additionally, to observe an unintentional movement of the ring, and thus detect invalid test results pictures are taken in interval mode. To capture possible stiffness degradation, an inductive displacement transducer is attached to the two brackets. The temperature rise of the specimen due to external and internal friction is monitored by a thermocouple and a thermal camera focused on the split. The set-up is shown in Figure 4.

5 Results

5.1 Quasistatic tests

A finite element model of the quasistatic tests was used to account for the influence of the resulting bending moment on the calculated strength. Under the ultimate load

Figure 4: Split-disk set-up for cyclic testing with mounted specimen. The displacement transducer is visible in the rear.

the maximum stress in fibre direction was calculated to Rt,II = 3260 MPa. Additionally, the mean strength was calculated according to the standard to Rt,II = 2460 MPa, taking hoop and helical layers into account. It can be stated that the influence of the bending moment on the calculated strength is significant. This is due to the high aspect ratio (w/t) of the specimen. In the present study, the fact that the highly stressed volume is very small cannot be neglected, and thus size effects have to be considered when estimating the effective strength via the finite element method. This effect is not covered by the present study. Thus for the estimation of the fatigue strength, the proceeding according to the standard is applied which is determination of the mean stress by the applied force divided by the loaded cross-section.

5.2 Cyclic tests

So far, the cyclic split-disk test has been conducted at different stress levels firstly for the reference material and secondly for the three most promising matrix modifications. The test results are presented in Figure 5. Despite the promising properties of the neat resin, the CNT modified matrix showed shorter lifetime compared to the reference material. Only for a very long lifetime CNT modification shows advantages. A significant improvement in the relevant region of about 100,000 cycles is obtained by modification with core-shell rubber (CSR) particles. A further improvement can be obtained by flexibilization of the resin used for CSR modification.

By the combination of stiffness degradation monitoring and imaging techniques such as microcomputed tomography ($\mu - C * T$) the damage mechanisms can be de-

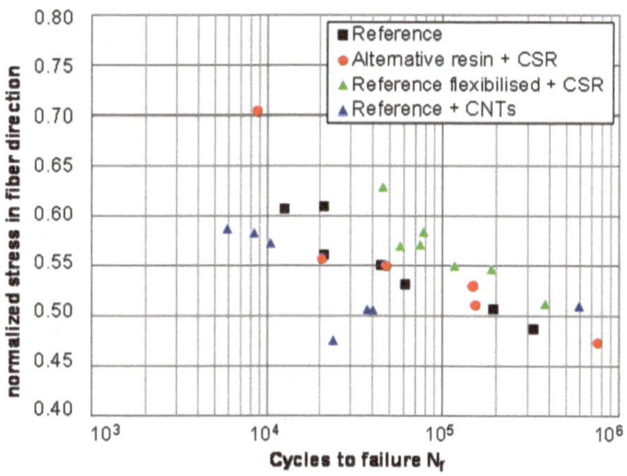

Figure 5: Results of the cyclic tests showing the maximum cyclic stress normalized to the quasi-static tensile strength of the reference material.

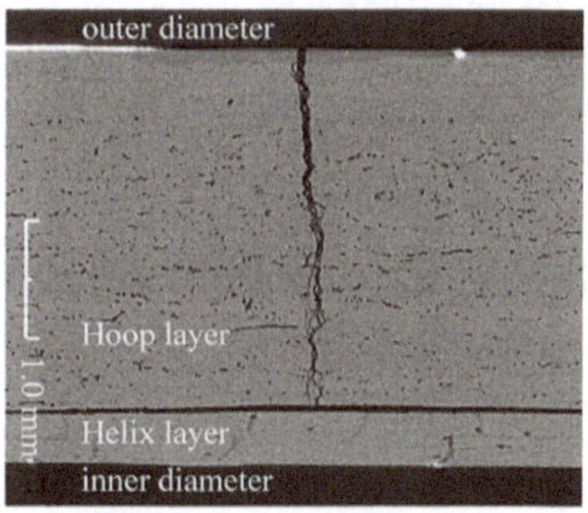

Figure 6: Cross-section of tested specimen obtained by x-ray computer-tomography within the region of the split.

tected. Figure 6 shows a $\mu - C * T$ scan of a specimen detail at the split. The specimen has been tested up to the point where a decreasing stiffness was observed. Two damage mechanisms can be detected: a delamination between helical and hoop layer and a radial crack in the hoop layer. It is not yet clear which damage mechanism developed first and caused the other one. Further tests can provide a more detailed insight into the damage mechanisms that lead to this stiffness evolution during the cyclic split-disk test.

6 Discussion

The application of the split-disk method for cyclic testing delivers convenient results, especially for the comparison of different materials. However, it may be not suitable to obtain absolute material properties, e. g., for the design of components or reliable lifetime prediction. Some side effects such as the friction between the disks and the specimen resulting in heating and abrasion need to be minimized for better test results. Despite the fact that the CNT modification showed good results in the properties of the matrix evaluation it reveals worse lifetime compared to the reference system. This is likely caused by the increased viscosity leading to a more difficult processability, and thus resulting in worse material quality. Eventually, optimization of the process parameters may eliminate this drawback. CSR modification improves the lifetime compared to the reference. The alternative resin shows very good mechanical properties, but similar to CNT modification, the viscosity is increased. This may be the reason for worse fatigue behavior compared to the flexibilized reference resin with CSR modification. Cross-sections obtained from specimens showed delaminations between the helical and the hoop layer as well as interfiber cracking of the hoop layer. The finite element analysis reveals an effort for fiber fracture of 0.5 while for interfiber fracture of about 0.3–0.4. Otherwise, interfiber fracture is much more sensitive to cyclic loading leading to premature interfiber cracking while cycling. Therefore, it is assumed, that cracking is initiated by axial stresses, related to the rotor axis, in the hoop layers. These are caused by the helical layers, which hinder a free contraction of the hoop layers in axial direction. Once the interfiber crack is initiated, it favors delamination in-between the helical and the hoop layer. When cyclic loading continues, disintegration of the structure and finally catastrophic failure of the component are the consequence.

7 Conclusions

The main focus of the present study was the development of a testing method to assess the lifetime performance of a flywheel made of CFRP. Therefore, the well-established split-disk method developed for quasi-static testing was adapted to cyclic loading conditions. The small scatter in the test results lead to the conclusion that this test method provides reproducible fatigue data and thereby is much less time-consuming than spin tests, which is the state-of-the-art for evaluation of fast rotating parts. Furthermore, a matrix modification using CSR-particles and CNTs leads to significantly improved mechanical properties of the epoxy resin. Cyclic split-disk tests on the laminate made of the modified resins showed the impact of the modification on the laminate fatigue properties. However, a low viscosity of the resin is mandatory to obtain improved matrix properties for the composite material as well. Therefore, CSR-particles in flexibilized epoxy resin performed superior in fatigue testing of the laminate. Lastly, a generic rotor was manufactured with the most promising matrix system. Cyclic spin tests of the component confirmed a significant lifetime improvement by the matrix modification with CSR and flexibilization.

Bibliography

[1] Nguyen, T. P., et al.: Improved fracture toughness and fatigue life of carbon fiber reinforced epoxy composite due to incorporation of rubber nanoparticles. J. Mater. Sci. 48, 6039–6047 (2013). https://doi.org/10.1007/s10853-013-7400-z..

[2] Ozdemir, N. G., et al.: Toughening of carbon fibre reinforced polymer composites with rubber nanoparticles for advanced industrial applications. eXPRESS Polym. Lett. 10, 394–407 (2016). https://doi.org/10.3144/expresspolymlett.2016.37.

[3] Klingler, A., Wetzel, B.: Fatigue Crack Propagation in Triblock Copolymer Toughened Epoxy Nanocomposites. Polym. Eng. Sci. (2017). https://doi.org/10.1002/pen.24558.

[4] Bleier, A.: Prüfverfahren zur Bestimmung exakter Werkstoffkennwerte einer unidirektionalen Schicht unter besonderer Berücksichtigung physikalischer Nichtlinearitäten. Shaker, Darmstadt (2011).

[5] ASTM D 2290: Standard Test Method for Apparent Hoop Tensile Strength of Plastic or Reinforced Plastic Pipe by Split Disk Method.

[6] Kinna, M. A.: NOL Ring Test Methods. No. NOLTR-64-156. Naval Ordnance Laboratory, White Oak, 1964.

[7] Krummenacker, J., Hausmann, J., Sorochynska, L., Klingler, A., Wetzel, B.: Development of a cyclic test method for ultra-fast rotating flywheels made of CFRP and improvement of their fatigue strength by matrix modifications, Proceedings of 21st International Conference on Composite Materials, Xi'an (China), 2017. http://www.iccm-central.org/Proceedings/ICCM21proceedings/papers/3693.pdf.

Enio P. Bandarra Filho and Letícia Raquel de Oliveira

Experimental study of thermal conductivity, viscosity and breakdown voltage of mineral oil-based TiO_2 nanofluids

Abstract: The nanofluids applied to transformer oil base can be considered the new insulating fluids for the next generation, since the potential to improve the dielectric strength and thermal performance of the transformer compared to pure mineral oils. In this study, performance tests of nanofluids with TiO_2 nanoparticles were performed to dielectric strength with various values of volume concentration and compared with the performance of the pure oil. Additionally, thermal conductivity and viscosity were experimentally measured. The dielectric strength (breakdown voltage AC) of nanofluids were obtained by measuring the breakdown voltage of the samples in a dielectric strength tester. The experimental results showed that the volume concentration of 0.05 % showed the largest increase of dielectric strength. Furthermore, it was obtained an increase in the thermal conductivity of 3–4 % and no significant change in the viscosity, showing good potential to application.

Keywords: Nanofluids, transformer oil, thermal conductivity, viscosity, breakdown voltage, titanium dioxide

1 Introduction

The dielectric oil in electrical transformer is a fundamental component responsible for cool down the temperature, act as an insulator between the windings, increasing the resistance, and avoid short circuit. In general, mineral oil is used and the characteristic of this oil is related to the low value of thermal conductivity, about 0.110 W/mK [1]. Because of this, the heat transfer is limited which can lead to overload and critical failures. In cases of failure in the electrical transformers, as these equipments are critical points in the energy transmission systems, their replacement may take time in these cases. They operate at high voltages and require insulating oil that has good di-

Acknowledgement: The authors acknowledge the support provided by CNPq, CAPES and FAPEMIG for this investigation, to the Laboratory of Electrical Transformers of Faculty of Electrical Engineering of the Federal University of Uberlândia (UFU) by taking measurements of breakdown strength and contributors LNMIS of Institute of Physics (UFU) for supplying the TiO_2 nanoparticles and their characterization.

Enio P. Bandarra Filho, Letícia Raquel de Oliveira, School of Mechanical Engineering, Federal University of Uberlandia (UFU), Av. João Naves de Ávila, 2121, Santa Monica, Uberlandia MG 38408-514, Brazil, e-mails: bandarra@ufu.br, leticia.raquel@ufu.br

https://doi.org/10.1515/9783110584455-019

electric properties, cooling capacity, as well as high resistance to ensure the normal operation [2].

The widespread use of insulating oil in high voltage transformers and cooling electric devices has led to an extensive investigation with the aim of improving the dielectric and thermal characteristics of the oil [3]. For this purpose, in the last two decades with the increasing advance of nanotechnology, the use of nanofluids has been a new alternative to modify and improve the fluid transport properties and also to intensify thermal efficiency [4]. Several different nanoparticle have been investigated with the main goal to improve as the same time the dielectric properties of the oil as well as the heat transfer [5–12]. The classification of nanoparticles can be done in three main groups, conducting nanoparticles (Fe_3O_4, Fe_2O_3, ZnO, SiC), semiconducting nanoparticles (TiO_2, CuO, Cu_2O) and ceramics nanoparticles (Al_2O_3, SiO_2, BN). Base oils are mineral, synthetic or vegetable.

Choi and Eastman [7] named as nanofluids the dispersions of nanoparticles sized between 1 to 100 nm in fluids. Since then many studies have been carried out describing the synthesis and potential applications of nanofluids. In this respect, many papers were published for thermal conductivity enhancement using nanofluids with water base, but for insulating fluids, such as transformer oil, research has been oriented mainly to breakdown strength and dielectric phenomena. Segal et al. [5] were the first researchers to study the addition of nanoparticles in the insulating oil and the effect on the dielectric properties and heat transfer. The mentioned authors evaluate the influence of magnetic fluids (ferrofluids) under different concentrations and magnetization values in a prototype transformer. They evaluated the dielectric strength and heat transfer of the ferrofluid in real conditions and concluded that the ferrofluid modified heat dissipation in a beneficial way transformer. Also, the addition of magnetic particles has improved dielectric strength during the lightning impulse test, contradicting the conventional standards that indicate the use of oil without any kind of particle to avoid degradation. Lee and Kim [3] tested nanofluids compose by oil-based and ferromagnetic particles with an average diameter of 10 nm to estimate the dielectric strength. They found an increase of more than twice for the value of dielectric strength using ferromagnetic nanoparticles. When they used an external magnetic field, an increase over than 30 % was obtained. The increase in dielectric strength by adding magnetic nanoparticles has been explained by the fact that nanoparticles act as electron entangling reducing the speed of electric current through the dielectric. The increase in dielectric strength by adding magnetic nanoparticles was also observed in [13] and [14].

A summary of some papers referenced in the literature on transformer oil based nanofluids with breakdown voltage is shown in Table 1.

From Table 1, it was observed that in some studies there was evident improvement of breakdown voltage, however, some others showed reduction with the addition of nanoparticles. Therefore, it is fundamental to study in detail the dielectric behavior of nanofluids regardless of the type of nanoparticle. Several experimental re-

Table 1: Papers related to nanofluids oil based in electrical transformer reported in the literature.

Nanofluid	Synthesis method	Average nanoparticles	Dispersant	Loading (%)	Percentage increase in breakdown
Fe_3O_4/mineral oil [5]	Two-step	–	-	Up to 1 % vol.	40 %
Ag-silica/mineral oil [1]	One-step	5.5	–	0.1 % to 0.6 wt%	60.7 % reduction in breakdown strength
Al_2O_3/mineral oil [15]	Two-step	<50 nm	–	5, 10, 20 and 40 wt%	13 % for concentration 20 % wt (Al_2O_3)
Fe_3O_4/mineral oil [15]	Two-step				8 % for concentration 10 % wt (Fe_3O_4)
TiO_2/mineral oil [15]	Two-step				4 % for concentration 10 % wt (TiO_2)
Fe_3O_4/mineral oil [16]	Two-step	–	-	–	16 %
Diamond/mineral oil [17]	Two-step	4-8	Sucrose and oleic acid	0.01–1 wt%	10 % for concentration 0.5 wt%
TiO_2/mineral oil [18]	Two-step	100	CTAB	0.06; 0.18; 0.24 and 0.36 g/L	127 % for concentration 0.06 g/L
TiO_2/mineral oil [19]	Two-step	35	–	0.05; 0.07 and 0.1 wt%	TiO_2 10.4 %
SiO_2/mineral oil [19]	Two-step	30			SiO2 8.2 %
Fe_3O_4/mineral oil [20]	Two-step	23.2	Oleic acid	0.1 to 0.6 vol%	16 % for concentration 0.3 vol%
TiO_2/mineral oil [24]	Two-step	<20	–	0.075 vol%	19 %
TiO_2/vegetal oil [29]	Two-step	20	–	0.00625 vol%	31 %
Diamond/mineral oil [4]	Two-step	3–6	–	0.05 to 0.05 vol%	reduction in breakdown strength
MWCNT/mineral oil [4]					
ZnO/mineral oil [21]	Two-step	34	–	0.0005 vol%	8.3 %
SiC/mineral oil [22]	Two-step	80	–	0.005 vol%	Breakdown strength reduction

sults, including [5], confirmed the good performance of magnetic nanoparticles. However, other conductive metal nanoparticles, semiconductor and ceramics can also be used and are being widely studied. One example is the application of titanium dioxide (TiO_2), which is an attractive material for producing nanofluids. It shows good stability when subjected to high electric fields, with its relatively high dielectric constant to be a semiconductor. It has been extensively used in research for not being toxic and exhibits a long-term stability. Among the three crystalline phases of titanium dioxide, anatase is the easiest to obtain and the most widely used, from paint pigments, electronic devices and currently nanofluids.

Du et al. [9] found that the addition of TiO_2 semiconductor nanoparticles to the insulating mineral oil significantly increased the dielectric strength under alternating. Lv et al. [27] investigated the effects of mechanical dispersion method (stirring, ultrasonic bath and probe processes) on the dispersibility of functionalized TiO_2 nanoparticles with oleic acid in the transformer oil for preparing stable nanofluid and its effects on AC breakdown strength. The results reveal that the combination of ultrasonic bath process and stirring method has the best dispersion efficiency and the obtained nanofluid possesses the highest AC breakdown strength current and on impulse, resistance to moisture content and properties of oil aging. However, they were not investigated the effect of adding TiO_2 nanoparticles in the transport properties such as thermal conductivity.

On the other hand, Fontes et al. [4] in their experimental studies with diamond nanoparticles and multiwalled carbon nanotubes dispersed in mineral insulating, evaluated thermal and transport properties measuring the thermal conductivity, dynamic viscosity and evaluated dielectric strengths through the measurements of the insulation breakdown voltage AC using a dielectric strength analyzer with brass electrodes. The authors noted the increase in thermal conductivity and dynamic viscosity with the increment in the concentration of nanoparticles. However, the addition of nanoparticles in the oil caused the breakdown voltage to decrease. In other work, Du et al. [2] added TiO_2 semiconductor nanoparticles of 10 nm to transformer oil for producing nanofluids semiconductors, with the aim to enhance the insulating characteristics. Performed breakdown voltage tests before and after the addition of the nanoparticles. As a result, they found that the breakdown voltage for the semiconductor nanofluid was 1.2 times higher than that of pure oil to a volumetric concentration of 0.075 % nanoparticles, and that the partial discharge resistance was also improved.

Fallah-Shojaie et al. [28] also experimentally investigate the influence of nanoparticles of titanium dioxide (TiO_2) related to the breakdown voltage in transformer insulating oil. Their results showed that the addition of TiO_2 semiconductor nanoparticles increased the dielectric strength of the insulating oil. They found that increasing the volumetric concentration of nanoparticles the breakdown voltage decreased. Although many researchers have found significant improvements in dielectric strength of the insulating oil with the addition of TiO_2 nanoparticles, theoretical and experimental results as Józefczak et al. [12] have demonstrated that the addition of nanoparticles (insulating, semiconducting or conducting) to a fluid base presented a negative effect on heat exchange by convection. However, it is important to highlight that the ferromagnetic nanoparticles have a higher agglomeration than that observed in titanium dioxide nanoparticles as already evidenced by Hwang et al. [13].

On the other hand, Kopcansky et al. [30] demonstrated that resistance to dielectric breakdown of nanofluids insulating oil base is strongly influenced by the effects of the nanoparticle aggregation and at lower levels are most appropriate for applying nanofluids as insulating fluid high-voltage transformer. Recently, Katiyar et al. [31] dispersed TiO_2 in the rutile and anatase crystalline forms in two different types of transformer oil to produce stable, homogeneous nanofluids with breakdown voltages greater than the pure oil. They observed an increase of approximately 79 % in dielectric strength for the anatase phase and 51 % for rutile with mass concentrations of 0.1 %. The main difference between the dynamic loads of dielectric nanoparticles within the oil was explained as being due to the trapping of electrons efficiently leading to a decrease of the electric current speed resulting in increased breakdown voltage. The difference between the crystalline phases was explained based on the electronic structure and the affinity for removing electrons. Still according to the authors, the presence of TiO_2 nanoparticles influence the distribution of the electric field lines within the oil due to its high permittivity and crystal semiconductor structure.

Based on these mentioned works, it can be stated that nanofluids are promising for use in electrical transformers, which can be considered as the next generation insulation fluids, since they offer new possibilities for improving the dielectric breakdown voltage, as well as heat transfer performance against pure oil. Therefore, due to the great importance of electrical transformers, both in industry and in residential power distribution, further research is needed in order to understand the physic mechanism and improve the thermal and insulating properties of the oil without compromising the performance. The objective of this experimental study was concerned with preparing nanofluids through dispersing TiO_2 in mineral oil, as well as evaluating electrical and thermal characteristics and rheological properties of pure oil and nanofluids. Thermal conductivity and viscosity of TiO_2 nanofluids were measured in a temperature range of 20° to 50 °C and volumetric concentrations of 0.0025 % to 0.05 %. Any surfactants or dispersion agents may affect the thermal performance and rheology of fluids.

2 Experimental methods and materials

2.1 Synthesis and characterization of nanoparticles

The TiO_2 nanoparticles were synthesized in aqueous solution at room temperature as showed in Oliveira et al. [25]. The X-ray diffraction (XRD) technique was used to determine the pase, crystalline structure and size of TiO_2 nanoparticles. The equipment used for measurements was a DRX-6000 (Shimadzu), with monochromatic radiation Cu-K_{α_1} ($\lambda = 1.54056A$).

The Raman spectroscopy technique wass used to identify the synthesized compound. The micro-Raman spectra were obtained sing an Ocean Optics spectrometer with excitation source of 780 nm. A transmission electron microscope (TEM) (Joel, JEM-2100) was used for structural and morphological analysis of the surface of the nanoparticle agglomerations.

2.2 Synthesis and characterization of the nanofluids

For the preparation of the nanofluids, the two-step method was used, which consisted of the dispersion of the nanoparticles (average size of 11 nm) in transformer mineral oil, in different volumetric concentrations: 0.0025 %, 0.005 %, 0.01 % and 0.05 %. The mineral oil was purchased from YPF S. A. manufacturer and it is a paraffinic mineral oil derived from petroleum, which has a natural feature of a high dielectric constant. It is an insulating medium suitable for use in voltage transformers, electrical systems and switches, whose operations are done in oil bath. Table 2 shows some properties of mineral oil and TiO_2 nanoparticles.

Table 2: Transport properties of the transformer mineral oil and nanoparticles.

Viscosity at 40 °C [cp]	12	Thermal conductivity [W m^{-1} K^{-1}]	6.531
Density at 20 °C [g cm^{-3}]	0.848	Density at 20 °C [g cm^{-3}]	3900
Flash point [°C]	140	cp [J kg^{-1} K^{-1}]	686
Pour point [°C]	−12		
Breakdown voltage [kV]	45		
Acidity [mg KOH g^{-1}]	0.40		
Oxidation inhibitor content [%wt]	0.30		

The dispersion of the nanoparticles in the oil was performed using a sonicator (Hielscher Ultrasound Technology, UP100H, 100W, 30 kHz) for 30 minutes. In Figure 1 are shown the detail of the sample preparation with nanofluid concentration of 0.0025 % and after 24 hours of its synthesis. The nanofluids samples presented visual stability for approximately 30 days.

Thermal conductivity was measured using a Linseis equipment, model THB-1. Figure 2 shows the schematic view of the equipment. The thermal conductivity measurement was conducted by inserting the probe in a 100 mL capacity container connected to the thermal bath. Ten measurements per cycle were performed automatically by the computer connected to the equipment.

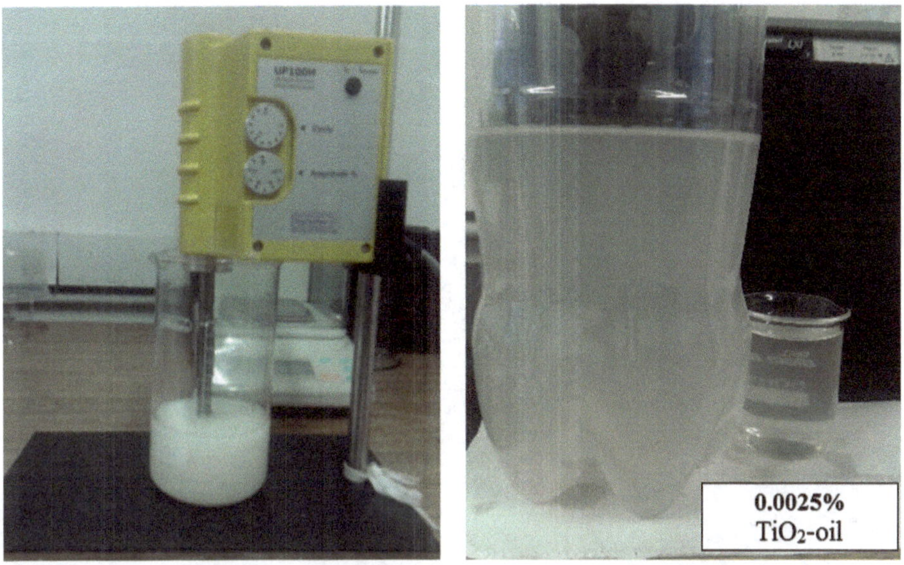

(a) (b)

Figure 1: Imagens (a) synthesis samples TiO_2-oil nanofluids and (b) 24 hours after synthesis.

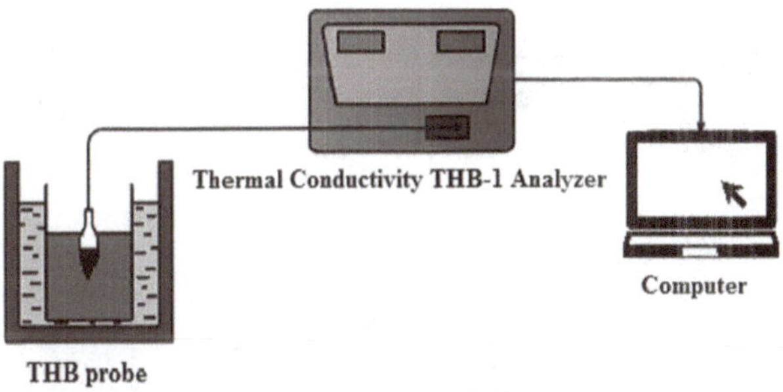

Figure 2: Scheme of the experimental apparatus used to measure thermal conductivity.

The dynamic viscosity has been measured in a rotational viscometer, model SVM 3000 from Anton Paar manufacturer, which has a cylindrical geometry. The temperature was controlled and the measurements were performed at temperatures of 20, 30, 40 and 50 °C. Figure 3 depicts the photo of the equipment.

The dielectric strength measurements have been performed in a dielectric strength tester produced by Triel 3E. The test fluid was loaded in a tub in such way to completely flood both electrodes, which were made of brass with diameter of 25.4 mm and distant of 2.5 mm. The voltage was ramped up at a rate of 3 kV/s. Figure 4 depicts the test basin image and it is possible to observe the electrodes submitting the nanofluid high voltage. Five measurements of the dielectric strength were performed for each sample, and between each measurement the sample was allowed to stand for 1 minute so

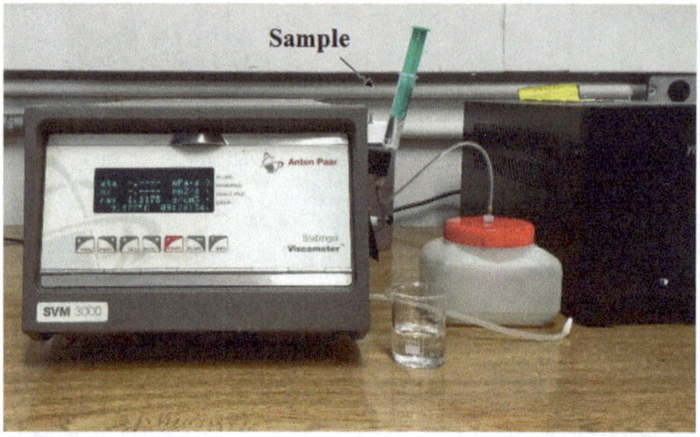

Figure 3: Picture of viscometer using for measurement viscosity of nanofluids.

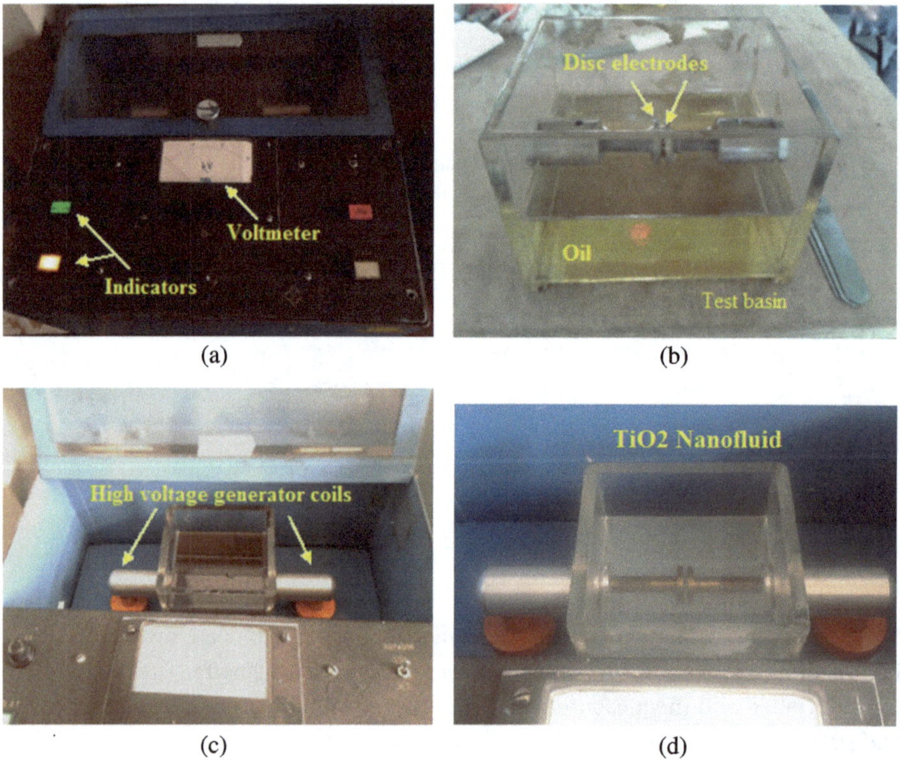

Figure 4: Tester components of dielectric breakdown: (a) Front view of the test setup, (b) Front view of test basin with the electrode assembly and pure oil sample. (c) Top view of the test setup showing the high voltage generator coils and (d) TiO_2 nanofluid sample in the test basin.

that the portion of the fluid between the electrodes, which was ionized be replaced by other of nonionized fluid, following standard ASTM D877/67. The value of the dielectric strength of each sample was obtained as the average of the five measured values.

3 Results and discussion

3.1 Characterization of the nanoparticles

Figure 5 presents X-ray diffraction (XRD) patterns, (a) and (b) Raman spectrum of TiO_2 nanoparticles synthesized. In Figure 5(a), the formation of the anatase phase of TiO_2 was confirmed due to the presence of Bragg diffraction peaks (JCPDS 21-1272). The average size of the TiO_2 NCs of 11 nm, estimated by the Debye–Scherrer equation as shown in [30]. In Figure 5(b), the micro-Raman spectrum of the nanoparticles in the anatase phase of TiO_2 was observed. The characteristics bands of vibration of this phase were evidenced according to [31, 32]. Figure 6 shows TEM images of nanoparticles where can be possible to observe the spherical shape.

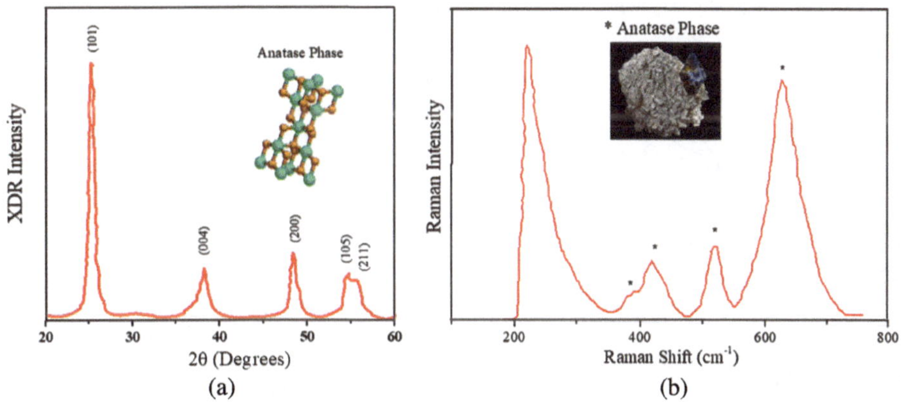

Figure 5: (a) X-ray diffraction patterns and (b) micro-Raman spectrum of TiO_2 nanoparticles synthesized.

3.2 Characterization of nanofluids

The electrical breakdown strength of transformer oil and nanofluids were measured and results are shown in Figure 7. It was observed that for the pure mineral oil, the dielectric strength was measured to be 42 kV, while the manufacturer is indicated 45 kV according to ASTM D-877. This small difference may be due to the presence of moisture or difference between the equipment used in the measurements. Figure 7 depicts the effect of volumetric concentration of nanoparticles the dielectric strength given

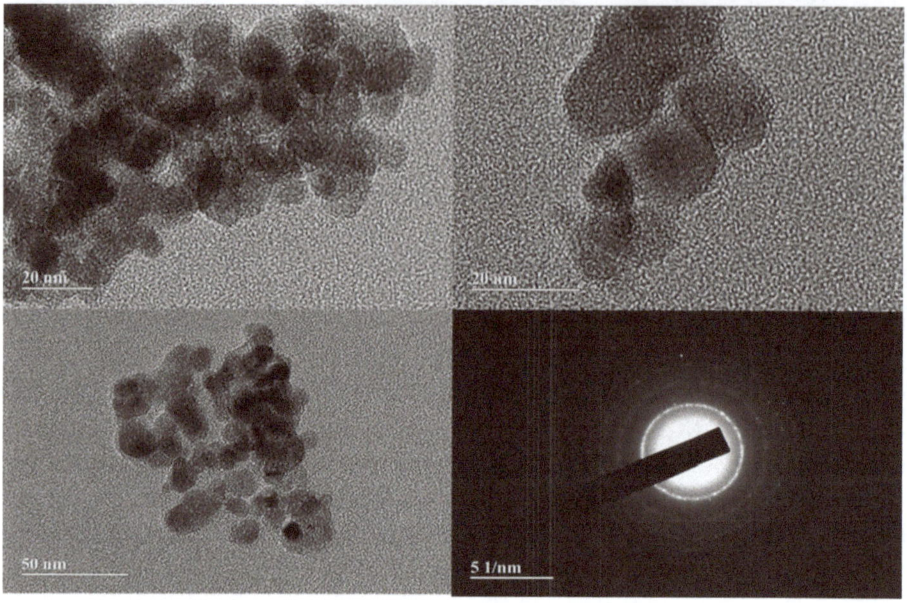

Figure 6: TEM images of TiO_2 nanoparticles and electron diffraction pattern.

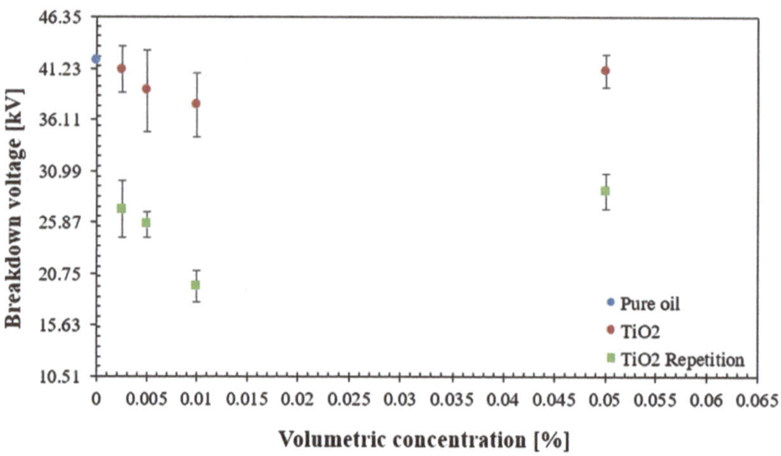

Figure 7: Variation of breakdown voltage with the volumetric concentration of nanoparticles.

in function of the breakdown voltage for nanofluids after manufacturing and after 2 months. It was observed that with the increase of the volumetric concentration of nanoparticles the dielectric strength of the mineral oil decreases, contrary to the literature consulted. When comparing the results of dielectric strength of nanofluids freshly prepared and two months after preparation, the decrease in dielectric strength is even greater. This can be related to oxidation of the components of the oil, even the degradation or agglomeration of nanoparticles. For a better visualization, the breakdown voltage values of pure oil and nanofluids are presented in Table 3.

Table 3: Experimental breakdown voltage data for TiO_2 in mineral oil after synthesis, after 2 months and deviation % on respect to measured mineral oil.

Volume concentration (%)	Breakdown voltage (kV)		Breakdown voltage deviation (%)	
	After synthesis	2 months after synthesis	$\Delta = \frac{BV_{nf}-BV_{oil}}{BV_{oil}}100$	
Pure oil	42		After synthesis	2 months after synthesis
0.0025	41.1	27.2	2.1	35.2
0.005	39	25.6	7.1	39.0
0.01	37.6	19.5	10.5	53.6
0.05	41	29	2.4	31.0

The breakdown voltage deviations found between nanofluids and pure oil may have occurred due to the presence of water, which is one of the main factors that contribute to the reduction of breakdown voltage, mainly observed after 2 months. Other factors such as oil degradation by oxidation and formation of by-products and higher temperatures may also contribute to this reduction. Moreover, the presence of nanoparticles and their clusters, unlike what has been reported by other authors [15–24, 26–29], can contribute to electrical phenomena that cause reduced breakdown voltage [1, 4, 22, 38]. Mergos et al. [35] said dielectric properties nanofluids based oil depend mainly on the size of the nanoparticles and their chemical composition; in addition, the polarization of the surface of nanoparticles is induced by the presence of adsorbed water, influencing the dielectric behavior. However, specific and in-depth studies should be done in an attempt to accurately explain these inconsistencies. Nevertheless, the addition of small volume fraction of the mineral insulating oil has been recommended by [36] to increase the breakdown voltage of mineral oils. Experimental results from Chiesa and Das [22] and Hwang et al. [9] also confirm these results.

Figuree 8 shows a summary of some works found in the literature of mineral oil based TiO_2 nanofluids. In most of these studies, it is possible to observe an increase of the breakdown voltage with the increase of the concentration of nanoparticles. It is interesting to note that there must be an optimum concentration, because after a certain concentration the voltage begins to fall. In addition, the increase in concentration contributes to the formation of agglomerations, which certainly causes the breakdown voltage of nanofluids to drop.

The thermal conductivity of TiO_2 nanofluids was measured, in order to evaluate its possible enhancement as a function of particles loading and temperature. Mean values of thermal conductivity measured in this study are shown in Figure 9 for different temperatures as a function of volume fraction. According to Maxwell [37], the

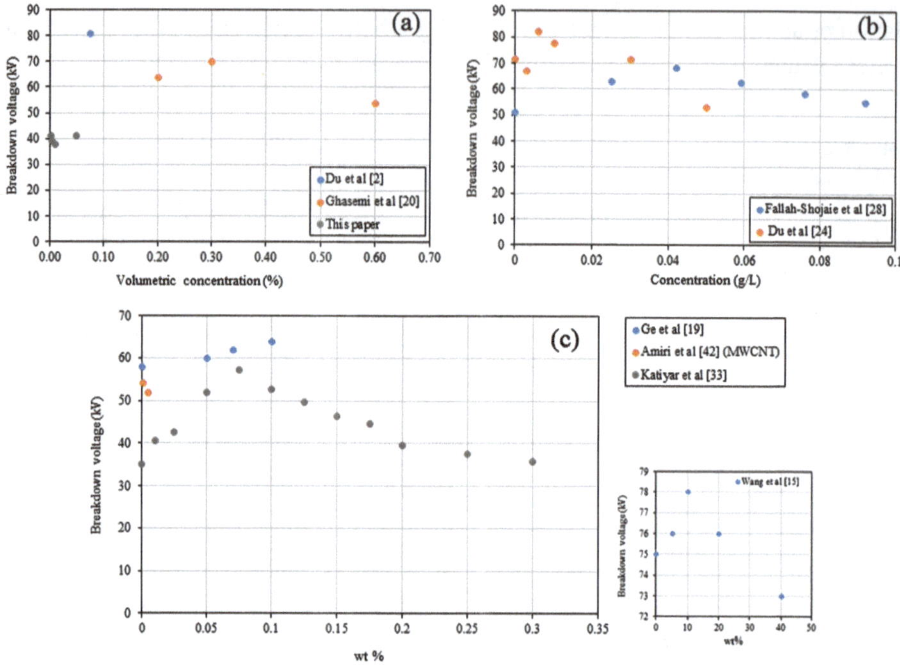

Figure 8: Summary of papers from breakdown voltage of TiO_2 nanofluids based mineral oil: (a) volume concentration, (b) concentration (g/L) and (c) Mass concentration.

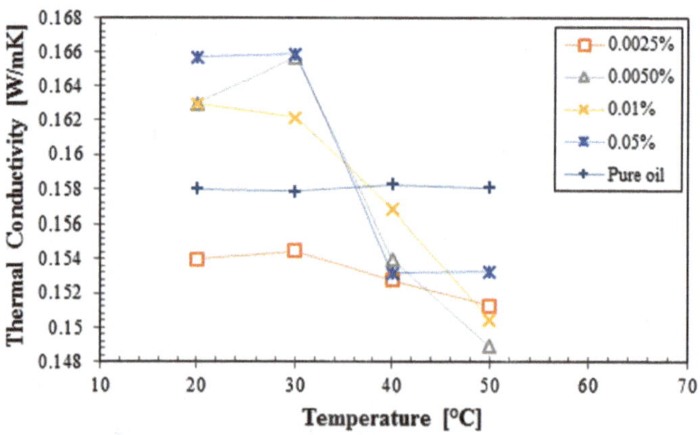

Figure 9: Thermal conductivity variation with temperature at different volumetric concentrations.

increase in the concentration of particles dispersed in a based fluid causes the enhancement of the thermal conductivity. It was observed that for the lowest concentration (0.0025 vol%) the thermal conductivity obtained was lower than that of pure oil. As the concentration increased, the conductivity increased satisfactorily, but with the increase in temperature there was a reduction in conductivity. This decrease in conductivity was certainly caused by nanoparticles agglomeration induced by increased temperature.

In general, comparing pure oil with nanofluids containing nanoparticles, an increase in thermal conductivity was observed, in the 2.85 to 5.12 % range. In Table 4, it is possible to observe the deviation of the conductivity values of nanofluids in relation to pure oil. The similar trend can be seen from previous literature [4, 7, 22].

Table 4: Percentage deviation on respect to measure mineral oil thermal conductivity and nanofluid.

Volumetric concentration (%)	Deviation (%) $\Delta\lambda = \frac{\lambda_{nf}-\lambda_{oil}}{\lambda_{oil}} 100$			
	20 °C	30 °C	40 °C	50 °C
0.0025	4.18	2.98	2.65	1.64
0.005	2.97	2.91	0.13	0.32
0.01	4.30	2.98	−2.08	−3.73
0.05	2.85	4.37	5.12	−4.05

The greater variation of the data collected for the nanofluids suspensions can be explained by the fact that this system was affected by some sedimentation issues. The thermal conductivity enhancement observed for volumetric concentration 0.01 % and 0.05 % were higher. Some papers, for example, [22] indicated that aggregation can explain the inconsistencies found experimentally for decrease in thermal conductivity.

Figure 10 shows the values of nanofluids viscosity for different temperatures. The addition of nanoparticles changed the viscosity of the nanofluid in relation to the pure mineral oil resulting in a higher viscosity. And with the increase in the concentration of nanoparticles, the greater the increase in viscosity. With the temperature increased, a decrease in the viscosity of the nanofluids was noticed. In addition, it was also observed that there was no significant increase in viscosity with the addition of varying concentrations of nanoparticles in relation to pure oil. Table 5 shows the values of viscosity deviations from pure mineral oil. The highest deviation was 1.25 % for 0.05 vol% at 40 °C.

Synthesis of a nanofluid is not simply a mixture of a liquid and nanoparticles, it requires an adequate methodology of dispersion and homogenization. The two-step method ensures good dispersion, but the sedimentation of nanoparticles is a natural phenomenon and difficult to control. Having sedimentation, it can directly affect the thermal conductivity and other properties of nanofluids. Therefore, sedimentation of nanofluids was controlled through photos enabling visual control of stability. Figure 11 shows pictures of the nanofluid samples captured after 24 h of preparation, 30 days and 120 days after.

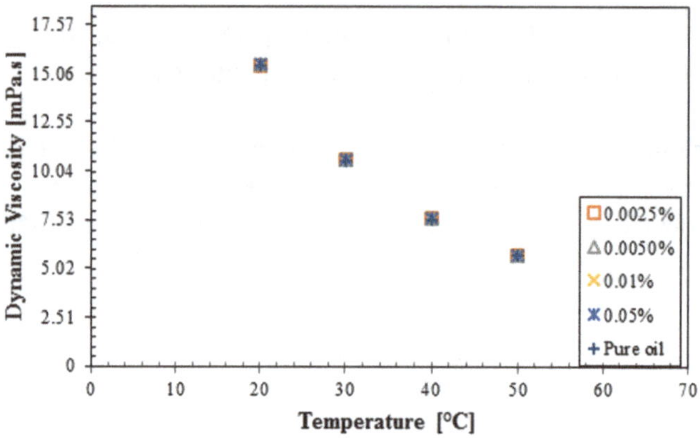

Figure 10: Viscosity of nanofluids and pure oil.

Table 5: Percentage deviation on respect to measure mineral oil viscosity and nanofluid.

Volumetric concentration (%)	Deviation (%) $\Delta\lambda = \frac{\mu_{nf}-\mu_{oil}}{\mu_{oil}}100$			
	20 °C	30 °C	40 °C	50 °C
0.0025	0.01	0.05	0.13	0.2
0.005	1.0	0.76	0.42	0.65
0.01	1.08	0.91	0.75	0.63
0.05	0.87	1.09	1.25	1.07

After preparation, it was carried out the sedimentation method in open containers and was observed visually sedimentation of nanoparticles. However, the samples remained visually stable for about 30 days before showing a thick layer of sediment on the bottom of the containers. To investigate the stability of nanofluids accurately and accurately, other techniques are required, such as, for example, measurement of zeta potential, analysis of nanoparticle size distribution or spectroscopy UV-visible. However, this research will be conducted in future work.

4 Conclusion

This paper reports the results of a recent investigation on the thermalphysical and dielectric properties of mineral oil based TiO_2 nanofluids. Experimental measurements of the breakdown voltage, thermal conductivity and viscosity of TiO_2 nanoparticles dispersed in transformer mineral oil were carried out using low volumetric concentrations.

Figure 11: Images of the samples: (a) 24h after preparation, (b) 30 days, (c) 120 days.

- Stable TiO_2 nanofluids were successfully prepared by the two-step method with ultrasound. According to the visual method of stability control, it was possible to observe that the nanofluids remained stable for approximately 30 days.
- The results of the dielectric strength measurements of nanofluids revealed that these nanofluids do not provide an increase in dielectric strength of the mineral oil as reported in the literature. However, still with the addition of the nanoparticles, the nanofluid could be used as insulating fluid since presented values of breakdown voltage lower than the recommended limit values. Furthermore, the use of such oil-based nanofluids may be prolonged the life of insulating fluid as it showed increases in thermal conductivity, which indicates a good heat exchange capacity and improved operation of the transformer.
- The viscosity of the measured nanofluids was satisfactory, with no increase greater than 1.25 %, indicating that the increased nanoparticle concentration did not significantly increase the viscosity.
- The thermal conductivity found increased with increasing nanoparticle concentration. However, with the increase in temperature there was a drop in the thermal conductivity. However, gains of 2.85 to 5.12 % were observed.

Additional studies should be done before use in transformer applications, however, it is expected that positive result using nanofluids generating lighter and smaller transformer design using the same power.

Bibliography

[1] Botha, S. S., Ndungu, P., Bladergroen, B. J.: Physicochemical properties of oil-based nanofluids containing hybrid structures of silver nanoparticles supported on silica. Ind. Eng. Chem. Res. 50(6), 3071–3077 (2011).

[2] Du, Y., Lv, Y., Li, C., Chen, M., Zhong, Y., Zhou, J., et al.: Effect of semiconductive nanoparticles on insulating performances of transformer oil. IEEE Trans. Dielectr. Electr. Insul. 770–776 (2012).

[3] Lee, J.-C., Kim, W.-Y.: Experimental study on the dielectric breakdown voltage of the insulating oil mixed with magnetic nanoparticles. Phys. Proc. 32, 327–334 (2012).

[4] Fontes, D. H., Ribatski, G., Bandarra Filho, E. P.: Experimental evaluation of thermal conductivity, viscosity and breakdown voltage AC of nanofluids of carbon nanotubes and diamond in transformer oil. Diam. Relat. Mater. 58, 115–121 (2015).

[5] Segal, V., Hjorstberg, A., Rabinovich, A., Nattrass, D., Raj, K.: AC and impulse breakdown strength of a colloidal fluid based on transformer oil and magnetite nanoparticles. In: IEEE Int. Symp. Electrical Insulation (ISEI), pp. 619–622 (1998).

[6] Lee, J.-C., Lee, W.-H., Lee, S.-H., Lee, S.: Positive and negative effects of dielectric breakdown in transformer oil based magnetic fluids. Mater. Res. Bull. 47(10), 2984–2987 (2012).

[7] Choi, S. U. S., Eastman, J. A.: Enhancing Thermal Conductivity of Fluids With Nanoparticles. In: ASME, International Mechanical Engineering Congress & Exposition, pp. 99–105. Argonne National Lab, IL, United States (1995).

[8] Tijerina, J. T., Narayanan, T. N., Gao, G. H., Rohde, M., Tsentalovich, D. A., Pasquali, M., Ajayan, P. M.: Electrically insulating thermal nano-oils using 2D fillers. ACS Nano 6, 1214–1220 (2012).

[9] Hwang, J. G., Zahn, M., O'Sullivan, F. M., Pettersson, L. A. A., Hjortstam, O., Liu, R. S.: Effects of nanoparticle charging on streamer development in transformer oil-based nanofluids. J. Appl. Phys. 107, 014310 (2010).

[10] Du, Y. F., Lv, Y. Z., Li, C. R., Chen, M. T., Zhou, J. Q., Li, X. X., Zhou, Y., Tu, Y. P.: Effect of electron shallow trap on breakdown performance of transformer oil-based nanofluids. J. Appl. Phys. 110, 104104 (2011).

[11] Choi, C., Yoo, H., Oh,
J.: Preparation and heat transfer properties of nanoparticle-in-transformer oil dispersions as advanced energy-efficient coolants. Curr. Appl. Phys. 8, 710–712 (2008).

[12] Józefczak, A.: Study of low concentrated ionic ferrrofluid stability in magnetic field by ultrasound spectroscopy. J. Magn. Magn. Mater. 321, 2225–2231 (2009).

[13] Hwang, J. G., Zahn, M., O'Sullivan, F. M., Pettersson, L. A. A., Hjortstam, O., Liu, R.: Electron Scavenging by Conductive Nanoparticles in Oil Insulated Power Transformers. In: 2009 Electrostatics Joint Conference, vol. 1.1, pp. 1–12 (2009).

[14] O'Sullivan, F., Hwang, J.: A model for the initiation and propagation of positive streamers in transformer oil. IEEE 2008, 210–214 (2008).

[15] Wang, Q., Rafiq, M., Lv, Y., Li, C., Yi, K.: Preparation of Three Types of Transformer Oil-Based Nanofluids and Comparative Study on the Effect of Nanoparticle Concentrations on Insulating Property of Transformer Oil. J. Nanotechnol. 2016, 1–6 (2016).

[16] Rafiq, M., Li, C., Lv, Y., Yi, K., Hussnain, S.: Preparation and Study of Breakdown Features of Transformer Oil Based Magnetic Nanofluids. In: 2017 International Conference on Electrical Engineering (ICEE). IEEE, (2017).

[17] Shukla, G., Aiyer, H.: IEEE Trans. Dielectr. Electr. Insul. 22, 2185–2189 (2015).

[18] Atiya, E. G., Mansour, D.-E. A., Azmy, A. M., Khattab, R. M.: Enhanced Dielectric Strength of Transformer Oil Using TiO2 Nanoparticles with Surfactant. In: 16th International Middle- East Power Systems Conference-MEPCON'2014, Ain Shams University, Cairo, Egypt (2014).

[19] Ge, X., Xia, Y., Cao, Z.: Tribological properties and insulation effect of nanometer TiO_2 and nanometer SiO2 as additives in grease. Tribol. Int. 92, 454–461 (2015).

[20] Ghasemi, J., Jafarmada, S., Nazari, M.: Effect of magnetic nanoparticles on the lightning impulse breakdown voltaje of transformer oil. J. Magn. Magn. Mater. 389, 148–152 (2015).

[21] Hanai, M., Hosomi, S., Kojima, H., Hayakawa, N., Okubo, H.: Dependence of TiO_2 and ZnO nanoparticle concentration on electrical insulation characteristics of insulating oil. In: Annu. Rep. Conf. Electrical Insulation and Dielectric Phenomena (CEIDP), pp. 780–783 (2013).

[22] Chiesa, M., Das, S. K.: Experimental investigation of the dielectric and cooling performance of colloidal suspensions. Insul. Media. Colloids Surf. A 335(1–3), 88–97 (2009).

[23] Lv, Y. Z., Wang, L. F., Li, X. X., Du, Y. F., Zhou, J. Q., Li, C. R.: Experimental investigation of breakdown strength of mineral oil-based nanofluids. In IEEE Int. Conf. Dielectric Liquids (ICDL), pp. 26–30 (2011).

[24] Du, Y. F., Lv, Y. Z., Wang, F. C., Li, X. X., Li, C. R.: Effect of TiO_2 nanoparticles on the breakdown strength of transformer oil. In: Electrical Insulation (ISEI), Conference Record of the 2010 IEEE International Symposium on, pp. 1–3. IEEE (2010).

[25] Oliveira, L. R., Silva, A. C. A., Dantas, N. O., Bandarra Filho, E. P.: Thermophysical properties of TiO2-PVA/water nanofluids. Int. J. Heat Mass Transfer, Part A 115, 795–808 (2017).

[26] Lv, Y. Z., Du, Y. F., Zhou, J. Q., Li, X. X., Chen, M. T., Li, C. R., Wang, G. L.: Nanoparticle effect on electrical properties of aged mineral oil based nanofluids, CIGRÉ 2012, Paper D1-106 (2012)..

[27] Lv, Y.-Z., Li, C., Sun, Q., Huang, M., Li, C. R., Qi, B.: Effect of Dispersion Method on Stability and Dielectric Strength of Transformer Oil-Based TiO_2 Nanofluids. Nanoscale Res. Lett. 11(1), 515 (2016).

[28] Fallah-Shojaie, A., Tavakoli, A., Ghomashpasand, M., Hoseinzadeh, S.: Experimental Evaluation on the dielectric breakdown voltage of fresh and used Transformer Oil mixed with Titanium dioxide nanoparticles in the Gilan Electrical distribution company. In: 21st Iranian Conference on Electrical Engineering (ICEE) (2013).

[29] Zhong, Y. X., Lv, Y. Z., Li, C. R., Du, Y. F., Chen, M. T., Zhang, S. N., Zhou, Y., Chen, L.: Insulating properties and charge characteristics of natural ester fluid modified by TiO_2 semiconductive nanoparticles. IEEE Trans. Dielectr. Electr. Insul. 20(1), 135–140 (2013).

[30] Kopcansky, P., Tomco, L., Marton, K., Koneracka, M., Timko, M., Potocova, I.: The DC dielectric breakdown strength of magnetic fluids based on transformer oil. J. Magn. Magn. Mater. 289, 415–418 (2005).

[31] Katiyar, A., Dhar, P., Nandi, T., Maganti, L. S., Das, S. K.: Enhanced dielectric breakdown performance of anatase and rutile titania based nano-oils. arXiv preprint (2016). arXiv:1511.09017v1.

[32] Guinier, A.: X-ray diffraction in crystals, imperfect crystals, and amorphous bodies. W.H. Freeman, San Francisco (1963).

[33] Ohsaka, T.: Temperature-Dependence of the Raman-Spectrum in Anatase TiO_2. J. Phys. Soc. Jpn. 48, 1661–1668 (1980).

[34] Zhang, W. F., He, Y. L., Zhang, M. S., Yin, Z., Chen, Q.: Raman scattering study on anatase TiO_2 nanocrystals. J. Phys. D Appl. Phys. 33, 912–916 (2000).

[35] Mergos, J. A., Athanassopoulou, M. D., Argyropoulos, T. G., Dervos, C. T.: Dielectric properties of nanopowder dispersions in paraffin oil. IEEE Trans. Dielectr. Electr. Insul. 19(5), 1502–1507 (2012).

[36] Dhar, P., Katiyar, A., Maganti, L. S., Pattamatta, A., Das, S. K.: Superior dielectric breakdown strength of graphene and carbon nanotube infused nano oils. arXiv preprint (2015). arXiv:1511.04540v1.
[37] Maxwell, J.: Univ. Oxford (1873).
[38] Beheshti, A., Shanbedi, M., Heris, S. Z.: Heat transfer and rheological properties of transformer oil-oxidized MWCNT Nanofluid. J. Therm. Anal. Calorim. 118, 1451–1460 (2014). https://doi.org/10.1007/s10973-014-4048-0.